Atmospheric Aerosols - Regional Characteristics - Chemistry and Physics

Atmospheric Aerosols - Regional Characteristics - Chemistry and Physics

Editor

Henri Duff

scitus academics

Atmospheric Aerosols - Regional Characteristics - Chemistry and Physics
Edited by **Henri Duff**

ISBN: 978-1-68117-316-0
Library of Congress Control Number: 2016937628

SCITUS Academics LLC,
www.scitusacademics.com
Box No. 4766, 616 Corporate Way,
Suite 2, Valley Cottage,
NY 10989

Printed in United States of America on Acid Free Paper

Preface

Aerosol is a colloidal system of solid or liquid gas particles. An aerosol contains both the particles and the suspended gas, normally air. Natural aerosols include fog, exudates from the forest and geyser steam. Artificial aerosols include haze, powder, particulate air pollutants, and smoke. Atmospheric aerosols are solid or liquid particles and are suspended in air between 0.002 μm and 100 μm in diameters. The size, source, chemical composition, quantity and distribution of aerosol particles in space and time and how long they survive in the atmosphere vary significantly. Primary atmospheric aerosols are particulates that are directly emitted into the atmosphere. Secondary atmospheric aerosols are particulates formed in the atmosphere through gas-to-particles processes.

This book offers characterization and impact of atmospheric aerosols on regional climate. There has also been detailed information on measuring techniques and atmospheric conditions, such as chemical and physical processes, dust particles emissions, and formation and transport of aerosol particles. The book should be an immensely useful resource to solve some of the problems related to the complex nature of the interaction between atmospheric aerosols and climate science.

Table of Contents

CHAPTER 1

Towards Optimal Aerosol Information for the Retrieval of Solar Surface Radiation Using Heliosat

Richard Mueller *, Uwe Pfeifroth and Christine Traeger-Chatterjee

Deutscher Wetterdienst, Frankfurter Str. 135, D-60387 Offenbach, Germany

ABSTRACT

High quality data of surface radiation is a prerequisite for climate monitoring (Earth radiation budget) and solar energy applications. A very common method to derive solar surface irradiance is the Heliosat method, a one channel approach for the retrieval of the effective cloud albedo (CAL). This information is then used to derive the solar surface irradiance by application of a clear sky model. The results of this study are based on radiative transfer modelling, visual inspection of satellite images and evaluation of satellite based solar surface radiation with ground measurements. The respective results provide evidence that variations in Aerosol Optical depth induced by desert storms and biomass burning events lead to a significant increase of the effective cloud albedo, thus, that certain aerosol events are interpreted as clouds by the method. For the estimation of the solar surface radiation aerosol information is needed as input for the clear sky model. As the aerosol effect is partly considered by

CAL, there is a need to modify external aerosol information for the use within the clear sky model, e.g., by truncation of high aerosol loads. Indeed, it has been shown that a modified version of the Monitoring Atmospheric Composition and Climate (MACC) aerosol information leads to better accuracy of the retrieved solar surface radiation than the original MACC data for the investigated 9 sites and time period (2006–2009). Further, the assumption of a constant aerosol optical depth of 0.18 provides also better accuracies of the estimated solar surface radiation than the original MACC data for the investigated sites and period. It is concluded that this is partly due to the consideration of scattering aerosols by the effective cloud albedo.

KEYWORDS

solar surface radiation; radiative transfer modeling; interactions with atmosphere; cloud albedo retrieval; aerosol effect; aerosol clouds

1. INTRODUCTION

The solar surface irradiance (*I*) is an important variable for climate monitoring and analysis (e.g., [1]), and is widely used, beside others, in agrarmeteorology and solar energy applications (e.g., [2]).

Satellite based solar surface radiation is the primary observational source of information in many regions of the world due to the low spatial coverage of well maintained ground measurements. Nowadays, the accuracy of satellite based solar surface radiation data is close to that of ground measurements (e.g., [3]).

Aerosols affect the solar surface radiation significantly (e.g., [4]). Uncertainties in aerosol information are therefore a main driver for errors in the retrieved solar surface irradiance. Thus, in [5] the performance of different aerosol climatologies have been investigated. In that study the aerosol information from the Monitoring Atmospheric Composition and Climate project (MACC) has shown the best performance compared to the Max-Planck-Institute Aerosol Climatology version 1 (MAC-v1) [6] and GADS/OPAC [7,8]. However, it has been also briefly discussed that high aerosol loads (e.g., desert storms, biomass burning) might be considered as clouds by the Heliosat method [2,3,9]. This would have significant consequences for the interpretation of the effective cloud albedo and the optimal aerosol information. This hypothesis is further investigated and discussed in this study.

The main items of the Heliosat method, which is used to retrieve the effective cloud albedo (CAL), are described inSection 2 in order to enable an understanding of the potential effect of aerosols on CAL. To the knowledge of the authors the "Heliosat" approach is one of the most established methods applied for the retrieval of CAL and solar surface radiation from visible images of geostationary satellites. Several key producer use either the Heliosat approach or the satellite derived cloud albedo as input to estimate the solar surface irradiance, e.g., the Climate Monitoring Satellite Application Facility (CM SAF) [10], Satellight [11], Solar Radiation Data (SODA) [12], PVGIS [13], Land Surface Analysis SAF (LSA SAF) [14] and the Ocean and Sea Ice SAF (OSI SAF) [15] to mention some examples.

2. HELIOSAT

Within this study the Heliosat method ([2,10]) is used to retrieve the effective cloud albedo (also referred to as cloud index) from visible images of geostationary satellites. As a first step of the retrieval procedure the effect of variations in the Sun-Earth distance and the solar zenith angle have to be corrected in order to normalize the illuminance conditions. Furthermore, the dark offset of the instrument has to be subtracted from the counts measured by the satellite instrument. Thus, the observed reflections are normalized by application of Equation (1):

$$\rho = \frac{D - D_0}{f \cdot cos(\theta)} \tag{1}$$

Here, D is the observed digital count including the dark offset of the satellite instrument. D0 is the dark offset, θ is the solar zenith angle and f corrects the variations in the Sun-Earth distance. The resulting ρ is the normalized reflection based on the satellite measurements.

The effective cloud albedo (CAL) is then derived from the normalized reflections by using Equation (2):

$$CAL = \frac{\rho - \rho_{cs}}{\rho_{max} - \rho_{cs}} \tag{2}$$

Here, ρ is the observed normalized reflection for each pixel and time. ρcs is the clear sky reflection, which is a monthly value derived for every pixel and time slot separately. This is done by iteration as follows. First, all reflection values ρ(within the time span covering usually 1 month) are used to calculate an average reflection ρaverage. This average reflection serves as initial threshold value ρthreshold. A new average is then calculated with all reflections that are smaller than the initial threshold value plus a small value σ, given as σ=0.035ρmax. This average is then the threshold for the next iteration, where again all reflection values below ρthreshold+σ are used to calculate a new value for ρthreshold. The iteration is proceeded until no reflection value is available that is higher than ρthreshold+σ, hence until the threshold does not change any more. ρcs is then set equal to the final ρthreshold. It is assumed that at this point all cloudy pixels are filtered out. ρmax is the "maximum" reflection. It is determined by the 95 percentile of all reflection values at local noon in a target region, characterized by high frequency of cloud occurrence for each month. In this manner changes in the satellite brightness sensitivity are accounted for. Further details are given in [16].

The effective cloud albedo (cloud index) is subsequently used to estimate the all sky irradiance, in general (CAL between 0 and 0.8) by:

$$G_{clear} \times (1 - CAL) = G \tag{3}$$

Hence, the application of a clear sky model enables the estimation of the all sky irradiance. For this step information about aerosols is needed, either directly or within turbidity information. For the calculation of the clear sky radiation the sophisticated SPECMAGIC method is used, which is described in detail in [16]. It is based on radiative transfer modeling and enables an appropriate treatment of the aerosol effect on solar surface irradiance.

3. RESULTS AND DISCUSSION

As described in Section 2, ρcs is a statistical minimum of the observed reflectances, which is retrieved from visible images. Clouds increase the reflection signal, hence ρ is usually higher than ρcs in the presence of clouds. ρcs represents therefore in a statistical manner the "clear sky" reflection. However, higher ρ values might also occur in cloud free situations during the occurrence of specific aerosol events like biomass burning or desert storms. These events lead typically to a temporal increase of aerosol optical depth relative to the background/climatological value. In order to investigate this possible effect theoretically radiative transfer calculations have been performed with libRadtran [17]. Figure 1 shows that the increase of scattering aerosols is coupled with the increase of the planetary reflection, and hence with the increase of the reflection signal received by the satellite. As the statistical method estimates the "minimum reflection" it has in turn to be expected that in these cases ρ is significantly higher than ρcs for a certain period. Thus, that aerosol events are at least partly treated as clouds, leading to increased values of CAL, even in cloud free situations.

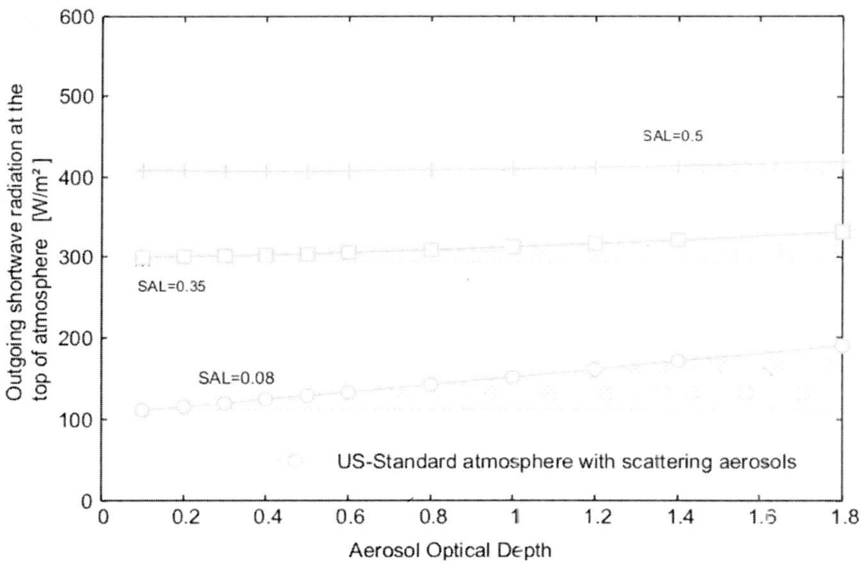

Figure 1. Results of the radiative transfer modeling with libRadtran for US-standard atmosphere, 335 DU ozone, 15 mm water vapor and scattering aerosols (ssa = 0.99). Increase of the aerosol optical depth leads to a significant increase for low to high surface albedo (SAL) values. Only for very high SAL values above 0.5 the increase is getting insignificant. For low to high SAL (up to *ca.* 0.45) it has to be expected that a temporary increase of AOD induced by specific aerosol events leads to an increase of cloud albedo (CAL) values.

Indeed comparison of CAL images with SeaWIFS (Sea-viewing Wide Field-of-view Sensor) ocean color images [18] and MODIS (Moderate-

Resolution Imaging Spectroradiometer) fire images [19] show evidence that aerosol clouds lead to enhanced CAL values even for cloud-free skies. Figure 2 and Figure 3 show respective examples.

In Sede Boker and Tamanrasset AERONET stations [20] are operated in addition to the BSRN stations. The respective AOD measurements at 500 nm/675 nm show high variability of AOD and regularly occurring high AOD values above 0.4. Therefore, the relation between AOD and CAL has been investigated at these sites for the period 2006–2009. Table 1 shows that on average CAL increases with increasing AOD, which provides further evidence that high AOD values induced by dust events leads to an increase in the observed CAL values.

In summary, empirical findings and RTM results give evidence that the aerosol effect on solar surface irradiance is partly considered by an increase of CAL. Hence, truncation of high aerosol values in the MACC aerosol climatology might lead to better results in the estimated solar surface irradiance than using the original MACC data. Further, also climatological background values might work well. Therefore, the performance of "original" MACC aerosol information is compared to a truncated MACC version and to constant aerosol approaches, in detail Aerosol Optical Depth of 0.18 and 0.0 in the troposphere, respectively. An AOD of 0.18 corresponds approximately to the constant visibility of 20 km, which has been proven to perform well within the SAF comparison study [15].

The "original" MACC aerosol information results from a data assimilation system for global reactive gases, aerosols and greenhouse gases. It consists of a forward model for aerosol composition and dynamics [22] and the data assimilation procedure described in detail in [23]. The data is generated by the European Centre for Medium Range Weather Forecasts.

The truncation of the MACC aerosol information, referred to as MACC modified (MACCmod), is performed as follows:

$$AOD_{Trunc} = \begin{cases} AOD_{MACC} & \text{for AOD} \leq 0.16 \\ 0.16 + 0.5 * (AOD_{MACC} - 0.16) & \text{for AOD} > 0.16 \end{cases} \quad (4)$$

Here, AODTrunc is the truncated AOD and AODMACC is the original MACC AOD. Monthly long term means on a 0.5 × 0.5 degree latitude-longitude grid are used. The pixel value is derived by spatial interpolation and assignment of the respective monthly mean.

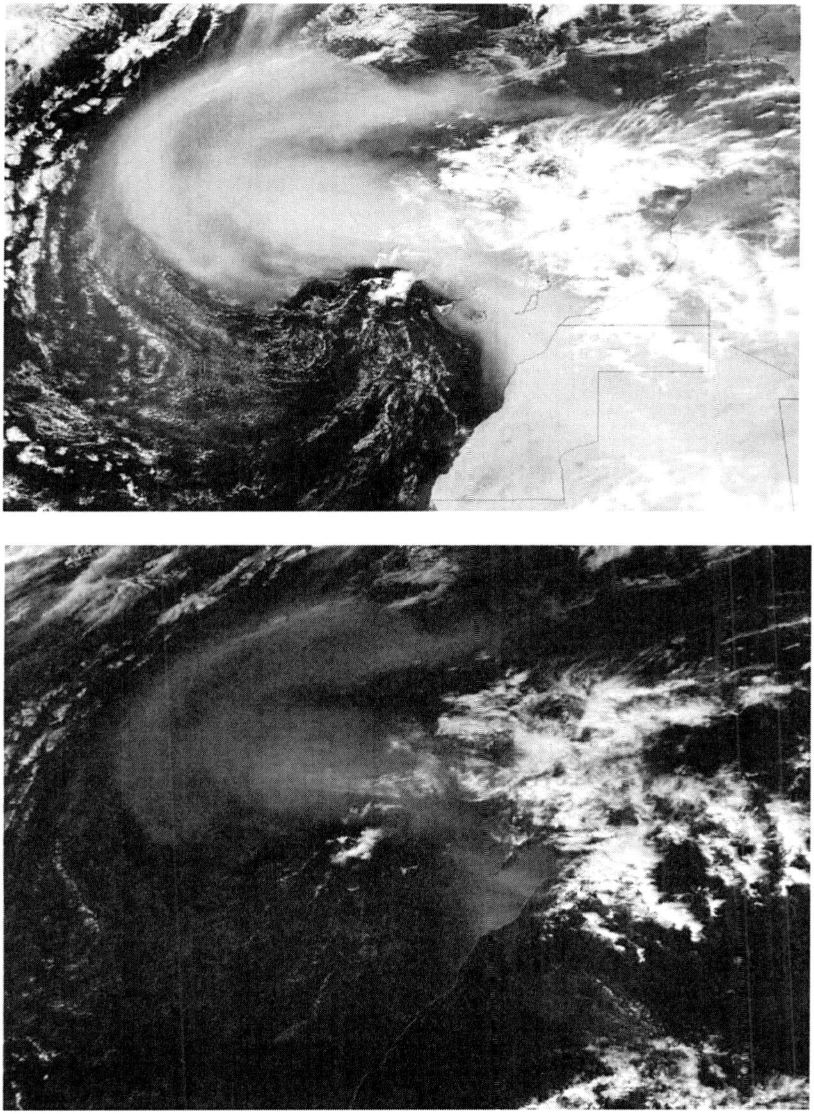

Figure 2. Example of the effect of aerosol events on CAL. The SesWIFS image from 26 February 2000 [18,21] shows clearly a dust storm over parts of the land and ocean (top, yellowish color). The "dust cloud" shows a different texture than the surrounding water clouds. This dust cloud leads to a significantly increased CAL value (bottom), which subsequently leads to significantly lower solar surface radiation, see Figure 4. The top figure has been provided by SeaWIFS project, NASA/GSFC and ORBIMAGE.

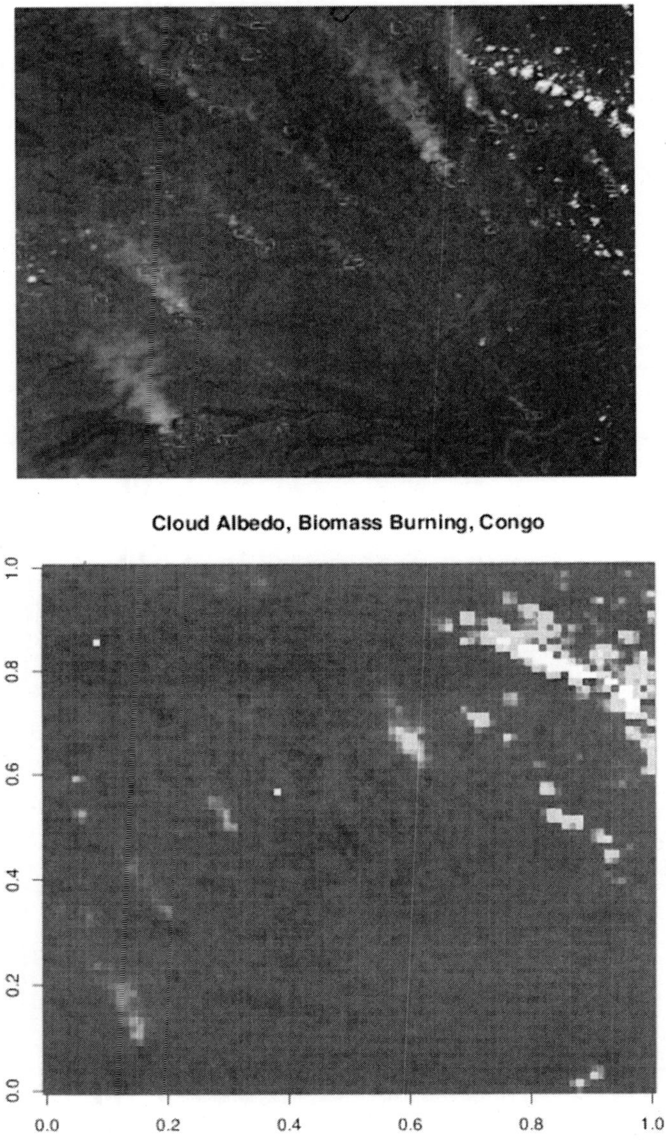

Figure 3. Biomass burning (red color) leads to biomass burning "clouds" in theMODIS [19,26] fire monitoring image, 13 June 2003, 12:30 (top). The thick fire clouds with large geographical coverage are also apparent as increased CAL values (pale blue) in the CAL image (same time). However, compared to the desert event images the effect seems to be less pronounced. This item will be discussed later on.

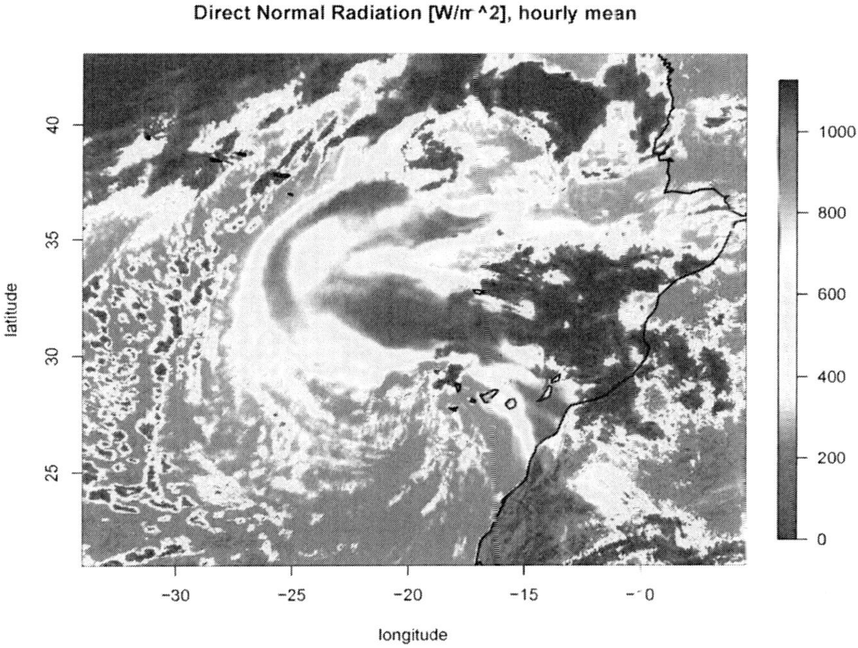

Figure 4. Example for the effect of high CAL values induced by aerosol events. The high CAL values induced by the dust storm (Figure 2) leads to a significant reduction of the direct normal irradiance relative to the surrounding area.

MACC is used as baseline for the comparison study as it has been evaluated to perform significantly better than the "Kinne" [24] GADS/OPAC [7,8] and MAC-v1 climatology [6], see [5] for further details.

For the estimation of the solar surface irradiance, and thus the comparison study, the same retrieval method than in [5] has been used. The method is referred to as SPECMAGIC and consists of the Heliosat approach for the treatment of the cloud effect (CAL) and a spectrally resolved clear sky model, the complete approach is described and discussed in detail in [3,16]. Monthly means of solar surface irradiance are calculated using the different aerosol information described above and subsequently compared to ground measurements of the Baseline Surface Radiation Network (BSRN) [25] for all sky conditions. The comparison is performed for the period of 2006 to 2009. The BSRN stations used for the validation are listed in Table 2. Only those stations were used that cover at least a period of 1 year. The stations and periods correspond to that used in [5].

Table 1. Comparison of CAL and AOD for the sites Tamanrasset (TAM) and Sede Boker (SBO). Only tuples where the observation matches within ±1 min have been used. On average the values of CAL increase with increasing AOD, hence with increasing frequency of dust storms. The CAL images as well as the AERONET AOD time series show that there is a relative high frequency of clouds, in particular of optical thin clouds. The satellite observes an area of about 3 × 3 km while the AERONET measurements look straight towards the sun. Thus, the cloud screening is different for the two observations, which explains that CAL is on average not zero for background aerosol, although, AOD is theoretically only observed in clear sky. The AOD at 500 nm has been used for Sede Boker. Tamanrasset does not provide data around 500 nm, thus, 675 nm has been used. SBO/TAM in the table means that the results for Sede Boker are on the left hand and that for Tamanrasset on the right hand of the "/" sign.

AOD Bin	AOD Mean SBO/TAM	CAL Mean SBO/TAM	Increase in CAL SBO/TAM	Number of Tuples SBO/TAM
<0.15	0.09/0.06	0.058/0.010	-	947/883
0.15–0.4	0.23/0.23	0.067/0.024	0.01/0.014	766/274
>0.4	0.64/0.68	0.145/0.080	0.087/0.07	109/113

Table 2. List of Baseline Surface Radiation Network (BSRN) stations used.

Station	Country	Code	Latitude $(DegN)$	Longitude $(DegE)$	Elevation (m)	Number of Months
Palaiseau	France	pal	48.73	2.21	156	16
Lerwick	UK	ler	60.13	−1.18	84	13
Lindenberg	Germany	lin	52.21	14.12	125	12
Carpentras	France	car	44.05	5.03	100	48
Payerne	Switzerland	pay	46.81	6.94	491	48
Sede Boker	Israel	sbo	30.9	34.78	500	48
Tamanrasset	Algeria	tam	22.78	5.51	1385	48
Toravere	Estonia	tor	58.25	26.46	70	48
Cabauw	Netherlands	cab	51.97	4.93	0	48

The stations cover different climates, but the majority is located in regions with weak to moderate aerosol loads and variations. Exceptions are Sede Boker and Tamanrasset which are located close to and in the Sahara desert, respectively. The limited number of stations does not allow to draw final conclusions on the global performance of the investigated aerosol information.

For the comparison only the aerosol information has been altered, all other input variables remained unchanged. Hence, the differences in the error measures can only result from the different aerosol information. As main error measure the Mean Absolute Bias (MAB) and Bias are used. The MAB and bias are defined by Equations (5) and (6).

$$MAB = 1/n * \sum_i |RADmm^i_{sat} - RADmm^i_{gbm}| \tag{5}$$

$$Bias = 1/n * \sum_i RADmm^i_{sat} - RADmm^i_{gbm} \tag{6}$$

RADmm is the monthly mean of radiation for satellite (sat) and ground based measurements (gbm), respectively, and n is the number of the monthly means. The absolute bias is preferred as, in contrast to RMSE, it is a linear quantity and avoids bias cancellation of errors across months or stations.

Figure 5 (absolute units) and Figure 6 as well as Table 3 show the results ofthe evaluation.

The use of the MACC modified climatology leads to a considerable increase in the accuracy of the estimated surface radiation compared to the original MACC data, in particular when the results at the stations Tamanrasset and Sede Boker are appropriately considered. These stations are located in, respectively, close to the Sahara desert region where desert storms occur regularly, and thus higher variability of AOD has to be expected. The use of MACC, MACCmod and an AOD of 0.18 result in better accuracies than the zero aerosol approach, which means that all these approaches provide useful information compared to the zero assumption. However, the accuracy in the retrieved solar surface irradiance is higher using the constant aerosol optical depth of 0.18 than using the MACC data. But, for the interpretation of this result it has to be considered that the majority of stations are located in Europe. There 0.18 is probably a good climatological value. This assumption is supported by the SAF surface radiation inter-comparison study [15], where a constant AOD value of around 0.18 (visibility of 20 km) has performed well over Europe [15]. Higher aerosol loads at Tamanrasset and Sede Boker induced by desert storms might be partly considered by the effective cloud albedo on the other hand. These arguments might explain the good performance of the 0.18 AOD approach.

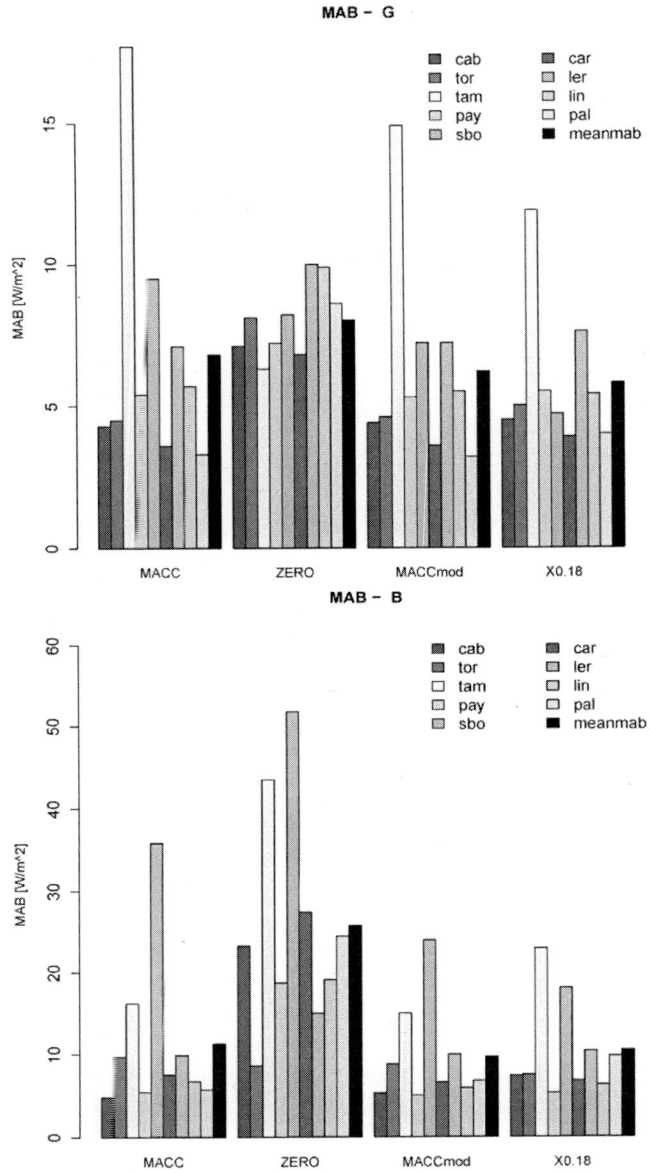

Figure 5. Mean Absolute Bias of the global irradiance (MAB-G, **top**) and the direct irradiance (MAB-B, **bottom**) for MACC, MACCmod and the zero and 0.18 AOD assumptions, evaluated at nine different Baseline Surface Radiation Network (BSRN) stations (colored bars). "meanmab" is the mean MAB over all nine stations. The effect of aerosols on direct irradiance is much more pronounced. As a result much higher errors are apparent for direct irradiance than for global irradiance.

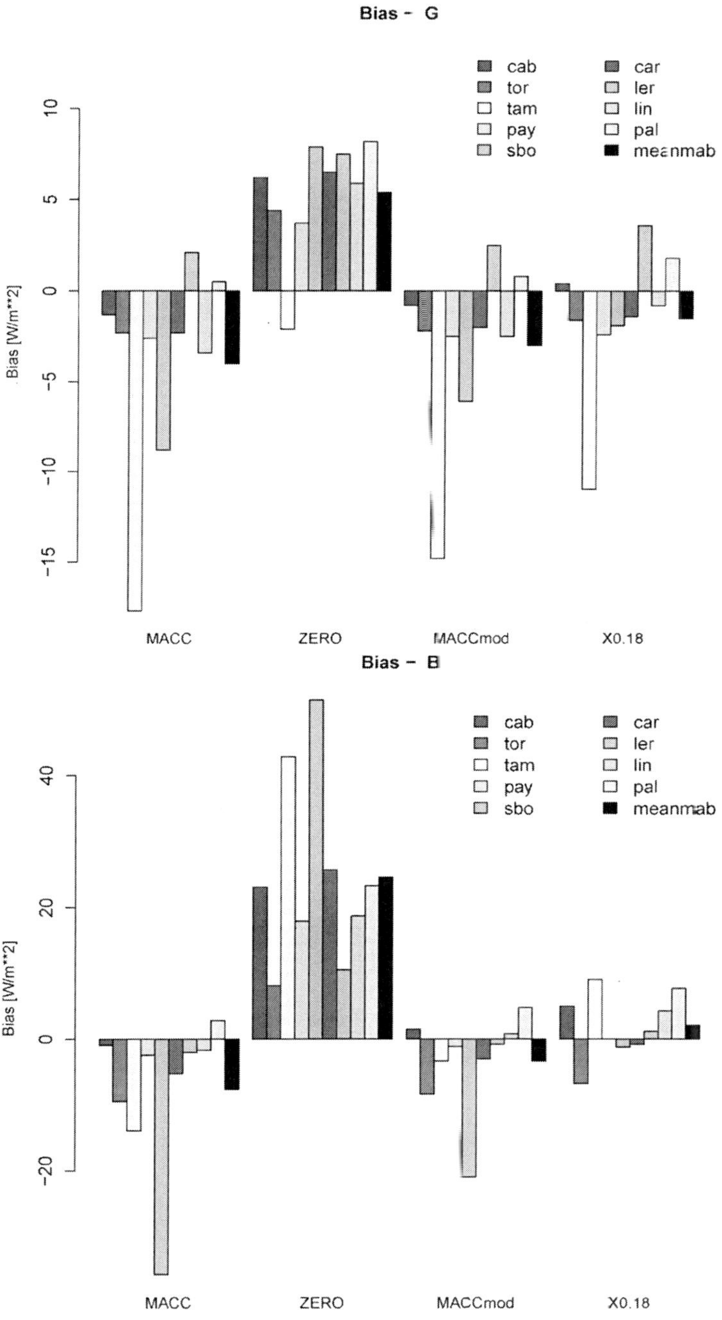

Figure 6. Bias for the global irradiance (Bias-G, **top**) and direct irradiance (Bias-B,**bottom**).

Table 3. Mean Absolute Bias (MAB) over all stations for global irradiance G and direct (beam) irradiance B in absolute units [W/m2] and relative units, MAB deviation relative to the mean [%]. The last two lines show the bias in W/m2. The AOD = 0 and AOD = 0.18 columns provide the results for zero and 0.18 AOD of tropospheric aerosol.

Variable	MACC	AOD = 0	MACCmod	AOD = 0.18
MAB G in W/m^2	6.8	8.0	6.2	5.8
in %	3.8	4.5	3.4	3.3
MAB B in W/m^2	11.3	25.8	9.7	10.5
in %	9.6	21.9	8.2	8.9
Bias G in W/m^2	−4.0	5.4	−3.0	−1.5
Bias B in W/m^2	−7.6	24.6	−3.3	2.1

The results show that aerosol information used within the clear sky model has to be modified in order to account for the aerosol effect on the CAL retrieval. Otherwise, the effect of scattering aerosols is considered twice, increasing the uncertainty of the retrieved solar surface irradiance. However, the best manner for the correction of aerosol information is not straightforward and needs further investigations, which in turn requires accurate information about the single scattering albedo (aerosol type) as pre-requisite. This is due to the fact that aerosols with high absorption rate (urban aerosols) behave different to scattering aerosols. For low to moderate surface albedo the planetary reflection still increases with increasing AOD, but the increase is much lower than for scattering aerosols. Yet, for higher surface albedo the reflection decreases with increasing AOD of urban aerosols, see Figure 7. Hence, the magnitude of the aerosol effect on CAL depends on the single scattering albedo, in other words, on the aerosol type. Indeed, by comparison of the biomass and desert storm events it seems that the effect of biomass burning is less pronounced. One reason could be that the single scattering albedo is expected to be significantly higher for dust than for biomass burning aerosols, which contains soot. Thus, for an optimal adjustment of the aerosol information applied in the clear sky model accurate knowledge of the single scattering albedo and the surface albedo is required.

A good potential to generate reliable single scattering albedo data might be given by the MACC project itself. MACC provides beside the total AOD also the black carbon AOD. The relation of total AOD and black carbon AOD can be used to estimate the single scattering albedo. However, using this information in order to achieve a better accuracy of solar surface radiation is not a trivial task, given that the accuracy of the solar surface radiation is already close to that of ground based measurements. It requires further and detailed RTM studies, which are out of the scope of this study, but which are intended as a next step towards optimal aerosol information for the retrieval of solar surface radiation using Heliosat.

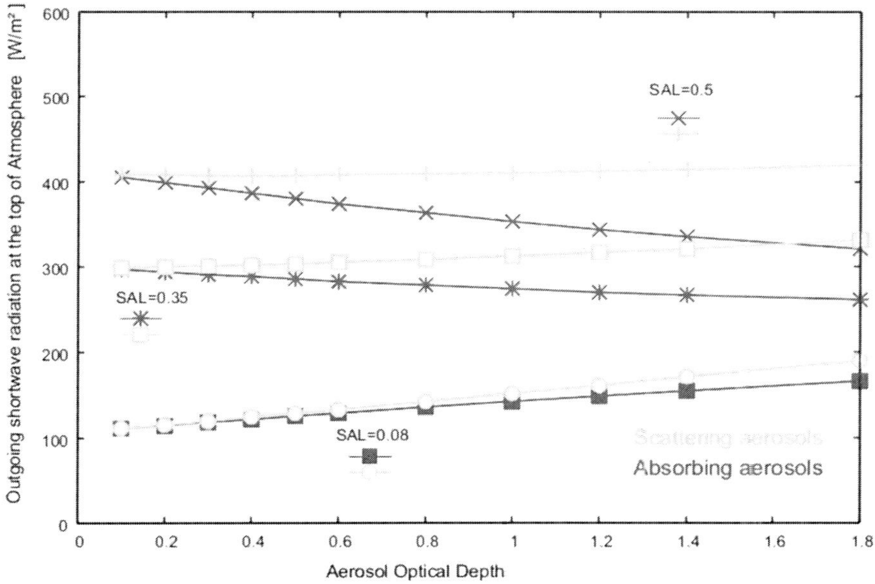

Figure 7. Results of the radiative transfer modeling with libRadtran for US-standard atmosphere, 335 DU ozone, 15 mm water vapor for scattering aerosols (ssa = 0.99) and urban (absorbing) aerosols (ssa *ca.* 0.8). For absorbing aerosols the increase of the aerosol optical depth leads to a significant decrease of the planetary reflection for bright surfaces. This behavior is quite contrary for scattering aerosols.

The results discussed here might be also relevant for the retrieval of cloud fraction and cloud properties using other methods that do not include an appropriate separation between water and aerosol clouds. Descriptively, the dust outbreak illustrated in Figure 2 leads also to an enhanced cloud fraction in the CLARA-A1 [27] data set.

Further, it has to be also considered that errors in the MACC aerosol information (overestimation of AOD) might contribute to the better performance of MACC modified and the 0.18 AOD assumption.

Howsoever, the results (e.g., better performance of 0.18 AOD approach) show that it can not be assumed a priori that higher spatial and temporal resolution of aerosol information applied within the clear sky model lead to higher accuracies in the retrieved solar surface irradiance.

4. MATERIAL AND METHODS

For the study radiative transfer modeling, visual inspection of satellite images and evaluation of satellite based solar surface radiation with ground measurements have been applied. For the latter BSRN ground measurements and the SPECMAGIC retrieval methods have been used, please see Section 3 for further information.

5. CONCLUSIONS

This study provides evidence that the effect of temporal and spatial aerosol variability on solar surface irradiance is partly considered by the retrieval of the effective cloud albedo performed within the Heliosat method [2,3,9], e.g., dust storms and biomass burning events are interpreted as clouds, leading to an increase of the effective cloud albedo. This finding is mainly a result of the one channel Heliosat approach. Separation between "aerosol clouds" and water clouds are not performed and thus "aerosol clouds" contribute to the retrieved CAL. Hence, aerosol information used within the clear sky model has to be modified in order to account for the aerosol effect on the CAL retrieval. Otherwise, the effect of scattering aerosols is considered twice increasing the error of the retrieved solar surface radiation. As a consequence modification of the aerosol information used within the clear sky model is required for the appropriate usage within the retrieval of solar surface irradiance. Within this scope it has been shown that truncation of high AOD values of the original MACC aerosol information leads to higher accuracies in the retrieved solar surface irradiance. Further, the assumption of a constant aerosol optical depth of 0.18 provides also better accuracies in the estimated solar surface irradiance than the original MACC data.

Overall, the results show that a modification of external aerosol information used within the clear sky model is advisable. Further, the results show that it can not be assumed a priori that a higher spatial and temporal resolution of aerosol information applied within the clear sky model lead to higher accuracies in the retrieved solar surface irradiance, but that the performance depends on the retrieval method and the quality of the aerosol information. Within this scope, it has to be also considered that uncertainties in the MACC aerosol information might contribute significantly to the better performance of MACC modified and the 0.18

AOD assumption for the investigated sites and period. However, the evaluation results are relevant within the scope of climate monitoring, but are of limited value for nowcasting and in particular for the forecasting of AOD or solar surface radiation. The results are therefore not applicable for these application areas.

ACKNOWLEDGMENTS

This work is partly funded by EUMETSAT within the SAF framework. We thank the European tax payers, as the funding is originally given by them. The authors thank the operation team of CM SAF for their support. Arnon Karnieli is acknowledged as PI of the Sede Boker Aeronet station and Emilio Cuevas-Agullo as PI of the Tamanrasset station. We thank the PIs of the BSRN stations for the BSRN data and the Alfred Wegener Institute for hosting and managing of the BSRN archive.

AUTHOR CONTRIBUTIONS

Uwe Pfeifroth checked the appearance of higher CAL values with desert storms and biomass burning events. Christine Treager-Chatterjee did the validation of the different aerosol runs. R. Mueller developed the SPECMAGIC algorithm, supervised the work and wrote large parts of the manuscript. All authors contributed to the writing and editing of the manuscript.

CONFLICTS OF INTEREST

The authors declare no conflict of interest.

REFERENCES

1 Trenberth, K.; Fasulo, J.; Kiehl, J. Earth's global energy budget. *Bull. Am. Meteorol. Soc.* **2009**, *90*, 311–323.

2 Hammer, A.; Heinemann, D.; Hoyer, C.; Kuhlemann, R.; Lorenz, E.; Mueller, R.; Beyer, H. Solar Energy Assessment Using Remote Sensing Technologies. *Remote Sens. Environ.* **2003**, *86*, 423–432.

3 Müller, R.; Pfeifroth, U.; Chatterjee, C.T.; Trentmann, J.; Cremer, R. Digging the METEOSAT treasure—3 decades of solar surface radiation and effective cloud albedo. *Remote Sens.* **2015**, *7*, 8067–8101.

4 Müller, R. Solar Irradiance, Global Distribution. In *Encyclopedia of Sustainable Science and Technology*; Meyers, R.A., Ed.; Springer: New York, NY, USA, 2012; Volume 1, pp. 9699–9729.

5 Mueller, R.; Träger-Chatterjee, C. Brief Accuracy Assessment of Aerosol Climatologies for the Retrieval of Solar Surface Radiation. *Atmosphere* **2014**, *1*, 9699–9729.

6 Kinne, S.; O'Donnel, D.; Stier, P.; Kloster, S.; Zhang, K.; Schmidt, H.; Rast, S.; Giorgetta, M.; Eck, T.F.; Stevens, B.; *et al*. MAC-v1: A new global aerosol climatology for climate studies. *J. Adv. Model. Earth Syst.* **2013**, *5*, 707–740.

7 Hess, M.; Koepke, P.; Schult, I. Optical properties of aerosols and clouds: The software package OPAC. *Bull. Am. Meteorol. Soc.* **1998**, *79*, 831–844.

8 Koepke, P.; Hess, M.; Schult, I.; Shettle, E. *Global Aerosol Data Set*; Technical Report, MPI Meteorologie Hamburg Report No. 243; MPI Meteorologie: Hamburg, Germany, 1997.

9 Cano, D.; Monget, J.; Albuisson, M.; Guillard, H.; Regas, N.; Wald, L. A method for the determination of the global solar radiation from meteorological satellite data. *Sol. Energy* **1986**, *37*, 31–39.

10 Posselt, R.; Mueller, R.; Stöckli, R.; Trentmann, J. Spatial and temporal homogeneity of solar surface irradiance across satellite generations. *Remote Sens.* **2011**, *3*, 1029–1046.

11 Fontoynont, M.; Dumortier, D.; Heinemann, D.; Hammer, A.; Olseth, J.; Skartveit, A.; Ineichen, P.; Reise, C.; Page, J.; Roche, L.; *et al.* Satellight: An European Programme Dedicated to Serving Daylight Data Computed from Meteosat Images. In Proceedings of the Lux Europa Conference, Amsterdam, The Netherlands, May 1997; Available online: http://www.satellight.com/indexgT.htm (accessed on 16 June 2015).

12 Wald, L.; Albuisson, M.; Best, C.; Delamare, C.; Dumortier, D.; Gaboardi, E.; Hammer, A.; Heinemann, D.; Kift, R.; Kunz, S.; *et al.* SoDa: A project for the integration and exploitation of networked solar radiation databases. In*Environmental Communication in the Information Society*; Part 2; Pillmann, W., Tochtermann, K., Eds.;

International Society for Environmental Protection: Vienna, Austria, 2002; pp. 713–720.

13 Huld, T.; Müller, R.; Gambardella, A. A new solar radiation database for estimating PV performance in Europe and Africa. *Sol. Energy* **2012**, *86*, 1803–1815.

14 Geiger, B.; Meurey, C.; Lajas, D.; Franchisteguy, L.; Carrer, D.; Roujean, J. Near real-time provision of downwelling shortwave radiation estimates derived from satellite observations. *Meteorol. Appl.* **2008**, *15*, 411–420.

15 Ineichen, P.; Barroso, C.; Geiger, B.; Hollmann, R.; Mueller, R. Satellite Application Facilities irradiance products: Hourly time step comparison and validation. *Int. J. Remote Sens.* **2009**, *30*, 5549–5571.

16 Mueller, R.; Behrendt, T.; Hammer, A.; Kemper, A. A New Algorithm for the Satellite-Based Retrieval of Solar Surface Irradiance in Spectral Bands. *Remote Sens.* **2012**, *4*, 622–647.

17 Mayer, B.; Kylling, A. Technical note: The libRadtran software package for radiative transfer calculationsdescription and examples of use. *Atmos. Chem. Phys.* **2005**, *5*, 1855–1877.

18 Acker, J.; Shen, S.; Leptoukh, G.; Serafino, G.; Feldman, G.C.; McClain, C.R. SeaWiFS ocean-color data archive and distribution system: Assessment of system performance. *IEEE Trans. Geosci. Remote Sens.* **2002**, *40*, 90–103.

19 Justice, C.O.; Giglio, L.; Korontzi, S.; Owens, J.; Morisette, J.; Roy, D.; Descloitres, J.; Alleaume, S.; Petitcolin, F.; Kaufman, Y.J.; *et al.* The MODIS fire products. *Remote Sens. Environ.* **2002**, *83*, 244–262.

20 Holben, B.N.; Eck, T.F.; Slutsker, I.; Tanré, D.; Buis, J.P.; Setzer, A.; Vermote, E.; Reagan, J.A.; Kaufman, Y.J.; Nakajima, T.; *et al.* AERONET—A Federated Instrument Network and Data Archive for Aerosol Characterization.*Remote Sens. Environ.* **1998**, *66*, 1–16.

21 SeaWIFS Ocean Color Image. Available online: http:// oceancolor. gsfc.nasa.gov/SeaWiFS/HTML/dust.html (accessed on 16 June 2015).

22 Morcrette, J.J.; Boucher, O.; Jones, L.; Salmond, D.; Bechtold, P.; Beljaars, A.; Benedetti, A.; Bonet, A.; Kaiser, J.W.; Razinger, M.; *et al.* Aerosol analysis and forecast in the European Centre for Medium-Range Weather Forecasts Integrated Forecast System: Forward modeling. *J. Geophys. Res.* **2009**, *114*.

23 Benedetti, A.; Morcrette, J.J.; Boucher, O.; Dethof, A.; Engelen, R.; Fisher, M.; Flentje, H.; Huneeus, N.; Jones, L.; Kaiser, J.; *et al.*

Aerosol analysis and forecast in the European Centre for Medium-Range Weather Forecasts Integrated Forecast System: 2. Data assimilation. *J. Geophys. Res.* **2009**, *114*.

24 Kinne, S.; Schulz, M.; Textor, C.; Guibert, S.; Balkanski, Y.; Bauer, S.E.; Berntsen, T.; Berglen, T.F.; Boucher, O.; Chin, M.; *et al.* An AeroCom initial assessment—Optical properties in aerosol component modules of global models. *Atmos. Chem. Phys.* **2006**, *6*, 1815–1834.

25 Ohmura, A.; Dutton, E.; Forgan, B.; Fröhlich, C.; Hegner, H.; Heimo, A.; König-Langlo, G.; Mc Arthur, B.; Müller, G.; Philipopon, R.; *et al.* Baseline Surface Radiation Network (BSRN/WCRP): New precision radiometry for climate research. *Bull. Am. Meteorol. Soc.* **1998**, *79*, 2115–2136.

26 Image Courtesy Jeff Schmaltz, MODIS Rapid Response Team, NASA GSFC. Available online: http:// earthobservatory.nasa. gov/ NaturalHazards/view.php?id=11417 (accessed on 16 June 2015).

27 Karlsson, K.G.; Riihelä, A.; Müller, R.; Meirink, J.F.; Sedlar, J.; Stengel, M.; Lockhoff, M.; Trentmann, J.; Kaspar, F.; Hollmann, R.; *et al.* CLARA-A1: A cloud, albedo, and radiation dataset from 28 year of global AVHRR data. *Atmos. Chem. Phys.* **2013**, *13*, 5351–5367.

CHAPTER 2

Aerosol Characteristics over the Indo-Gangetic Basin: Implications to Regional Climate

A.K. Srivastava[1], Sagnik Dey[2] and S.N. Tripathi[3]

[1] Indian Institute of Tropical Meteorology (Branch), Prof. Ramnath Vij Marg, New Delhi, , India

[2] Centre for Atmospheric Sciences, Indian Institute of Technology Delhi, New Delhi, , India

[3] Department of Civil Engineering and Centre for Environmental Science and Engineering, Indian Institute of Technology Kanpur, Kanpur, India

1. INTRODUCTION

The climatic and environmental effects of atmospheric aerosols are the critical issues in global science community because aerosols, derived from variety of natural and man-made (or anthropogenic) emission sources, are well known to affect the air quality, human health and radiation budget [1]. While comparing the third and fourth assessment report of the

Intergovernmental Panel on Climate Change (IPCC) as shown in Figure 1, the level of scientific understanding for the role of green house gases (GHGs) in projected temperature changes is higher relative to that of aerosols [2,3]. This is because of inadequate measurements of aerosols, their microphysical and optical properties and poor understanding of their role in the Earth's radiation budget.

Aerosols influence the Earth's climate directly by scattering and absorbing the solar and terrestrial radiations and indirectly by modifying the cloud macro- and micro-physical properties [4]. The direct and indirect effects of atmospheric aerosols are shown in the schematics in Figure 2a and 2b, respectively. Varity of aerosols present in the atmosphere from natural and anthropogenic emission sources can influence our Earth's atmosphere directly by absorbing/scattering the incoming solar radiations (Figure 2a). It can also absorb and re-radiated the outgoing radiations emitted from the Earth. On the other hand, aerosols indirectly affect the climate system by acting as cloud condensation nuclei (CCN) and ice nuclei (IN) and thereby modify the cloud properties and their impacts depending upon the environment like polluted or un-polluted regions (Figure 2b). In a recent study, reported in [5], they have investigated the indirect aerosol effect during the successive contrasting monsoon seasons over Indian subcontinent. However, a different study reported in [6] was carried out to investigate the intensity and the spatial extent of the indirect effect over the Indo-Gangetic Basin (IGB) region.

The uncertainty in quantifying these impacts have no doubt improved over the years due to assimilation of observations (especially after global observations of aerosols by EOS-Terra started in 1999), but not up to the desired level, particularly at regional scale [3]. The Indian subcontinent is one such region, where heterogeneity in aerosol optical and microphysical properties over a wide range of spatial and temporal scales continues to hinder in improving the estimates of aerosol-induced climate forcing. Thus, it is important to improve aerosol characterization with high spatio-temporal resolutions; particularly over the IGB region, which supports nearly 70% of the country's population and is one of the highly polluted regions in the world. The problem is also critical due to lack of adequate long-term measurements of aerosol properties and large uncertainty in emission factors, leading to poor representation of aerosol distribution by General Circular Model (GCM) [7].

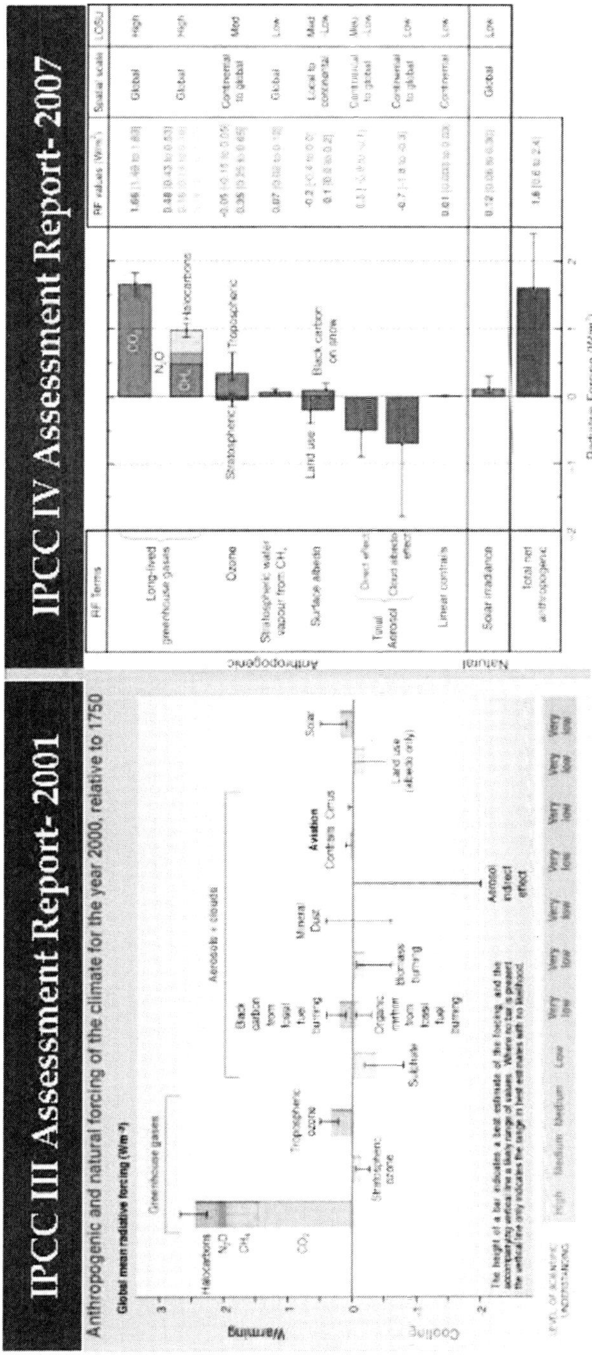

Figure 1. A comparison between III and IV assessment reports of IPCC for 2001 and 2007 (*Adopted from [2,3]*).

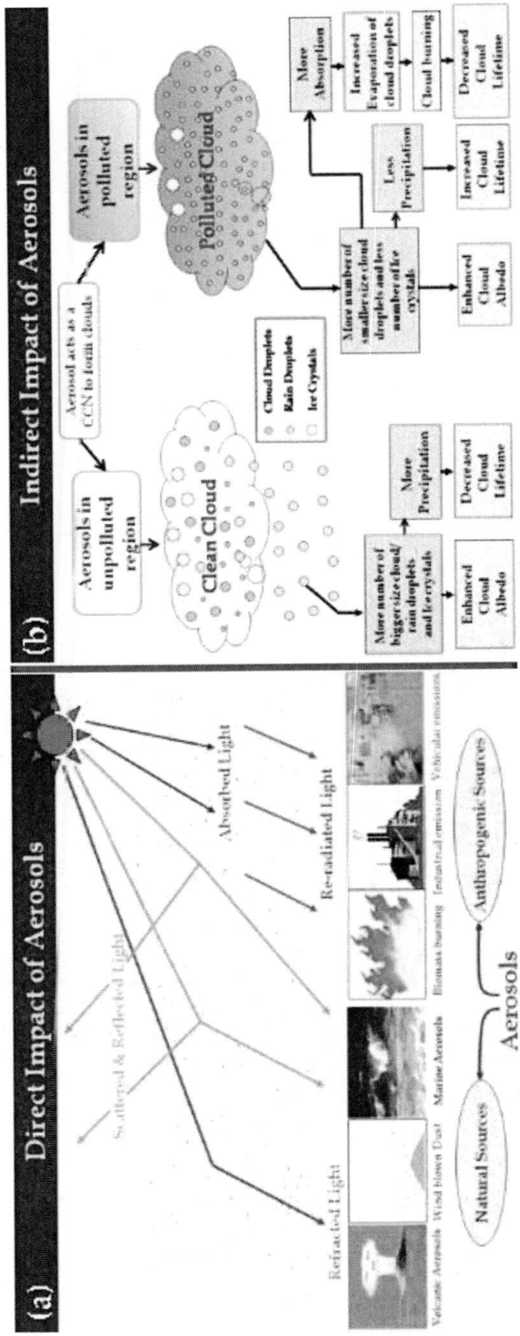

Figure 2. Schematics showing (a) Direct and (b) Indirect impacts of atmospheric aerosols.

Although, aerosol properties have been measured at many sites in India in continuous and campaign modes (the *in-situ* observations have summarized in [8]) in the last two decades, only few of them have fairly long-term data of aerosol microphysical properties [9-11]. Satellites are proved to be a good tool to understand the broad spatio-temporal characteristics of aerosols and associated effects from global to local scales [12-15]; but they are unable to provide an in-depth view of aerosol properties on a local scale and pose higher uncertainties as compared to the ground-based instruments [16]. In the IGB, aerosols of natural and anthropogenic origins mix with each other during dust loading season [17,18]. As a result, aerosol properties change, leading to even larger uncertainty in satellite retrievals [14,19] as this is not considered in the aerosol retrieval algorithm [20]. National Aeronautics and Space Agency (NASA) has setup ground-based aerosol monitoring network under the Aerosol Robotic Network (AERONET) program [21], in which automatic sun/sky radiometers are deployed at various places around the world. As per India, particularly in the northern part, is concerned, the routine measurements of aerosols under this network were started initially by the deployment of the sun/sky radiometer at Kanpur over the IGB region in year 2001 [22]. At a later stage, it was deployed at other places in the IGB, considering the region as crucial for aerosol measurements where significant aerosol loads of pollution mostly from the combustion of biomass, bio-fuel/fossil fuel emissions and the transported mineral dust have led to one of the largest regional TOA energy losses worldwide (For Example see [11,23-25]). The measurements by individual instruments can particularly be useful for the quantification of the regional impact of aerosols on the radiative energy balance. Provided a parallel individuation of aerosol types is possible, these local studies can also help to reduce the uncertainties on the effect of individual aerosol species, which is necessary because the direct radiative forcing by individual aerosol species is less certain than the total direct radiative forcing by all aerosols [3]. The complex mixture of aerosols over the IGB has been evaluated in the literature over the last decade starting with Indian Ocean Experiment (INDOEX) [26], and the research has been continued with the AERONET data [27 and references therein].

Understanding and quantifying the aerosol effects are important in the IGB region due to several pathways have been hypothesized to explain the possible impacts of aerosols on the regional hydrological cycle. The region is of great research interest due to its unique topography surrounded by the Himalayas to the north, moderate hills to the south, Thar Desert and Arabian Sea in the west, and Bay of Bengal to the east. The IGB region is dominated by the urban/industrial aerosols [28-30], which demonstrate significant seasonal variability based on the complex mixture of these aerosols with the naturally produced aerosols, particularly during the pre-monsoon and monsoon seasons. In addition to the urban-industrial pollution, desert dust is one of the other major natural sources of the aerosols over the Ganga basin

[24,27,31-36], transported frequently from the neighboring desert regions, mostly during pre-monsoon periods. Dust storms are often experienced in northern and northwestern part of India, including over different parts of the IGB region during the pre-monsoon season, when dust aerosols are transported by southwesterly summer winds from the western Thar Desert [37,38]. High dust loading over the IGB region during the pre-monsoon period has been established by remote sensing data [39,40]. These dust storms apparently deposited silty materials in the downwind directions, as observed on the quartzite ridges in the Delhi area [41]. The wind also carries heavy metals to the IGB during the summer season [42] along with the dusts, causing severe air pollution and degradation in the visibility. On contrary, the spatial distribution of aerosols (in terms of AOD) during the winter season also revealed high aerosol loading over the IGB and its outflow to the northern Bay of Bengal due to high anthropogenic emission sources, which was observed by satellite [8,12,13,15] and ground-based measurements [22,43-46].

2. FACTORS AFFECTING AEROSOL CHARACTERISTICS OVER IGB

Complex nature of aerosols over the IGB is mainly because the region is very diverse in topography, population distribution, meteorology and emission sources. Figure 3 shows unique topography of the IGB region, surrounded by the variety of aerosol emission sources and thus making it hotspot for aerosol research.

2.1. Topography

The IGB region, world's most populated river basin having more than 700 million populations, stretches from Pakistan in the west to Bangladesh in the east, encompassing most of the northern part of India. The region is bounded by the Himalayas to the north, and by Vindhyan and Satpura range of mountains in the south. Due to its unique topography, this region can be summarized as a type of region, where, both anthropogenic and natural, aerosols show distinct seasonal characteristics and mixing [13,17,22,25,28,30,47]. General seasonal abundance shows that the winter months are dominated by the fine-mode aerosols, produced by various anthropogenic sources from the IGB region, and pre-monsoon or summer months are dominated by the coarse-mode mineral dust, primarily from the Thar Desert region in the western Rajasthan and its frequent transportation over the IGB region. Further details regarding geography, climate, regional sources and emissions of these aerosols over the IGB as well as over the other Indian region, however, can be found in [15]. This region also provides favorable climate for the agricultural activities due to its fertile soils and abundant water supply from the southwest monsoon and the rivers originating from the Himalayan glaciers such as the Ganges. Consequently, the cultivable land forms a major fraction of the total geographical area in the IGP region (~76%) as compared to the rest of India (~50%) (http://dacnet.nic.in/).

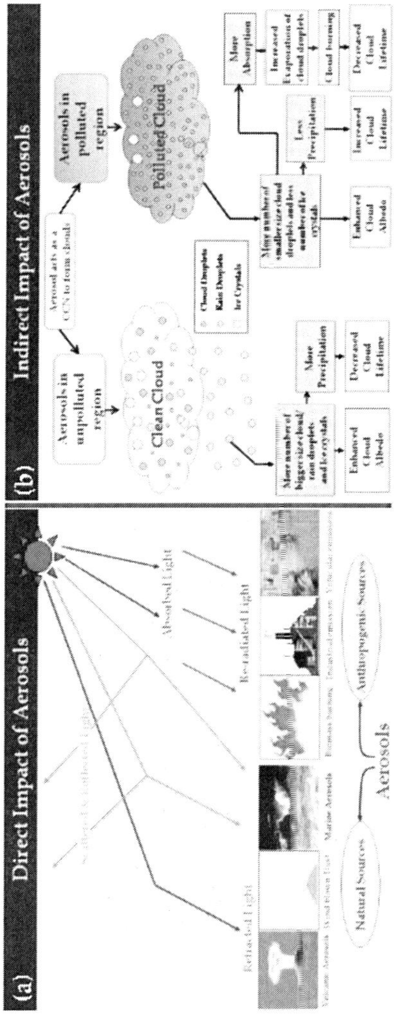

Figure 3. IGB region, showing unique topography surrounded by the variety of aerosol emission sources (Adopted after modification from personal presentation of William K. M. Lau on Aerosol, Monsoon Rainfall Variability and Climate Change).

2.2. Synoptic Conditions And Aerosol Characteristics Over Igb

Synoptic meteorology (e.g. wind pattern, air temperature and specific humidity) over the IGB region along with its surroundings is shown in Figure 4 for (a) winter and (b) summer seasons for the period of 2007-2008. The European Centre for Medium-Range Weather Forecasts (ECMWF) reanalysis monthly data of weather parameters such as wind, air temperature and specific humidity at 850 hPa pressure level were used to study the synoptic meteorological conditions over the region. In both the figures, winds are shown with arrows pointing towards the wind direction, where length of arrows defines the magnitude of wind speed (in ms^{-1}), line contour represents air temperature (in °C) and shaded color contour represents specific humidity (in kg kg^{-1}), showing in dark blue color for low and red color for high magnitude of specific humidity. Results reveal that the IGB region during the winter period is relatively drier than during the summer. The persistence of low temperature and the westerlies (with low intensity) can be seen over the region during the winter whereas during the summer, relatively high temperature with intense southwesterly winds was observed to dominate. These winds are found to pass through arid regions of the western India (particularly from the Thar Desert) and bring dry air masses over the region [24,27,48].

The general aerosol characteristics over the entire IGB region are shown in Figure 5 as mean AOD values at 550 nm for (a) winter and (b) summer seasons for the period of 2007-2008 in color codes, obtained from the Moderate Resolution Imaging Spectroradiometer (MODIS). Large spatial heterogeneity in AOD can readily be noticed over the IGB region during both winter and summer periods, which has also been confirmed through various ground-based measurements, discussed in the later section of this chapter. Relatively large magnitude of AOD was observed throughout the IGB region during the summer, which is mainly due to frequent occurrence of dust storms over the Thar Desert region that caused large amount of dust particles to be transported over the station (showing the highest AOD). However, large AOD during the winter is confined mostly over the eastern part of IGB. During the winter, the IGB region is often enveloped by thick fog and haze [49]. The prevailing winds over the region are westerly to northwesterly with relatively low wind speeds (<5 ms−1) as compared to the summer (as can be seen in Figure 4a) and the eastern parts of the IGB are impacted by a localized area of strong subsidence in winter [8,12,13,50]. These conditions tend to trap the pollution at low altitudes and responsible for the higher AOD along the eastern part of IGB. Results obtained, although, include the impacts of aerosol emissions from various natural and anthropogenic sources and the prevailing meteorology over the region, it also encourages to further investigate the plausible causes and impacts over the radiation budget as well as on weather and climate.

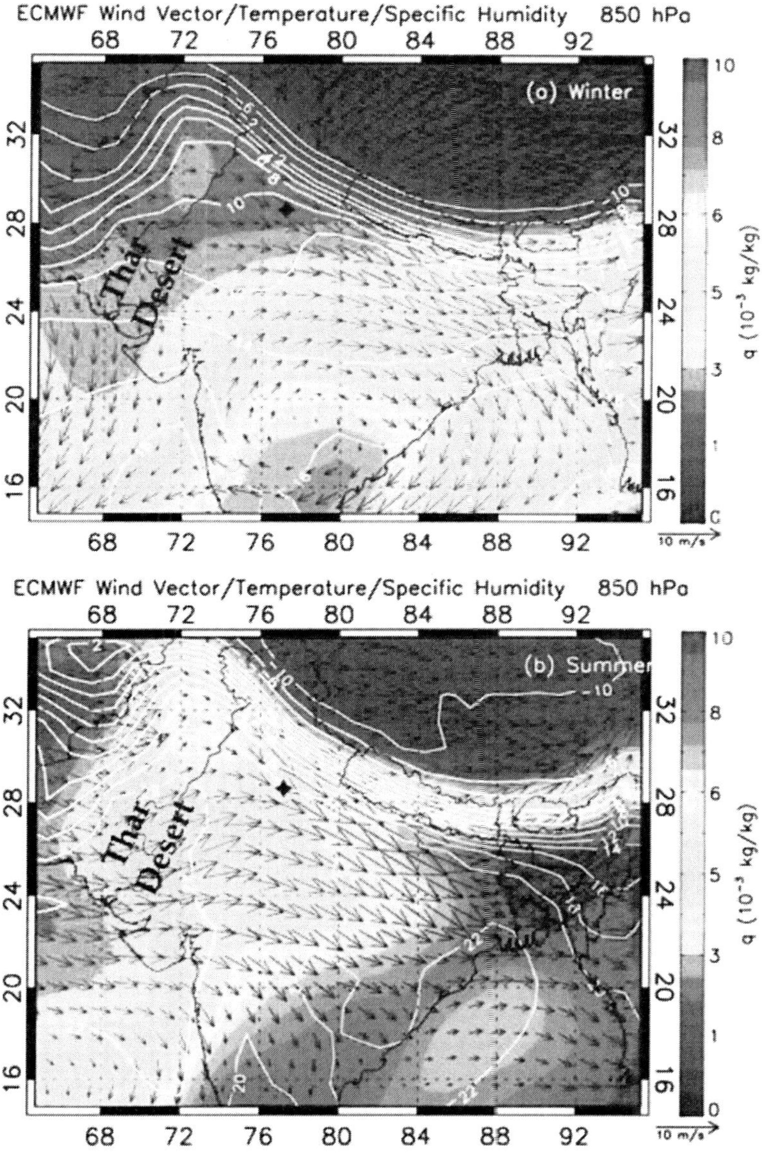

Figure 4. Synoptic meteorological conditions over the entire IGB region derived from ECMWF at 850 hPa pressure level during (a) winter and (b) summer periods (*Adopted from [46]*).

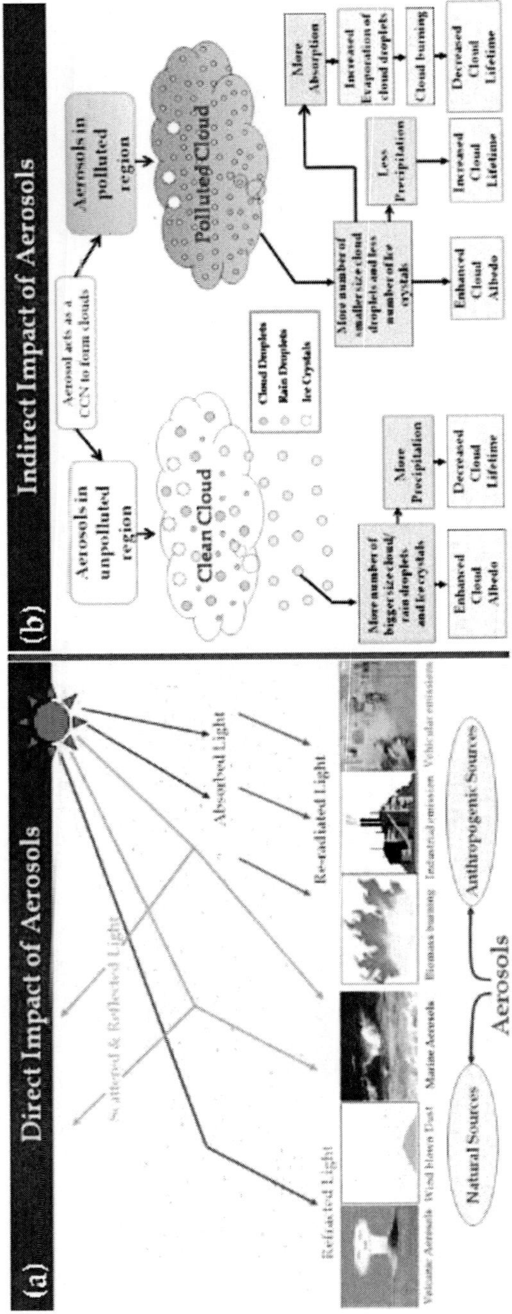

Figure 5. General aerosol characteristics in terms of AOD (550 nm) over the entire IGB region derived from MODIS during (a) winter and (b) summer periods (*Adopted from [46]*).

2.3. Emission Sources

The IGB region, apart from being a major source region for aerosols, is bordered by densely populated and industrialized areas on the west and eastern sides from where different aerosol species such as mineral dust, soot, nitrate, sulfate particles and organics are produced and transported to this region and thus making it an aerosol hotspot, as can also be seen in Figure 3. The region itself has both, rural and urban population and various kinds of emission sources such as natural and industrial. In rural areas, bio-fuels burning such as wood, dung cake and crop waste, predominantly contribute to the major aerosols loading [51]. However, in urban areas, aerosol emissions from fossil fuels burning such as coal, petrol and diesel oil dominate [52,53]. Large fluxes of absorbing aerosol emissions (black carbon and inorganic oxidized matter, which is mostly fly ash from coal-based power plants and particles from open burning of crop waste/forest-fires) were reported over the IGB [51]. Apart from the dust emissions from the Thar Desert, predominantly during the pre-monsoon months, the influence of emissions from the forest-fires and open burning of crop waste from the central India were also found over IGB during these months as biomass aerosol contribution [15,27].

3. MEASUREMENTS

3.1. Ground-Based

Aerosol measurements in the Indian sub-continent started as early as the 1960s, when [54] studied Angstrom turbidity from solar radiance measurements. Later, a multi-wavelength radiometer was developed by the Indian Space Research Organization (ISRO) to monitor spectral AOD at Trivandrum in the year 1985 [55] and in the same year, aerosol vertical distribution measurement by ground-based lidar was initiated at Pune [56]. Further, NASA has setup ground-based aerosol monitoring network in India under the Aerosol Robotic Network (AERONET) program [21], in which automatic sun/sky radiometers are deployed at various places, particularly in the northern part of India, including the Himalayan foothills. The routine measurements of aerosols under this network were started initially by the deployment of the automatic sun/sky radiometer at Kanpur over the IGB region in year 2001 [22]. At a later stage, it was deployed at other places in the IGB, considering the region as crucial for aerosol measurements [36].

Using ground-based radiometric measurements, [22] have reported for the first time the seasonal characteristics of aerosol optical properties and the spectral behavior of AODs over Kanpur, an urban-industrial city, situated in the central part of the IGB. They showed pronounced seasonal influence of various aerosol properties, with maximum dust loading during the pre-monsoon season. The increase of pollution has a direct impact on climatic conditions, especially the increase of haze, fog, and cloudy conditions, which decrease the visibility especially during the winter season. On the other hand, *in-situ*aerosol measurements by Central Pollution Control Board (CPCB) have also showed very high annual average concentrations (>210 μg/m^3, in the critical range compared to the air quality standard in India) of particulate matter of diameter less than 10 μm (PM_{10}) in the atmosphere of the major cities of the Ganga basin like Delhi, Kanpur and Kolkata (http://www.cpcb.nic.in). These high PM_{10}concentrations provide an opportunity for SO_4 formation on the particulate surface, leading to very high concentration of sulfate aerosols in the atmosphere, which is the case observed over the IGB and reported in [29]. Several studies indicate strong seasonal variability in aerosol loading and changes in aerosol properties over the IGB [14,22,25,31,33-36,48,57,58]. In the recent studies, [23] and [24] have demonstrated the distribution of aerosols and associated optical and radiative properties in the IGB region and its further expansion to the foothills of Himalayas during the pre-monsoon period. The pre-monsoon period is of particular interest because this is the key period when locally generated and regionally transported aerosol loading peaks over the IGB region and spread up to the foothills of Himalayas [23,35,59-62], which has been linked to influence the monsoon circulation in India [63,64]. Significant gradient in the magnitude of most of the aerosol characteristics was observed over the IGB, which may be due to the gradual changes in weather parameters and/or emission sources. Such gradient is, ultimately, found to impact the Earth-atmosphere system by negative radiative forcing, thus causing cooling, at the surface, and positive aerosol forcing, thus causing heating in the atmosphere for the study period. Such gradient in heating rate raises several climatic issues, and is needed to be answered on the basis of longer period investigations at several stations to improve the scientific understanding of the regional climate in inter-annual as well as intra-seasonal scale.

The first simultaneous measurements of chemical composition (carbonaceous and inorganic species) and optical properties (absorption coefficient and mass absorption efficiency) of ambient aerosols ($PM_{2.5}$ and PM_{10}) have been recently reported in [58] at an urban site (Kanpur) in the IGB region. The study provides important information on the temporal variability in the abundance of organic matter and mineral dust over the IGB region, which has large implications to the large temporal variability in the atmospheric radiative forcing due to these aerosols. Based on the measured aerosol chemical composition, other studies have been carried out to understand the characteristics of anthropogenic aerosols and their quantification to the total radiative forcing over the IGB region, which are limited only at Kanpur [11] and Delhi [25]. Figure 6 shows seasonal variability of optical properties of composite aerosols estimated over Delhi (a typical urban station at the western part of the IGB near to the Thar Desert) during the winter, summer and post-monsoon seasons; however, the same for anthropogenic aerosols are shown in Figure 7. The anthropogenic components were found to be contributing ~72% to the composite aerosol optical depth ($AOD_{0.5}$ ~0.84) at Delhi. The contribution was found to be more during the winter (~84%) and post-monsoon (~78%) periods and less during the summer (~58%). On the other hand, mean SSA for composite aerosols was found to be ~0.70 (ranging from 0.63 to 0.79). However, SSA for anthropogenic aerosols was found to be slightly less (by ~1%) than that for composite aerosols, and the difference may be due to the mixing of natural dusts with anthropogenic aerosols in the region (for composite aerosols).

The resultant atmospheric forcing due to composite and anthropogenic aerosols at Delhi is shown inFigure 8. The anthropogenic contributions to the composite aerosol were found to be ~93%, 54%, and 88%, respectively during the winter, summer, and post-monsoon seasons (with a mean contribution of ~75%). However, the anthropogenic fraction of ~73% is responsible for the composite aerosol atmospheric heating rate (2.42±0.72 $Kday^{-1}$) at Delhi. On the other hand at Kanpur, another typical urban station at central part of the IGB region, the heating rate due to anthropogenic aerosols was reported to be ~65% to the heating due to composite aerosols [11]. Relatively higher heating rate at Delhi may be caused by the large contribution of transported mineral dust aerosols due to the proximity of the station to the Thar Desert region as compare to Kanpur and their probable mixing with the other absorbing aerosol species like black carbon (BC) [27].

Large atmospheric heating rate of the order of more than 2 $Kday^{-1}$ is quite significant. Moreover, the large surface cooling due to negative forcing at the surface and strong heating due to positive forcing in the atmosphere, particularly for the anthropogenic aerosols, can strongly affect the atmospheric dynamics over the region. The warmer atmosphere close to the surface (due to high atmospheric absorption) and the colder surface during winter and post-monsoon periods over Delhi would create low-level inversions and strengthen the boundary layer stability [11], which restrict the mixing and dispersion of aerosols into the atmosphere. On the other hand, during summer, the observed large heating in the atmosphere, which is probably due to the mixing of anthropogenic aerosols with abundance of natural dusts, may supply excess energy to be trapped in the atmosphere during dry season and can have significant impact on regional climate and monsoon circulation systems [65,66]. Anomalous atmospheric heating due to absorbing aerosols (dust and BC) over the northern part of India during pre-monsoon season has been reported in [63]. A comparative study of aerosol direct radiative forcing was made with the available estimates from the literatures at various regions and given in Table 1. Various regions, characterized by different kinds of aerosol sources and prevailing meteorological conditions, are associated with different values of aerosol forcing and have provided an understanding of the aerosol radiative effect on regional scales, which are significantly different from the global mean radiative effect (indicating slightly cooling of the atmosphere).

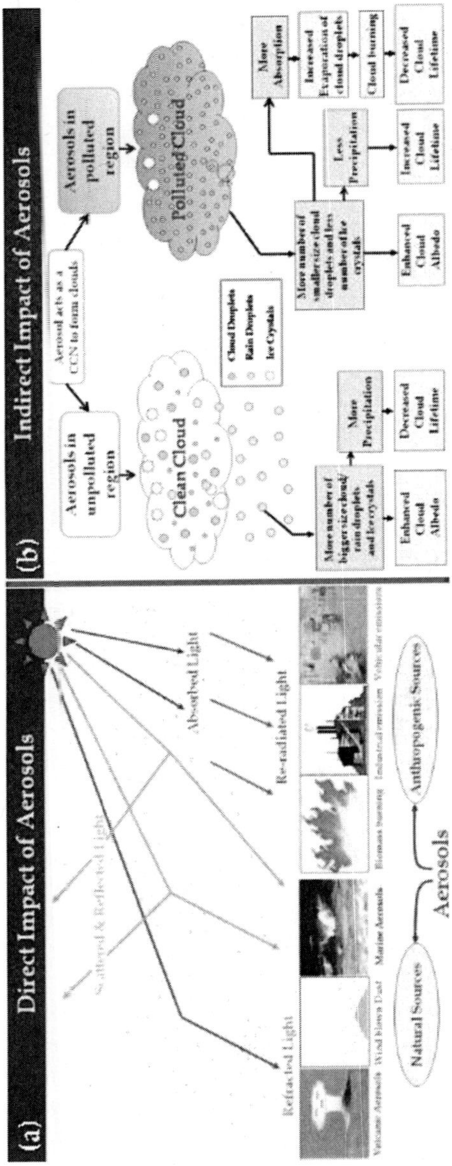

Figure 6. Seasonal mean spectral variation of (a) AOD and (b) SSA for composite aerosols over Delhi (*Adopted from [25]*).

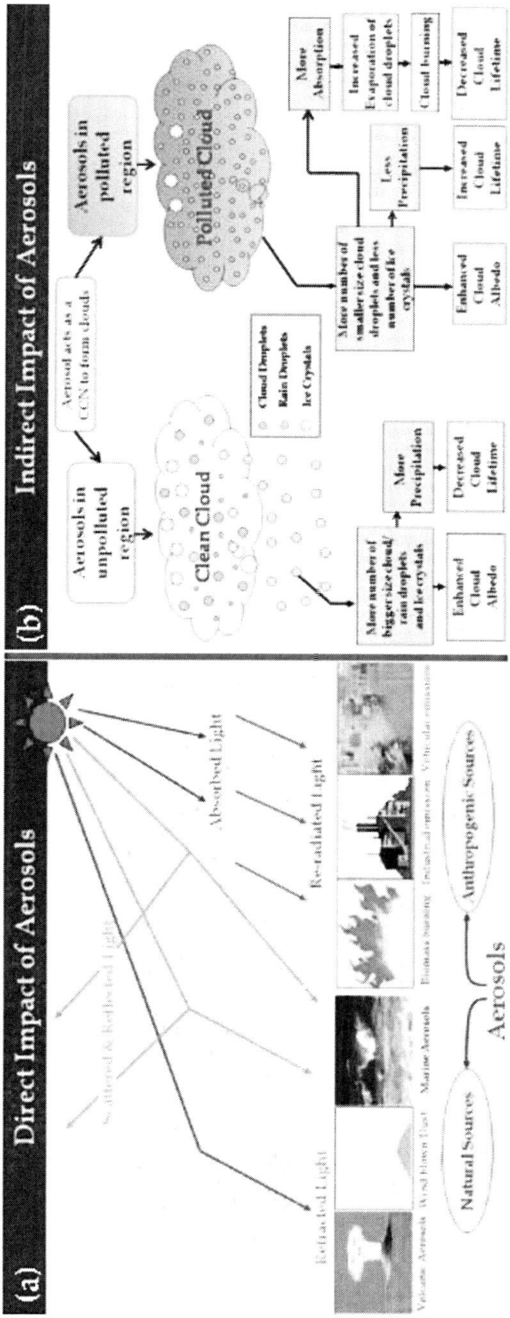

Figure 7. Same as figure 6, except for anthropogenic aerosols (*Adopted from [25]*).

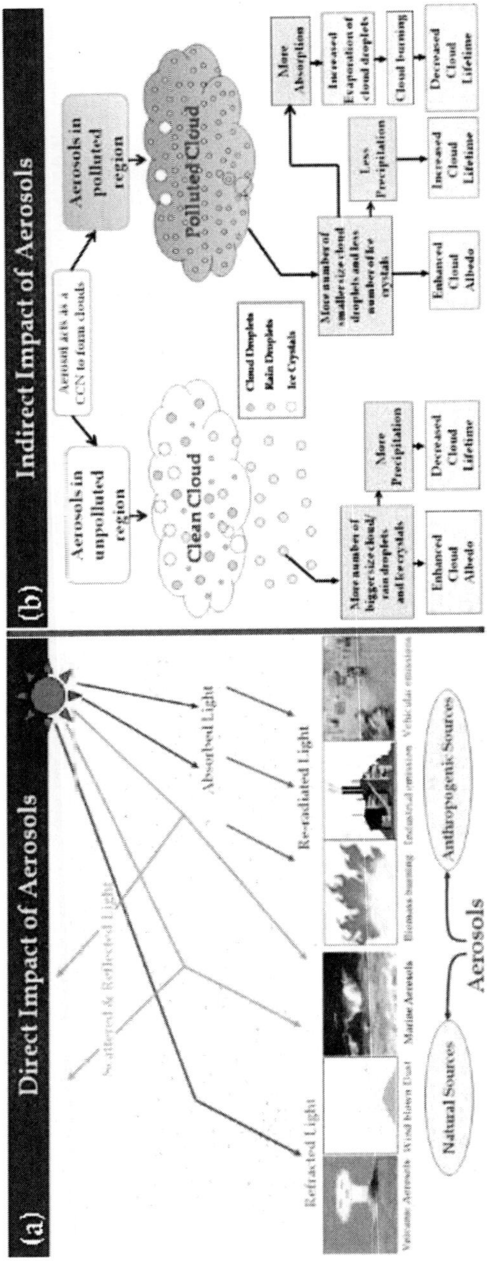

Figure 8. Monthly variation of atmospheric forcing for composite and anthropogenic aerosols over Delhi. The corresponding heating rate values for respective aerosols are given in the parenthesis (*Adopted from [25]*).

Table 1. Aerosol direct radiative forcing (DRF) at different locations in India.

Location	Type of Location	Period	Aerosol DRF (Wm^{-2}) at		References
			Surface	Atmosphere	
New Delhi	Urban	Annual	-67	71	[57]
Kanpur	Urban	Annual	-32	28	[11]
Ahmedabad	Urban	Annual	-49	44	[131]
New Delhi	Urban	Jan-Nov 2007	-79	87	[25]
New Delhi	Urban	Mar-Jun 2006	-77	80	[48]
Kanpur	Urban	Apr-Jun 2009	-26.1 to -29.2	19.5 to 16.1	[24]
Gandhi College	Rural	Apr-Jun 2009	-29.7 to -31.9	20.9 to 16.6	[24]
New Delhi	Urban	Winter (Dec 2004)	-66	67	[132]
Hissar	Urban	Winter (Dec 2004)	-[illegible]0 (before fog) -20 to -25 (during fog)	15 (before fog) 25 to 40 (during fog)	[128]
Pune	Urban	Nov-Apr 2001 and 2002	-33	33	[133]
Hyderabad	Urban	Jan-May 2003	-33	42	[134]
Bangalore	Urban	Oct-Dec 2001	-23	28	[135]
Central India (multiple stations)	Urban and Rural	Winter (Feb 2004)	-15 to -40	16 to 29	[72]
Nainital	Rural (high-altitude)	Winter (Dec 2004)	-4.2	0.7	[77]
Nainital	Rural (high-altitude)	July 2006-May 2007	-14	14	[62]
Kathmandu	Urban (high-altitude)	Winter 2003	-25	25	[136]
Chennai	Urban	Feb-Mar	-19	13	[137]
Arabian Sea	Polluted Marine	Mar-Apr	-27	15	[70]
Bay of Bengal	Polluted Marine	Mar	-27	23	[138]
Indian Ocean	Polluted Urban	Feb-Mar	-29	19	[139]
Global mean	Natural and Anthropogenic			-0.5±0.4	[3]

Due to high level of anthropogenic emissions, aerosol distribution in terms of type and loading undergo strong variability associated with the episodic yet strong influence of dust transport and biomass burning during the pre-monsoon period [23]. Dust was found to be one of the major components of aerosol composition (apart from other species) over the region [32], which significantly affects the region during pre-monsoon period due to enhanced surface convection activities [24,25,31,33-36,48,67], and thus essential to quantify its contribution over the region. Aerosol composition was measured with the chemical analysis method over IGB during different periods of time as reported in [44,52,53]. Retrieval of columnar black carbon and organic carbon has been carried out over IGB using the AERONET data [60,68]. Moreover, [7] have integrated AERONET and Cloud-Aerosol Lidar with Orthogonal Polarization (CALIOP) data into atmospheric GCM to infer aerosol types at two AERONET sites in the IGB. However, in a recent study, [27] have discriminated the major aerosol types over the IGB region during pre-monsoon period using multi-year AERONET measured aerosol products associated with the size of aerosols (mainly fine mode fraction, FMF) and radiation absorptivity (mainly single scattering albedo, SSA). Figure 9a shows density plot of SSA versus FMF at Kanpur (KNP, a typical urban AERONET site over the central IGB region) and Gandhi College (GC, a typical rural AERONET site over the central IGB region) for different aerosol types. High dust enriched aerosols (i.e. polluted dust, PD) were found to contribute more over the central IGB station at Kanpur (~62%) as compared to the eastern IGB station at Gandhi College (~31%) whereas vice-versa was observed for polluted continental (PC) aerosols, which contain high anthropogenic and less dust aerosols. Contributions of carbonaceous particles having high absorbing (mostly black carbon, MBC) and low absorbing (mostly organic carbon, MOC) aerosols were found to be 11% and 10%, respectively at Gandhi College, which was ~46% and 62% higher than the observed contributions at Kanpur; however, very less contribution of non-absorbing (NA) aerosols was observed only at Gandhi College (2%). The mean SSA and FMF based on cluster analysis of daily-averaged data at Kanpur and Gandhi College, associated with the different aerosol categories is also shown inFigure 9b. The horizontal and vertical lines indicate the standard deviations of SSA and FMF from their respective means, indicating the variability of these parameters for different aerosol types. Although similar magnitude of SSA was observed for PD, PC and MBC type aerosols, they are further distinguished based on FMF thresholds following Lee et al. (2010), i.e. FMF<0.4 indicates dominantly coarse-mode and hence is assigned to PD aerosols, FMF>0.6 indicates dominantly fine-mode and hence is assigned to MBC aerosols, and PC aerosols are considered for 0.4FMF0.6. MOC and NA type aerosols have similar FMF, but higher scattering relative to the other aerosol types.

Spectral information of SSA for each aerosol type was also shown and discussed in [27], which clearly discriminates the dominance of natural dust (SSA increases with increasing wavelength) with anthropogenic aerosols (SSA decreases with increasing wavelength) at Kanpur and Gandhi College over the IGB. As expected, SSA for PD and PC aerosols was found to have spectrally increased, suggesting relative importance of dust. PD has higher spectral trend relative to PC due to larger fraction of dust, which was found to be dominated at Kanpur as compared to Gandhi College. On the contrary to PD and PC type aerosols at Gandhi College, SSA for NA aerosols was found to have spectrally decreased with relatively larger magnitude at all the wavelengths. However, relatively less spectral dependence in SSA was seen for MBC and MOC aerosols, which shows slight decrease in spectral SSA at Gandhi College and opposite at Kanpur.

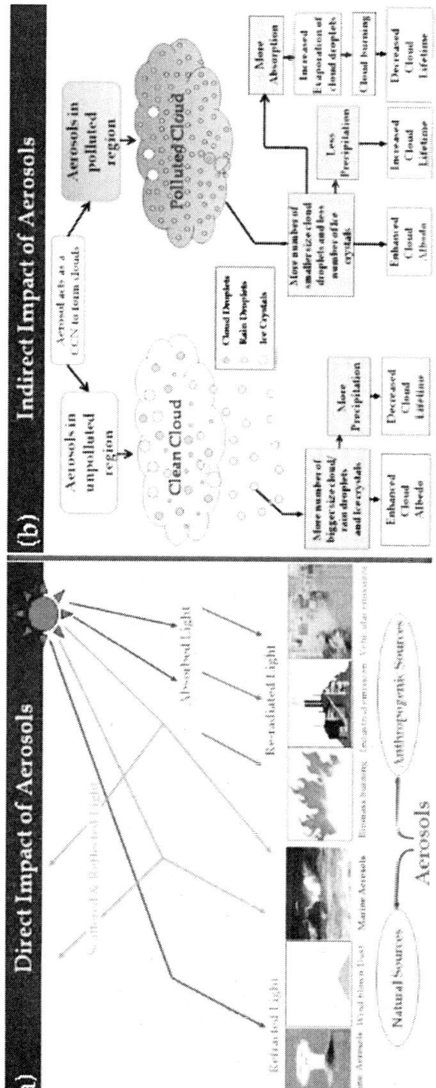

Figure 9. a) Density plot and corresponding (b) cluster plot of AERONET-derived SSA vs. FMF for two stations over the IGB region (showing different aerosol types) during pre-monsoon period (*Adopted from [27]*).

Further, the absorption aerosol optical depth (AOD*abs*) at different wavelengths (λ) can be obtained, suggested in [69] as

$$AOD_{abs}(\lambda) = [1 - SSA(\lambda)] \times AOD(\lambda) \quad (1)$$

(1) The absorption Ångström exponent (AAE) has been computed as negative of the slope of fitted line of the natural logarithm of AOD*abs* vs. natural logarithm of the respective wavelengths and used to substantiate the inferred aerosol types over IGB, as shown in Figure 10. The magnitude of AAE near to 1.0 (marked by dotted line in Figure 10) represents a theoretical AAE value for black carbon as reported in [69]. AAE values for PD and PC aerosol types were found to be 1.70 and 1.43, respectively at Kanpur and 1.30 and 1.18, respectively at Gandhi College. However, for MBC and MOC type aerosols, AAE values were relatively higher at Kanpur (~20%) than at Gandhi College, where values were found to be closer to the theoretical AAE value for black carbon (i.e. AAE≈1.0), thus indicating the presence of fresh BC at Gandhi College, which can be expected from the potential source of combustion of fossil fuel and biomass burning used for domestic purposes. On the other hand, aged BC or mixed BC can be expected at Kanpur (mostly from biomass burning and urban/industrial sources), which is favorable scenario during the summer periods [17-36]. The estimated AAE values over the IGB thus suggest relative dominance of absorbing type aerosols over the central part of IGB (due to dominant dust mixed with other absorbing aerosols) as compared to the eastern part during pre-monsoon period.

Apart from these continuous measurements, various field campaigns have also been conducted regionally to study and improve the aerosol remote sensing measurements as well as provide data for atmospheric prediction over the past decade. Campaigns conducted in or near India, which used space-based, airborne, and surface-based instrumentations to observe high aerosol loading over the Indian subcontinent and the surrounding Oceanic regions, included the Indian Ocean Experiment (INDOEX) reported in [26], Arabian Sea Monsoon Experiment (ARMEX) reported in [70], Indian Space Research Organization Geosphere Biosphere Programme (ISRO-GBP) Land Campaign reported in [43-45]. The first phase of Land Campaign (LC-I) was conducted during February to March 2004, to understand the spatial distribution of aerosols and trace gases over central/peninsular India. The details of these campaigns and the major findings have been reported in literatures in [71-74]. As a continuation of this experiment, second phase of land campaign (LC-II) was conducted during December 2004, to characterize the regional aerosol properties and trace gases across the entire IGB region including the Himalayan foothills. This phase of the campaign provided a comprehensive database on the optical, microphysical and chemical properties of aerosols over the IGB and the foothills of Himalayas and reported in [44,45,75-80]. All these studies showed the persistence of high aerosol loading (in terms of high AOD) and black carbon mass concentrations over the region.

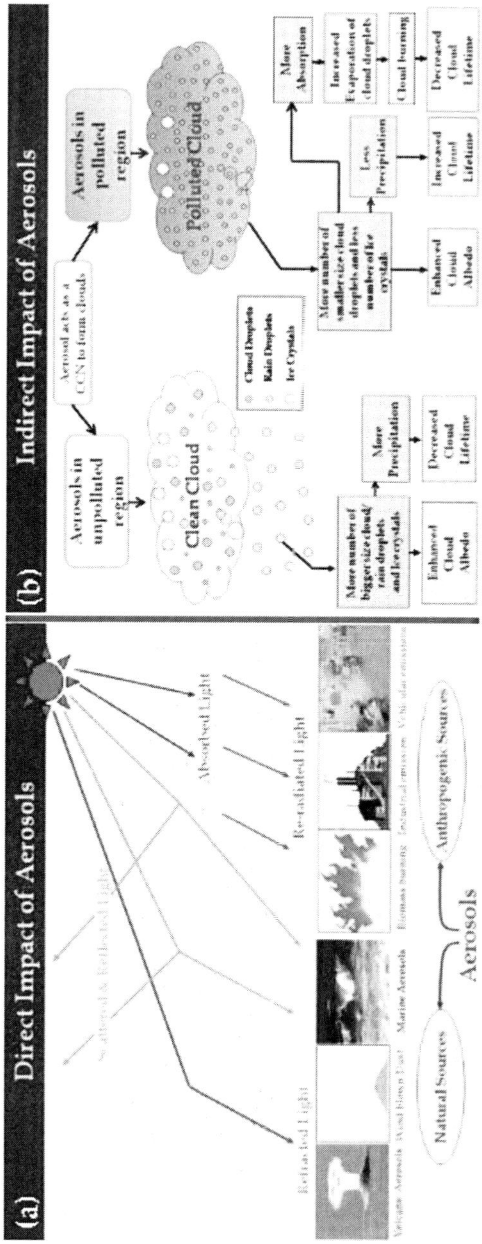

Figure 10. Mean AAE values for each aerosol type at Kanpur and Gandhi College over the IGB (*Adopted from [27]*).

Further, an International TIGERZ experiment was conducted by the NASA AERONET group within the IGB region around the industrial city of Kanpur during the pre-monsoon period [36]. The major objectives of TIGERZ include the spatial and temporal characterization of columnar aerosol optical, microphysical and absorption properties; the identification of aerosol particle types/mixtures; and the validation of remotely sensed aerosol properties from satellites. In a recent past, Ramanathan group from the Scripps Institution of Oceanography, University of California, USA conducted a field measurement (from November 2009–September 2010), called *Project Surya*, in a rural area over the IGB region. Studies were focused on to establish the role of both solid biomass based cooking in traditional stoves and diesel vehicles in contributing to high BC and organic carbon (OC), and solar absorption [81,82]. In continuation to this, [83] have studied the link between local scale aerosol properties and column averaged regional aerosol optical properties and atmospheric radiative forcing.

Apart from ground-based aerosol measurements, vertical distribution of aerosols were carried out for the first time over Kanpur in the IGB region during the winter and summer of the year 2005, reported in [84-85]. Vertical measurements of aerosols up to 1.5 km provided useful information during the winter because aerosols were mostly confined to the boundary layer; however, during summer, aerosols get convected and reached up to the higher altitudes. The Integrated Campaign for Aerosols, gases, and Radiation Budget (ICARB) was initiated to address these issues with multi-institutional, multi-instrumental, multi-platform field campaign, where integrated observations and measurements of aerosols with special emphasis on black carbon, radiation and trace gases along with other complementary measurements on boundary layers and meteorological parameters were made simultaneously [86]. The ICARB was conducted during February-May period of 2006 as an integrated campaign, comprising three segments namely the land, ocean and air, to assess the regional radiative impact of aerosols and trace gases, and to quantify the effect of the long-range transport of aerosols and trace gases over the Indian mainland, Arabian Sea, Bay of Bengal and the tropical Indian Ocean. The details of this campaign and the major findings have been reported in different literatures in [86-90]. ICARB was covered only the eastern part of the IGB (Bhubaneswar) [91] while focusing mostly on the peninsular India and surrounding oceans. Continental Tropical Convergence Zone (CTCZ) campaign, focused on the aerosol distribution in the pre-monsoon and monsoon (June–September) seasons was initiated in the year 2008, and covering the continental part of the more common tropical convergence zone over India, including the IGB [92]. During the campaign, Aircraft and ground-based measurements together were carried out over the IGB and Central part of India to quantify the aerosol indirect effect. The details of the campaign and the major findings have been reported in a recent publication in [61].

Even though all these national and international field experiments and campaigns have greatly improved our understanding on aerosol optical, physical as well as chemical properties and have indeed reduced the uncertainty in regional aerosol direct radiative forcing at various parts of India including the IGB region, they are limited to a certain period or location due to their specific goals. In this perspective, long-term experiments with a high spatio-temporal scale can add advantages of understanding aerosol influences on a longer time scale, thereby helping to infer the signs of anthropogenic impacts. This is where satellite data become very useful and can complement the ground-based and/or *in-situ* measurements.

3.2. Space-Borne

Satellite retrievals of aerosol properties over land have only been available in recent years and a few studies have been done using these data over the Indian subcontinent, focusing on the IGB region. In reference [12], they were the first to study the spatial distribution of AOD over India using Multiangle Imaging Spectroradiometer (MISR) during the winter period from 2001 to 2004, where they were able to explain the enormous pollution observed over the IGB based on meteorology, topography and potential aerosol emission sources. Further, subsequent studies using Moderate Resolution Imaging Spectroradiometer (MODIS) data have confirmed this observation [13,15,93] with additional information on the seasonal variability of AOD and fine mode fraction, to some extent. In continuation to that recently in reference [8], they have presented a detailed analysis of a 9 year (2000–2008) seasonal climatology of size- and shape-segregated aerosol properties over the Indian subcontinent derived from the MISR. The spatial heterogeneity of the aerosol parameters are shown in Figure 11 for each season.

The spatial distribution of AOD in the winter season reveals high AOD over the IGB and its outflow to the northern Bay of Bengal because of high anthropogenic emission sources, as previously observed from satellite [12,13,15,93] and ground-based [22,43,45] measurements. It is well known that the IGB region is often enveloped by thick fog/haze during this period, which is typically associated with high aerosol loading over the region [43,49,94-96]. AOD averaged over eight winter seasons is highest (>0.4) over the eastern part of IGB, which was referred to as the 'Bihar pollution pool' in [12]. As pointed out in [12], this observation has strong implications for the large population residing in this area and thus calls for further work. In [97], they have used CO retrieved from MOPITT (version 3 data) and found a corresponding pool of high CO mixing ratios at 850 hPa level in the same area in winter. In continuation of this, in [50], they have further demonstrated the extensive pollution along the eastern parts of the IGB during winter months using the improved version 4 CO data from MOPITT and the new version 3 height resolved aerosol data from CALIPSO as well as the tropospheric column ozone from two different data products. Both the CO and aerosol data from this study confirm the trapping of pollution at low altitudes by subsidence. Aerosols across the IGB was found to be transported from west to east by northwesterly winds, encounter a narrowing valley floor and are trapped efficiently within the atmospheric column in the eastern part of the IGB by subsiding air [8]. Relatively low AE (<0.8) in the eastern IGB than the other parts, suggests high concentration of coarse dust particles emitted possibly by rural activities (e.g., agriculture, etc.) from the densely populated rural population.

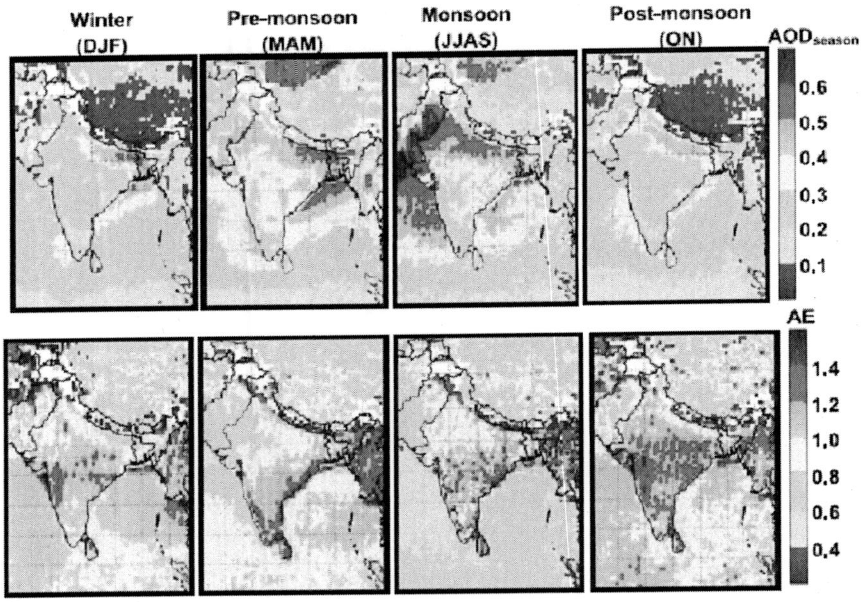

Figure 11. Spatial distribution of climatological mean mid-visible AOD (first panel) and AE (second panel) during the winter, pre-monsoon, monsoon and post-monsoon seasons over the Indian subcontinent (including IGB region) for the period from March 2000 to November 2008 (*Adopted from [8]*).

In the pre-monsoon season, aerosol spectral optical properties change significantly from the preceding winter season because of enhancement in dust loading, particularly over the IGB region [15,31,24]. Large emission of small particles from open biomass burning compensates the relative influence of dust on spectral AOD in the eastern part of the IGB [27] as indicated by an increase of AE during pre-monsoon season compared to the preceding winter season (Figure 11). This also leads to an overall increase in AOD in this region compared to the winter season. Thus, winter to pre-monsoon changes in aerosol properties are not just dominated by an increase in dust, as previously thought, but also by an increase in anthropogenic components, particularly in the regions where biomass combustion is in the common practice during this period. Changing atmospheric aerosol properties caused by anthropogenic activities carries serious implications for climate change and human health [98].

The anthropogenic emissions, particularly BC and sulfate aerosols are present throughout the year in northern India over the IGB [99,100]. Such aerosols form thick layers of haze in winter, termed as Atmospheric Brown Clouds (ABC), which block the solar radiation reaching to the surface [101]. In [102], they have reported in their study over India that the AODs derived from TOMS data from 1979 to 2000 increased by ~11% per decade during the winter with large values over the IGB region, which consequently affect the surface reaching solar radiation, known as "solar dimming" [103-105]. The average solar dimming observed over India is about -0.86 W m^{-2} yr^{-1}, while during winter, pre-monsoon and monsoon seasons the same was observed to be about -0.94, -1.04 and -0.74 Wm^{-2}, respectively [104]. The significant reduction in ground-reaching solar radiation can directly be correlated with the increased aerosol loading in the atmosphere due to enhancement in industrialization, vehicular pollution, biomass burning and dust storm activities over the region [106,107]. Apart from solar dimming effect due to variety of aerosols in terms of haze/fog conditions, our understanding about the role of secondary organic aerosols (particulate organic matters produced by gas-to-particle conversion process), particularly in climate change and its connection to health effects is very limited by numerous uncertainties. In a recent study in [108], they have observed an enhanced production of secondary organic aerosols over Kanpur in the IGB during the foggy periods of winter, which was hypothesized that the aqueous phase chemistry in fog drops is responsible for increased production of secondary organic aerosols.

During the monsoon, stronger westerly winds were found to be transport greater components of dust from the Arabian Peninsula to the Indian subcontinent [8]. In general, the spatial distribution of AOD (Figure 11) in this season is largely influenced by monsoon precipitation. Suppressed precipitation in the monsoon break phase allows for a rapid buildup of aerosols in the high anthropogenic source regions (e.g. IGB), while particles are being washed out by the precipitation in the active monsoon phase. This also leads to very high intra-seasonal and inter-annual variability in aerosol characteristics. Aerosol regional mean climatology in the post-monsoon season is very similar to that for the winter season (Figure 11), but the spatial distribution differs in several regions. For example, the wintertime high AOD zone in the IGB shows a larger spread and higher inter-annual variability across the basin in this season, owing to a stronger peak in crop waste burning in the western part of IGB than the eastern part [109] and weaker subsidence in the eastern part of IGB compared to the winter season. As a result, IGB is the region with highest aerosol absorption and thus occurred large discrepancy in MISR and MODIS derived AODs [20].

4. COUPLING OF IGB AEROSOLS TO THE HIMALAYAN REGION AND THEIR POSSIBLE IMPACTS

Due to combined effects of IGB topography and the Himalayan orography, aerosols over the IGB region are lifted up quite often and found to be extended up to the Himalayan foothills and also to the other high-altitude regions [23,61,110-113]. Absorbing aerosols in the elevated regions heat the mid-troposphere by absorbing solar radiation, and produce an atmospheric dynamical feedback called as elevated heat pump (EHP) effect. Consequently, this can lead to an increase in the summer monsoon rainfall over India [63] and enhancement in the rate of snow melting in the Himalayan regions [64], which is one of the potential themes for global scientific community and need to be addressed to improve scientific understanding of the regional climate on inter-annual as well as intra-seasonal scales. In particular, the main emphasis of the IGB region coupled with the Himalayan foothills is due to the highest AOD values in this region among the South Asia regions, which are persistent throughout the winter and spring seasons [114].

In a recent study in [115], they have shown a possible influence of desert dust aerosols originated and transported from the Thar Desert region to the high-altitude station at Manora Peak, Nainital in the central Himalayas (Figure 12). The high values of aerosol index (AI) derived from the Ozone Monitoring Instrument (OMI) attest to the presence of absorbing aerosol particles over the region; however, air mass back-trajectory analysis over the station shows different pathways for the transport of air masses from the source region to the experimental site over different time periods (Figure 12). In this study, [115] observed a thick aerosol layer at ~1500 m altitude (Figure 13), above the station level, which was substantiated by the air mass back-trajectory analysis (Figure 14).

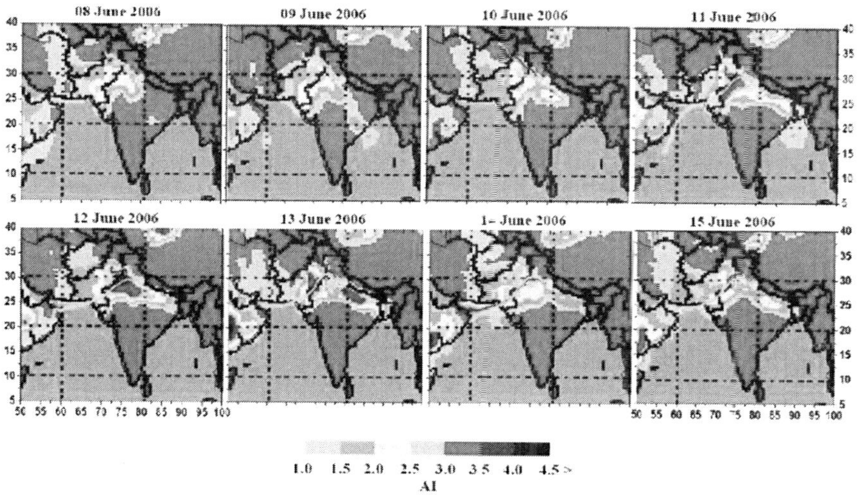

Figure 12. OMI AI images showing the source and the progressive movement of the dust aerosols occurred in June 2006. Five-day air mass back-trajectories at Manora Peak for different time periods are superimposed on respective days AI images (*Adopted from [115]*).

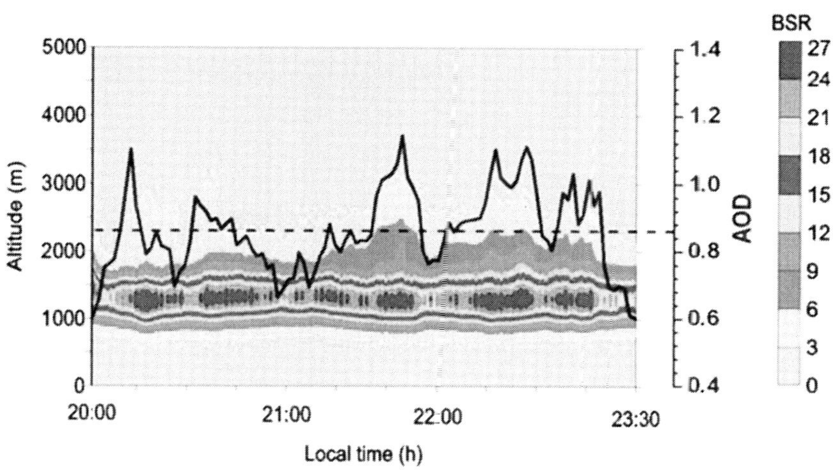

Figure 13. Altitude versus time variations in BSR on a dust day (12 June 2006, with color contour) along with the variations in lidar-derived AOD at 532 nm (with solid line, average AOD = 0.83 ± 0.12) (*Adopted from [115]*).

Apart from dust transport from the Desert regions, recent study in [116], they have also demonstrated significant impact of north Indian biomass burning on aerosols and trace gases and the resultant radiation budget over the central Himalayas during the spring period through air mass back-trajectory analysis coupled with fire counts (Figure 15). The same has also been reported in [117] to be one of the major sources of BC over the same station in the central Himalayas, which was observed to be much lower (in terms of magnitude) as compared to the urban location in the IGB, but was found to have significant contribution to the total aerosol optical depth (~17%) and the resultant atmospheric forcing (~70%) at Manora Peak [62]. Based on BC measured at two different wavelengths at ultra-violet (370nm) and near-infrared (880nm), [117] have distinguished the potential sources of BC at Delhi (one of the densely populated and industrialized urban megacities in Asia and typically represents the plains of Ganga basin) and Manora Peak (one of the high-altitude and sparsely inhabited clean site in the Indian Himalayan foothills situated in the central Himalayas). Based on the analysis, [117] have found the major contribution of BC at Manora Peak is from biomass burning while fossil fuel is found to be the dominating contributor at Delhi.

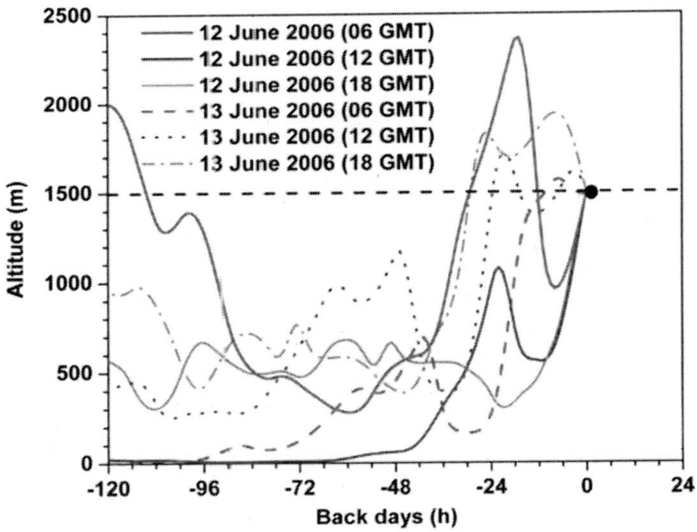

Figure 14. Temporal evolution of air masses at 1500 m altitude for three different time intervals on 12 and 13 June 2006 (*Adopted from [115]*).

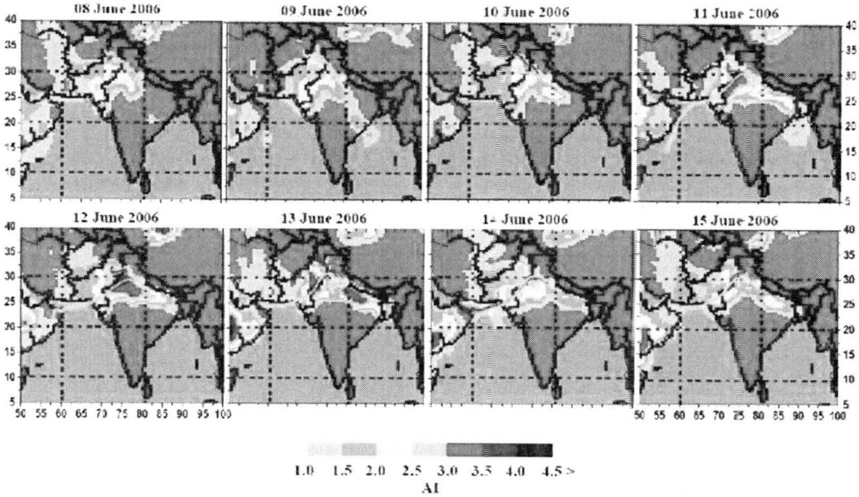

Figure 15. Three day back-air trajectories arriving at Manora Peak during the fire-impacted periods in 2007, 2008 and 2009 (triangle represents the observation site) (*Adopted from [116]*).

5. SUMMARY AND FUTURE DIRECTIONS

The study over IGB region revealed different aerosol characteristics over the region from western to central and to the eastern parts, which show significant gradient in magnitude of most of the aerosol characteristics. Such gradient can be explained due to the gradual changes in weather parameters and/or emission sources apart from geographical heterogeneity. Such gradient is, ultimately, found to have impact on the Earth-atmosphere system by negative radiative forcing, thus causing cooling, at the surface, and positive aerosol forcing, thus causing heating in the atmosphere. Such gradient in heating rate raises several climatic issues, and is needed to be answered on the basis of longer period investigations at several stations to improve the scientific understanding of the regional climate in inter-annual as well as intra-seasonal scale.

Due to large uncertainty in satellite derived aerosol products over the IGB during pre-monsoon dust periods, long-term ground-based measurements during different seasons can indeed provide useful information of the characteristics of aerosol types over the region on seasonal and inter-annual basis, which are meager and crucial for the regional climate models. Further, the mixing of natural dust with anthropogenically produced aerosol particles, has been hypothesized in [17] over the IGB region, mostly during the pre-monsoon period and corroborated with the AERONET data [36], suggested the complication of the satellite retrieval of aerosol characteristics and quantifying the climatic effects [118]. Hence, it is also one of the important research areas in understanding aerosol characteristics over the IGB region to make realistic assessments of aerosol-hydro-climate interplay.

The issue of black carbon or soot particles and its relationship with climate change has gained enormous scientific and popular interest over the last few years. The knowledge and understanding on aspects such as vertical distribution and mixing of black carbon with other aerosols, effects of cloud cover and monsoon still remains uncertain and incomplete. Few studies have shown that when sulphate or organics is coated over black carbon aerosols, its absorption effects are enhanced by 50% [119]. In case of black carbon mixed with large dust particles, absorption of the composite dust-black carbon aerosol system is enhanced by a factor of two to three compared to sum of black carbon and dust absorption [120]. However, we have no information on the state of mixing of black carbon. The proper assessment of mixing and/or coating of various aerosol species and their impacts on various aerosol characteristics have not been well quantified [121], which makes the investigation a real challenge [122]. IGB, being in proximity to the Thar Desert region, is found to be affected predominantly by the enhanced dust aerosols, mostly during the pre-monsoon period. As a result, the probability of this interaction (i.e. mixing) was suggested to be more over the region during this period [17,36] and is one of the future perspectives. To better understand these crucial issues, National Carbonaceous Aerosol Program (NCAP) was recently launched in India, focusing on the measurement of black carbon; their role in atmospheric stability and the consequent effect on cloud formation, monsoon and retreating of Himalayan glacier [123].

Based on recent observations using aircraft [61] and satellite measurements ([34,46], it has been reported that during pre-monsoon season, IGB region is characterized by the elevated aerosol layers extended up to the altitude from about 3 to 5 km. When the amount of absorbing aerosols such as black carbon and dust, are significant in the atmosphere, the aerosol optical depth and chemical composition are not the only determinants of aerosol radiative effects, but the altitude of the aerosol layer and its altitude relative to clouds (if present) are also essential. Thus, it is also essential to gather information on vertical distribution of aerosols over this region.

Further, fog over the IGB region is observed to be a common feature, occurs mostly during the winter period. The number of foggy days has been increasing in recent years as compared to earlier decades [124], with strong increasing trends of anthropogenic pollution in the IG plains [125]. Fog formation usually begins in the latter half of December and continues till the end of January, thus blanketing some regions for more than a month [126]. The low topography of the IGB, adjacent to the Himalayan range, favors formation of fog and provides high concentration of air pollutants in the plains which serve as additional CCN for nucleation. Fog affects day to day lives of millions of people living in this region, resulting in poor visibility down to less than 100 meters causing frequent flight and train delays and even a significant number of deaths from vehicular accidents in many severe events [127]. Though few studies were done focusing on fog-induced aerosol characteristics over the IGB region and their impacts to the aerosol radiative forcing [49,128], detailed studies of aerosol composition and inter-annual variation of aerosols are required to better understand the interaction of winter haze with the formation of fog over the IGB.

Apart from the measurements for various aerosol characteristics through different ground-based and space-born instrumentations, a 1-D aerosol optical model named as optical properties of aerosols and clouds (OPAC) has been developed by [129], estimating crucial optical properties of aerosols such as AOD and SSA, under the assumption of spherical aerosol particles and external mixing. In [130], they have shown that the optical depth and SSA of aerosol particles have strongest sensitivity on the direct radiative forcing, and these optical properties have found to be large deviation with shape and composition [13]. Further, with the model studies reported in [47,118], the particle composition (i.e. mixing state) and shape (i.e. morphology) attributes to more cooling at both top of the atmosphere and surface, and the combined effect is ~6% more warming than the spherical particles. The significance of consideration of particle shape is more in the regions where black carbon mixes with pure mineral dust, which are the most probable case over the IGB in northern India, because enhancement in the atmospheric warming will be under-estimated if particle morphology is not considered [47]. Thus, there is an urgent need for modeling studies over the IGB region to examine quantitatively the influence of particle morphology along with their mixing states on optical and radiative characteristics of aerosols with their size distribution.

ACKNOWLEDGEMENT

Authors are thankful to the various Journals for allowing the use of published materials for this chapter. Figures used in this chapter are considered after taking permission from the respective Journals. SNT acknowledges the support under the program, Changing Water Cycle funded jointly by the Ministry of Earth Sciences, India and Natural Environment Research Council, UK. AKS thanks to Prof. B. N. Goswami, Director, IITM, Pune and Dr. P. C. S. Devara for their encouragement and support.

REFERENCES

1. Pöschl 2005 (2005) Atmospheric aerosols: composition, transformation, climate and health effects. Angew. Chem. Int. Ed. 44 75207540 .

2. Intergovernmental Panel on Climate Change (IPCC) 2001 Climate Change 2001: The Scientific Basis, edited by J. T. Houghton et al., Cambridge Univ. Press, New York, 881 p.

3. Intergovernmental Panel on Climate Change (IPCC) 2007 Climate change 2007: The physical science basis: Contribution of Working Group I to the Fourth Assessment Report of the Intergovernmental Panel on Climate Change, Chapter 2, Cambridge Univ. Press, New York, 129 p.

4. Schwartz SE, et al. 1995 Group Report: Connections between aerosol properties and forcing of climate, John Wiley, Hoboken, N. J, 251280 p.

5. 2010, Pandithurai G, Dipu S (2010) Aerosol indirect effect during successive contrasting monsoon seasons over Indian subcontinent using MODIS data. Atmos. Environ. 44 19371943 .

6. Tripathi, S. N. Pattnaik, A. Dey, S. , 2007a Aerosol indirect effect over Indo-Gangetic plain. Atmos. Environ. 41 70377047 .

7. Ganguly, D. Ginoux, P. Ramaswamy, V. Dubovik, O. Welton, J. Reid, E. A. Holben, B. N. , 2009 Inferring the composition and concentration of aerosols by combining AERONET and MPLNET data: Comparison with other measurements and utilization to evaluate GCM output. J. Geophys. Res. 114: D16203.

8. Dey, S. , Di Girolamo, 2010 (2010) A climatology of aerosol optical and microphysical properties over the Indian subcontinent from 9 years (2000-2008) of Multiangle Imaging Spectroradiometer (MISR) data. J. Geophys. Res. 115: D15204

9. Moorthy KK, Babu SS, Satheesh SK 2007 Temporal heterogeneity in aerosol characteristics and the resulting radiative impact at a tropical coastal station-Part 1: Microphysical and optical properties. Ann. Geophys. 25 22932308 .

10. Devara, P. C. S. Raj, P. E. Dani, K. K. Pandithurai, G. Kalapureddy, M. C. R. Sonbawne, S. M. Rao, Y. J. Saha, S. K. , 2008 Mobile lidar profiling of tropical aerosols and clouds. J. Atmos. Oceanic Technol. 25 12881295.

11. Dey, S. , S. N. Tripathi, 2008 Aerosol direct radiative effects over Kanpur in the Indo-Gangetic basin, northern India: Long-term (2001-

2005) observations and implications to regional climate. J. Geophys. Res. 113: D04212.

12. Di Girolamo, L. et, al, 2004 Analysis of Multi-angle Imaging Spectro-Radiometer (MISR) aerosol optical depths over greater India during winter 20012004 . Geophys. Res. Lett. 31: L23115.

13. Jethva, H. Satheesh, S. K. Srinivasan, 2005ethva H, Satheesh SK, Srinivasan J (2005) Seasonal variability of aerosols over the Indo-Gangetic basin. J. Geophys. Res. 110: D21204.

14. Prasad AK and Singh RP 2007 Changes in aerosol parameters during major dust storm events (2001-2005) over the Indo-Gangetic Plains using AERONET and MODIS data. J. Geophys. Res. 112: D09208.

15. Ramachandran, S. Cherian, 2008amachandran S, Cherian R (2008) Regional and seasonal variations in aerosol optical characteristics and their frequency distributions over India during . J. Geophys. Res. 113: D08207.

16. El -Metwally, M. Alfaro, S. C. Abdel, Favez. O. Wahab, Z. Mohamed, Chatenet, 2011 (2011) Aerosol properties and associated radiative effects over Cairo (Egypt). Atmos. Res. 99 263276 .

17. Dey, S. Tripathi, S. N. Mishra, S. K. , 2008 Probable mixing state of aerosols in the Indo-Gangetic Basin, northern India. Geophys. Res. Lett. 35: L03808.

18. Mishra SK, Tripathi SN 2008 Modeling optical properties of mineral dust over the Indian Desert. J. Geophys. Res. 113: D23201.

19. Tripathi, S. N. Dey, S. Chandel, A. Srivastva, S. Singh, R. P. Holben, B. , 2005a Comparison of MODIS and AERONET derived aerosol optical depth over the Ganga basin, India. Ann. Geophys. 23 10931101 .

20. Kahn RA, et al. 2009 MISR aerosol product attributes and statistical comparisons with MODIS. IEEE Transactions on Geoscience and Remote Sensing 47 12 40954114 .

21. Holben BN, et al. 1998 AERONET-A federated instrument network and data archive for aerosol characterization. Remote Sens. Environ. 66 116.

22. Singh, R. P. Dey, S. Tripathi, S. N. Tare, V. Holben, 2004 (2004) Variability of aerosol parameters over Kanpur, northern India. J. Geophys. Res. 109: D23206.

23. Gautam, R. Hsu, N. C. Tsay, S. C. Lau, K. M. Holben, B. Bell, S. Smirnov, A. Li, C. Hansell, R. Ji, Q. Payra, S. Aryal, D. Kayastha, R. Kim, K. M. , 2011 Accumulation of aerosols over the Indo-Gangetic

plains and southern slopes of the Himalayas: distribution, properties and radiative effects during the 2009 pre-monsoon season. Atmos. Chem. Phys. 11 1284112863 .

24. Srivastava, A. K. Tiwari, S. Devara, P. C. S. Bisht, M. K. Srivastava, S. N. Tripathi, P. Goloub, B. N. Holben, 2011a Pre-monsoon aerosol characteristics over the Indo-Gangetic Basin: Implications to climatic impact. Ann. Geophys. 29 789804 .

25. Srivastava, A. K Singh, S. Tiwari, S. Bisht, , 2012a Contribution of anthropogenic aerosols in direct radiative forcing and atmospheric heating rate over Delhi in the Indo-Gangatic Basin. Environ. Sci. Pollut. Res. 19 11441158.

26. Ramanathan, V. et, al, 2001 Indian Ocean Experiment: An integrated analysis of the climate forcing and effects of the great Indo-Asian haze. J. Geophys. Res. 106(D22): 28371-28398.

27. Srivastava, A. K. Tripathi, S. N. Dey, S. Kanawade, V. P. Tiwari, S. , 2012b Inferring aerosol types over the Indo-Gangetic Basin from ground based sunphotometer measurements. Atmos. Res. 109-110: 64-75.

28. Guttikunda, S. K. Carmichael, G. R. Calori, G. Eck, C. Woo, J. H. , 2003 The contribution of megacities to regional sulfur pollution in Asia. Atmos. Environ. 37 1122 .

29. Sharma, M. Kiran, Y. N. V. M. Shandilya, K. K. , 2003 Investigations into formation of atmospheric sulfate under high PM10 concentration. Atmos. Environ. 37 20052013 .

30. Monkkonen P, et al. (2004) Relationship and variations of aerosol number and PM10 mass concentrations in a highly polluted urban environment –New Delhi, India. Atmos. Environ. 38: 425– 433.

31. Dey, S. Tripathi, S. N. Singh, R. P. Holben, 2004 (2004) Influence of dust storm on the aerosol parameters over the Indo-Gangetic basin. J. Geophys. Res. 109: D20211.

32. Chinnam, N. Dey, S. Tripathi, S. N. Sharma, 2006 (2006) Dust events in Kanpur, northern India: chemical evidence for source and implications to radiative forcing. Geophys. Res. Lett. 33: L08803 EOF.

33. Gautam, R. Liu, Z. Singh, R. P. Hsu, N. C. , 2009 Two contrasting dust-dominant periods over India observed from MODIS and CALIPSO data. Geophys. Res. Lett. 36: L06813.

34. Gautam, R. Hsu, N. C. Lau, K. M. , 2010 Pre-monsoon aerosol characterization and radiative effects over the Indo-Gangetic Plains:

Implications for regional climate warming. J. Geophys. Res. 115: D17208.

35. Eck, T. et, al, 2010 Climatological aspects of the optical properties of fine/coarse mode aerosol mixtures. J. Geophy. Res. 115: D19205.

36. Giles DM, et al. 2011 Aerosol properties over the Indo-Gangetic Plain: A mesoscale perspective from the TIGERZ experiment. J. Geophys. Res. 116: D18203 EOF.

37. Middleton NJ 1986 A geography of dust storms in southwest Asia. Int. J. Climatol. 6 183196 .

38. Sikka DR 1997 Desert climate and its dynamics. Curr. Sci. 72 1 3546 .

39. Prospero, J. M. Ginoux, P. Torres, O. Nicholson, S. E. Gill, T. E. , 2002 Environmental characterization of global sources of atmospheric soil dust identified with the Nimbus 7 Total ozone Mapping Spectrometer (TOMS) absorbing aerosol product. Rev. Geophys. 40(1): 1002.

40. Washington, R. Todd, M. Middleton, N. J. Goudie, , 2003 Dust storm source areas determined by the Total Ozone Monitoring Spectrometer and surface observations. Ann. Assoc. Am. Geogr. 93 297313 .

41. Tripathi JK, Rajamani 1999 Geochemistry of the loessic sediments on Delhi ridge, eastern Thar Desert, Rajasthan: Implications for exogenic processes. Chem. Geol. 155: 265-278.

42. Yadav, S. Rajamani, 2003 Aerosols of NW India-A potential Cu source. Curr. Sci. 84(3): 278- 280.

43. Tripathi SN, et al. 2006 Measurements of atmospheric parameters during Indian Space Research Organization Geosphere Biosphere Programme Land Campaign II at a typical location in the Ganga basin: 1. Physical and optical properties. J. Geophys. Res. 111: D23209 EOF.

44. Tare, V. et, al, 2006 Measurements of atmospheric parameters during Indian Space Research Organization Geosphere Biosphere Program Land Campaign II at a typical location in the Ganga Basin: 2. Chemical properties. J. Geophys. Res. 111: D23210 EOF.

45. Nair VS, et al. 2007 Wintertime aerosol characteristics over the Indo-Gangetic Plain (IGP): Impacts of local boundary layer processes and long-range transport. J. Geophys. Res. 112: D13205.

46. Srivastava, A. K. Singh, S. Tiwari, S. Kanawade, V. P. Bisht, , 2012c Variation between near-surface and columnar aerosol characteristics during winter and summer at Delhi in the Indo-Gangatic Basin. Journal of Atmospheric and Solar-Terrestrial Physics 77 5766 .

47. Mishra, S. K. Dey, S. Tripathi, S. N. , 2008 Implications of particle composition and shape to dust radiative effect: A case study from the Great Indian Desert. Geophys. Res. Lett. 35: L23814.

48. Pandithurai, G. Dipu, S. Dani, K. K. Tiwari, S. Bisht, P. C. S. Devara, R. T. Pinker, 2008 Aerosol radiative forcing during dust events over New Delhi, India. J. Geophys. Res. 113: D13209.

49. Gautam, R. Hsu, N. C. Kafatos, M. Tsay, S. C. , 2007 Influences of winter haze on fog/low cloud over the Indo-Gangetic plains. J. Geophys. Res. 112: D05207.

50. Kar, J. Deeter, M. N. Fishman, J. Liu, Z. Omar, A. Creilson, J. K. Trepte, C. R. Vaughan, D. M. Winker, 2010 Wintertime pollution over the Eastern Indo-Gangetic Plains as observed from MOPITT, CALIPSO and tropospheric ozone residual data. Atmos. Chem. Phys. 10 1227312283 .

51. Habib, G. Venkataraman, C. Chiapello, I. Ramachandran, S. Boucher, O. Reddy, , 2006 Seasonal and interannual variability in absorbing aerosols over India derived TOMS: Relationship to regional meteorology and emissions. Atmos. Environ. 40 19091921 .

52. Tiwari S, Srivastava AK, Bisht DS, Bano T, Singh S, Behura S, Srivastava MK, Chate DM, Padmanabhamurty B (2009) Black carbon and chemical characteristics of PM10 and PM2.5 at an urban site of North India. J. Atmos. Chem. 62(3): 193–209.

53. Ram, K. Sarin, , 2010 Spatio-temporal variability in atmospheric abundances of EC, OC and WSOC over Northern India. J. Aerosol Sci. 41 8898 .

54. Mani, A. Chacko, O. Hariharan, 1969 (1969) A study of Ångström turbidity parameters from solar radiation measurements in India. Tellus 21 829843.

55. Moorthy KK, et al. 1999 Aerosol climatology over India: ISRO GBP MWR network and database. Rep. ISRO GBP SR-0399 , Indian Space Res. Org., Bangalore, India.

56. Devara, P. C. S. Maheskumar, R. S. Raj, P. E. Pandithurai, G. Dani, K. K. , 2002 Recent trends in aerosol climatology and air pollution as inferred from multiyear lidar observations over a tropical urban station. Int. J. Climatol. 22 435449.

57. Singh, S. Soni, K. Bano, T. Tanwar, R. S. Nath, S. Arya, B. C. , 2010 Clear-sky direct aerosol radiative forcing variations over mega-city Delhi. Ann. Geophys. 28 11571166 .

58. Ram K, Sarin MM, Tripathi SN (2012) Temporal trends in atmospheric PM2.5, PM10, EC, OC, WSOC and optical properties: Impact of biomass burning emissions in the Indo-Gangetic Plain. Environ. Sci. and Tech. 46: 686-695.

59. Venkataraman, C. Habib, G. Eiguren, A. Fernandez, A. H. Miguel, S. K. Friedlander, 2005 Residential biofuels in South Asia: carbonaceous aerosol emissions and climate. Science 307 14541456 .

60. Arola, A. Schuster, G. Myhre, G. Kazadzis, S. Dey, S. Tripathi, S. N. , 2011 Inferring absorbing organic carbon content from AERONET data. Atmos. Chem. Phys. 11 215225 .

61. Devi, J. J. Tripathi, S. N. Gupta, T. Singh, B. N. Gopalakrishnan, V. Dey, 2011N, Gupta T, Singh BN, Gopalakrishnan V, Dey S (2011) Observation-based 3 -D view of aerosol radiative properties over Indian Continental Tropical Convergence Zone: implications to regional climate. Tellus 63B: 971-989.

62. Srivastava, A. K. Ram, K. Pant, P. Hegde, P. Joshi, H. , 2012d Black carbon aerosols over central Himalayas: implications to climate forcing. Environ. Res. Lett. 7: 014002.

63. Lau KM, Kim MK, Kim KM 2006 Asian summer monsoon anomalies induced by aerosol direct forcing: The role of the Tibetan Plateau. Clim. Dyn. 26(7-8): 855 EOF864 EOF.

64. Lau KM, Kim MK, Kim KM, Lee WS 2010 Enhanced surface warming and accelerated snow melt in the Himalayas and Tibetan Plateau induced by absorbing aerosols. Environ. Res. Lett. 5 110 .

65. 2007ilewskie 2007 Climate change: Aerosols heat up. Nature 448: 541-542, doi:10.1038/448541a.

66. Ramanathan, V. Ramana, M. V. Roberts, G. Kim, D. Corrigan, C. Chung, C. Winker, 2007, Corrigan C, Chung C, Winker D (2007) Warming trends in Asia amplified by brown cloud solar absorption. Nature 448 575578 .

67. Srivastava, A. K. Tiwari, S. Bisht. P. C. S. Devara, P. Goloub, Z. Li, M. K. Srivastava, 2011c Aerosol characteristics during the coolest June month over New Delhi, northern India. Int. J. Remote Sens. 32 23 84638483 .

68. Dey, S. Tripathi, S. N. Singh, R. P. Holben, 2006 (2006) Retrieval of black carbon and specific absorption over Kanpur city, Northern India during 2001-2003 using AERONET data. Atmos. Env. 40 3 445456 .

69. Russell, P. B. Bergstrom, R. W. Shinozuka, Y. Clarke, A. D. De -Carlo, P. F. Jimenez, J. L. Livingston, J. M. Redemann, J. Dubovik, O. Strawa, 2010D, De-Carlo PF, Jimenez JL, Livingston JM, Redemann J, Dubovik O, Strawa A (2010) Absorption Angstrom Exponent in AERONET and related data as an indicator of aerosol composition. Atmos. Chem. Phys. 10 11551169.

70. Moorthy KK, Babu SS, Satheesh SK 2005a Aerosol characteristics and radiative impacts over the Arabian Sea during the inter-monsoon season: Results from ARMEX Field campaign. J. Atmos. Sci. 62 192206 .

71. Ganguly, D. Jayaraman, A. Gadhavi, H. Rajesh, T. A. , 2005a Features in wavelength dependence of aerosol absorption observed over central India. Geophys. Res. Lett. 32: L13821.

72. Ganguly, D. Gadhavi, H. Jayaraman, A. Rajesh, T. A. Mishra, A. , 2005b Single scattering albedo of aerosols over the central India: implications for the regional aerosol radiative forcing. Geophys. Res. Lett. 32: L18803.

73. Moorthy KK, et al. 2005b Wintertime spatial characteristics of boundary layer aerosols over peninsular India. J. Geophys. Res. 110: D08207.

74. Singh S, Singh B, Gera BS, Srivastava MK, Dutta HN, Garg SC, Singh R (2006) A study of aerosol optical depth in the central Indian region (17.3–8.6 oN) during ISRO-GBP field campaign. Atmos. Environ. 40: 6494–6503.

75. Ganguly, D. Jayaraman, A. Rajesh, T. A. Gadhavi, 2006 (2006) Wintertime aerosol properties during foggy and non-foggy days over urban center Delhi and their implications for shortwave radiative forcing. J. Geophys. Res. 111: D15217.

76. Niranjan, K. Sreekanth, V. Madhavan, B. L. Moorthy, K. K. , 2006 Wintertime aerosol characteristics at a north Indian site Kharagpur in the Indo-Gangetic plains located at the outflow region into Bay of Bengal. J. Geophys. Res. 111: D24209.

77. Pant, P. Hegde, P. Dumka, U. C. Sagar, R. Satheesh, S. K. Moorthy, K. K. Saha, A. Srivastava, M. K. , 2006 Aerosol characteristics at a high-altitude location in central Himalayas: optical properties and radiative forcing. J. Geophys. Res. 111: D17206.

78. Ramachandran, S. et, al, 2006 Aerosol radiative forcing during clear, hazy, and foggy conditions over a continental polluted location in north India. J. Geophys. Res. 111: D20214 EOF .

79. Srivastava, M. K. Singh, S. Saha, A. Dumka, U. C. Hegde, P. Singh, R. Pant, 2006, Singh R, Pant 2006 Direct solar ultraviolet irradiance over

Nainital, India, in the central Himalayas for clear-sky day conditions during December 2004. J. Geophys. Res. 111: D08201.

80. Rengarajan, R. Sarin, A. K. Sudheer, 2007 Carbonaceous and inorganic species in atmospheric aerosols during wintertime over urban and high-altitude sites in North India. J. Geophys. Res. 112: D21307.

81. Ramanathan, V. Rehman, I. H. Ramanathan, 2010 (2010) Project Surya Prospectus. University of California, San Diego, USA, 14 p.

82. Rehman, I. H. Ahmed, T. Praveen, P. S. Kar, A. Ramanathan, 2011 Black carbon emissions from biomass and fossil fuels in rural India. Atmos. Chem. Phys. 11: 7289-7299.

83. Praveen, P. S. Ahmed, T. Kar, A. Rehman, I. H. Ramanathan, 2012 Link between local scale BC emissions in the Indo-Gangetic Plains and large scale atmospheric solar absorption. Atmos. Chem. Phys. 12: 1173-1187.

84. Tripathi, S. N. Dey, S. Tare, V. Satheesh, S. K. Lal, S. Venkataramni, S. , 2005b Enhanced layer of black carbon in a north Indian industrial city. Geophys. Res. Lett. 32(12): L12802.

85. Tripathi, S. N. Srivastva, A. K. Dey, S. Satheesh, S. K. Krishnamoorthy, K. , 2007b The vertical profile of atmospheric heating rate profile due to black carbon at Kanpur (Northern India). Atmos. Env. 41 32 69096915 .

86. Satheesh SK, et al. 2009 Vertical structure and horizontal gradients of aerosol extinction coefficients over coastal India inferred from airborne lidar measurements during the Integrated Campaign for Aerosol, Gases and Radiation Budget (ICARB) field campaign. J. Geophys. Res. 114: D05204.

87. Moorthy KK, Satheesh SK, Babu SS, Dutt CBS 2008 Integrated Campaign for Aerosols, gases and Radiation Budget (ICARB): An Overview. J. Earth. Sys. Sci. 117 243262 .

88. Nair VS, Babu SS, Moorthy KK 2008 Aerosol characteristics in the marine atmospheric boundary layer over the Bay of Bengal and Arabian Sea during ICARB: Spatial distribution and latitudinal and longitudinal gradients. J. Geophys. Res. 113: D15208.

89. Satheesh, S. K. Moorthy, K. K. Babu, S. S. Vinoj, V. Dutt, C. B. S. , 2008 Climate implications of large warming by elevated aerosol over India. Geophys. Res. Lett. 35: L19809.

90. Satheesh, S. K. Vinoj, V. Moorthy, K. K. , 2010 Radiative effects of aerosols at an urban location in southern India: Observations versus model. Atmos. Environ. 44 52955304 .

91. Babu SS. et al. 2008 Aircraft measurements of aerosol black carbon from a coastal location in the north-east part of peninsular India during ICARB. J. Earth System Science 117 (Sp. Iss. 1): 263-271.

92. Department of Science and Technology (DST), Continental Tropical Convergence Zone (CTCZ) Programme 2008 Science Plan, Indian Clim. Res. Programme, Gov. of India, New Delhi, 167 p.

93. Prasad 2006K, Singh RP, Singh A (2006) Seasonal climatology of aerosol optical depth over the Indian subcontinent: Trend and departures in recent years. Int. J. Remote Sens. 27 12 23232329.

94. Jenamani RK 2007 Alarming rise in fog and pollution causing a fall in maximum temperature over Delhi. Curr. Sci. 93 314322 .

95. Badarinath KVS, Kharol SK, Sharma AR, Roy PS 2009 Fog over Indo-Gangetic Plains-A study using multi-satellite data and ground observations. IEEE J. Selec. Topics Appl. Earth Obs. Remote Sens. 2 3 185195.

96. Eck, T. et, al, 2012 Fog- and cloud-induced aerosol modification observed by the Aerosol Robotic Network (AERONET). J. Geophys. Res. 117: D07206.

97. 2008, Drummond JR, Attie JL, Liu J, Zou J, Nichitiu F, Seymour MD, Edwards DP, Deeter MN, Gille JC, Richter A (2008) Measurement of low altitude CO over the Indian subcontinent by MOPITT. J. Geophys. Res. 113: D16307.

98. Dey, S. , Di Girolamo, 2011 (2011) A decade of change in aerosol properties over the Indian subcontinent. Geophys. Res. Lett. 38: L14811.

99. Reddy, , C. Venketaraman, 2002a Inventories of aerosols and sulphur dioxide emissions from India: I. Fossil fuel combustion. Atmos. Environ. 36 677697 .

100. Reddy, , C. Venketaraman, 2002b Inventories of aerosols and sulphur dioxide emissions from India: II. Biomass combustion. Atmos. Environ. 36 699712 .

101. Ramanathan, V. et, al, 2005 Atmospheric brown clouds: Impacts on South Asian climate and hydrological cycle, Proc. Natl. Acad. Sci., U.S.A. 102 53265333 .

102. Massie, S. T. Torris, O. Smith, S. J. , 2004 Total Ozone Mapping Spectrometer (TOMS) observations of increases in Asian aerosol in winter from 1979 to 2000. J. Geophys. Res. 109: D18211.

103. Wild, M. Gilgen, H. Roesch, A. Ohmura, A. Long, C. N. Dutton, E. G. Forgan, B. Kallis, A. Russak, V. Tsvetkov, 2005, Ohmura A, Long CN, Dutton EG, Forgan B, Kallis A, Russak V, Tsvetkov A (2005) From dimming to brightening: Decadal changes in surface solar radiation. Science 308 847850.

104. Kumari, B. P. Londhe, A. L. Daniel, S. Jadhav, D. B. , 2007 Observational evidence of solar dimming: Offsetting surface warming over India. Geophys. Res. Lett. 34: L21810.

105. Badarinath KVS, Sharma AR, Kaskaoutis DG, Kharol SK, Kambezidis HD 2010 Solar dimming over the tropical urban region of Hyderabad, India: Effect of increased cloudiness and increased anthropogenic aerosols. J. Geophys. Res. 115: D21208 EOF

106. Streets, D. G. Wu, Y. Chin, 2006 (2006) Two-decadal aerosol trends as a likely explanation of the global dimming/brightening transition. Geophys. Res. Lett. 33: L15806.

107. Porch, W. Chyleka, P. Dubeya, M. Massie, 2007 (2007) Trends in aerosol optical depth for cities in India. Atmos. Environ. 41 75247532 .

108. Kaul, T. Gupta, S. N. Tripathi, V. Tare, J. L. Collett Jr, 2011 Secondary Organic Aerosol: A comparison between foggy and non-foggy days. Environ. Sci. Technol. 45 73077313 .

109. Venkataraman, C. Habib, G. Kadamba, D. Shrivastava, M. Leon, J. F. Crouzille, B. Boucher, O. Streets, D. G. , 2006 Emissions from open biomass burning in India: Integrating the inventory approach with high resolution Moderate Resolution Imaging Spectroradiometer (MODIS) active-fire and land cover data. Global Biogeochem. Cycles 20: GB2013.

110. Beegum, I. N. Moorthy, K. K. Babu, S. S. Satheesh, S. K. Vinoj, V. Badarinath, K. V. S. Safai, P. D. Devara, P. C. S. Singh, S. Vinod, U. C. Dumka, Pant, 2009D, Devara PCS, Singh S, Vinod, Dumka UC, Pant 2009 Spatial distribution of aerosol black carbon over India during pre-monsoon season. Atmos. Environ. 43: 1071-1078.

111. Bonasoni, P. et, al, 2010 Atmospheric Brown Clouds in the Himalayas: first two years of continuous observations at the Nepal Climate Observatory-Pyramid (5079 m). Atmos. Chem. Phys. 10 75157531 .

112. Decesari S, et al. (2010) Chemical composition of PM10 and PM1 at the high-altitude Himalayan station Nepal Climate Observatory-Pyramid (NCO-P) (5079m a.s.l.). Atmos. Chem. Phys. 10: 4583–4596.

113. Gobbi, G. P. Angelini, F. Bonasoni, P. Verza, G. P. Marinoni, A. Barnaba, 2010, Bonasoni P, Verza GP, Marinoni A, Barnaba F (2010) Sunphotometry of the 2006-2007 aerosol optical/radiative properties at the Himalayan Nepal Climate Observatory-Pyramid (5079ma.s.l.). Atmos. Chem. Phys. 10 1120911221.

114. Ramanathan, V. Ramana, M. V. , 2005 Persistent, widespread, and strongly absorbing haze over the Himalayan foothills and the Indo-Gangetic plains. Pure Appl. Geophys. 162 16091626 .

115. Srivastava, A. K. Pant, P. Hegde, P. Singh, S. Dumka, U. C. Naja, M. Singh, N. Bhavanikumar, Y. , 2011d Influence of south Asian dust storm on aerosol radiative forcing at a high-altitude station in central Himalayas. Int. J. Remote Sens. 32 22 78277845 .

116. Kumar, R. et, al, 2011 First ground based observations of influences of springtime Northern Indian biomass burning over the central Himalayas. J. Geophys. Res. 116: D19302.

117. Srivastava, A. K. Singh, S. Pant, P. Dumka, U. C. , 2012e Characteristics of black carbon over Delhi and Manora Peak-a comparative study. Atmos. Sci. Let. (in press).

118. Mishra, S. K. Tripathi, S. N. Aggarwal, A. Arola, 2012ggarwal A, Arola A (2012) Optical properties of accumulation mode polluted mineral dust: Effects of particle shape, hematite content and semi-external mixing with carbonaceous species. Tellus (accepted).

119. Bond, T. C. Streets, D. G. Yarber, K. F. Nelson, S. M. Woo, J. Klimont, 2004 (2004) A technology-based global inventory of black and organic carbon emissions from combustion. J. Geophys. Res. 109: D14203.

120. Chandra, S. Satheesh, S. K. Srinivasan, 2004 (2004) Can the mixing state of black carbon aerosols explain the mystery of 'excess' atmospheric absorption?. Geophys. Res. Lett. 31: L19109.

121. Xue, M. J. Yan, P. Pan, Xue M, Ma J, Yan P, Pan X (2011) Impacts of pollution and dust aerosols on the atmospheric optical properties over a polluted rural area near Beijing city. Atmos. Res. 101: 835-843.

122. Das SK, Jayaraman 2011 (2011) Role of black carbon in aerosol properties and radiative forcing over western India during pre-monsoon period. Atmos. Res. 102 320334 .

123. National Carbonaceous Aerosols Programme (NCEP), Science Plan, Indian Network for Climate Change Assessment 2011 Gov. of India, New Delhi, 44 p.

124. Singh S, Singh R, Rao VUM (2004) Temporal dynamics of dew and fog events and their impact on wheat productivity in semi-arid region of India, Third International Conference on Fog, Fog Collection and Dew, NetSys Int. (Pty) Ltd., Cape Town, South Africa, 11- 15 Oct.(http://www.up.ac.za/academic/geog/meteo/EVENTS/fogdew2003/PAPERS/C65.pdf).

125. S. Sarkar, R. Chokngamwong, G. Cervone, R. P. Singh, Kafatos, 2006 (2006) Variability of aerosol optical depth and aerosol forcing over India. Adv. Space Res. 37 12 21532159 .

126. K. Ali, G. A. Momin, S. Tewari, P. D. Safai, D. M. Chate, P. S. P. Rao, 2004 Fog and precipitation chemistry at Delhi, north India. Atmos. Environ. 38 42154222 .

127. S. Hameed, M. I. Mirza, Siddiqui. Z. R. Ghauri, R. Javed, A. R. Khan, O. V. Rattigan, S. Qureshi, Husain, 2000 (2000) On the widespread winter fog in Northeastern Pakistan and India. Geophys. Res. Lett. 27 18911894 .

128. S. K. Das, A. Jayaraman, Mishra, 2008, Mishra A (2008) Fog-induced variations in aerosol optical and physical properties over the Indo-Gangetic Basin and impact to aerosol radiative forcing. Ann Geophys. 26 13451354 .

129. Hess. M., P. Koepke, Schultz, 1998 (1998) Optical properties of aerosols and clouds: the software package OPAC. Bull. Am. Meteorol. Soc. 79 831844 .

130. A. Comiskey, S. E. Schwartz, B Schmid, H. Guan, E. R. Lewis, P. Ricchiazzi, J. A. Ogren, 2008 Direct aerosol forcing: Calculation from observables and sensitivities to inputs. J. Geophys. Res. 113: D09202.

131. D. Ganguly, Jayaraman, 2006 (2006) Physical and optical properties of aerosols over an urban location in western India: implications for shortwave radiative forcing. J. Geophys. Res. 111: D24207.

132. D. Ganguly, A. Jayaraman, T. A. Rajesh, Gadhavi, 2006 (2006) Wintertime aerosol properties during foggy and non-foggy days over urban center Delhi and their implications for shortwave radiative forcing. J. Geophys. Res. 111: D15217.

133. G. Pandithurai, R. T. Pinker, T. Takamura, P. C. S. Devara, 2004 Aerosol radiative forcing over a tropical urban site in India. Geophys. Res. Lett. 31: L12107.

134. Badarinath KVS, Latha KM 2006 Direct radiative forcing from black carbon aerosols over urban environment. Adv. Space. Res. 37 12 21832188 .

135. Babu SS, Satheesh SK, Moorthy KK 2002 Aerosol radiative forcing due to enhanced black carbon at an urban site in India. Geophys. Res. Lett. 29(18): 1880.

136. M. V. Ramana, V. Ramanathan, I. A. Podgorny, B. B. Pradhan, Shrestha, 2004B, Shrestha B (2004) The direct observations of large aerosol radiative forcing in the Himalayan region. Geophys. Res. Lett. 31: L05111.

137. Ramachandran 2003 (2003) Aerosol radiative forcing over Bay of Bengal and Chennai: Comparison with maritime, continental, and urban aerosol models. J. Geophys. Res. 110: D21206 EOF.

138. Satheesh SK 2002 Radiative forcing by aerosols over Bay of Bengal region. Geophys. Res. Lett. 29(22): 2083.

139. S. K. Satheesh, V. Ramanathan, B. N. Holben, K. K. Moorthy, N. G. Loeb, H. Maring, J. M. Prospero, Savoie, 2002 (2002) Chemical, microphysical, and radiative effects of Indian Ocean aerosols. J. Geophys. Res. 107: 4725.

CHAPTER 3

Mie LIDAR Observations of Tropospheric Aerosol over Wuhan

Wei Gong [1,2], Boming Liu [1,2,*], Yingying Ma [1,2] and Miao Zhang [1,2]

[1] State Key Laboratory of Information Engineering in Surveying, Mapping and Remote Sensing (LIESMARS), Wuhan University, Wuhan 430079, China

[2] International Research Center of Satellite Remote Sensing and Atmospheric Monitoring, Wuhan University, Wuhan 430079, China

ABSTRACT

Wuhan is a rapidly developing large city in central China. To analyze the aerosol characteristics over Wuhan, the optical properties of the nocturnal aerosol layers in the lower troposphere were observed using a ground-based LIDAR(Light Detection And Ranging) located in the Laboratory of Information Engineering in Surveying, Mapping and Remote Sensing (LIESMARS) from Wuhan University, China (114°21′E, 30°32′N) in January 2013–January 2015. The vertical distribution and temporal variation of tropospheric aerosols over Wuhan were summarized. The atmospheric boundary layer height (ABLH) was mainly at an altitude of 1.5–2 km (~33.1% probability), with an annual average of 1.66 km. The ABLH was higher in spring–summer (~2 km) and lower in autumn–winter (~1.2 km). The aerosol optical depth (AOD) was higher in spring–autumn than in summer–winter. The highest AOD was about 0.79 in October and the lowest was about 0.08 in January. The annual average was about 0.3. To study the relationship between the AOD and the particulate matter ≤2.5 μm in the aerodynamic diameter ($PM_{2.5}$) in the lower troposphere, a typical haze event from 9–14 October 2014 was investigated. The results showed a correlation coefficient of 0.5165 between these two variables.

KEYWORDS

LIDAR; aerosol extinction coefficient; atmospheric boundary layer height; aerosol optical depth; $PM_{2.5}$

1. INTRODUCTION

Aerosols play a major role in the assessment of climate, Earth's radiation, the formation of clouds and precipitation, and environmental problems [1]. The Intergovernmental Panel on Climate Change (IPCC) has reported that large uncertainties on climate change are attributed to both the direct and indirect effect of aerosols [2]. They have also been considered to be the greatest contributors to the total uncertainty in radiative forcing because of their direct and indirect effects. Thus, investigation of the spatiotemporal distribution of tropospheric aerosols is very important for understanding aerosol radiative forcing [3].

The current primary means of detecting aerosols is the AERONET monitoring network [4], which can provide aerosol column information but is unable to obtain its accurate vertical distribution. However, LIDAR (Light Detection And Ranging) detection systems have been developed and widely used for this purpose. The European Aerosol Research LIDAR Network (EARLINET) was the first aerosol LIDAR network mainly aimedat providing a comprehensive, quantitative, and statistically significant database for the aerosol distribution on a continental scale [5]. Matthias *et al.* studied the vertical distribution of aerosols over Europe [6]. In the US, the Micro Pulse LIDAR Network (MPLNET) was initiated by NASA (National Aeronautics and Space Administration), covering more area and supporting a wide range of atmospheric aerosols. Welton *et al.* [7] monitored global changes in clouds and aerosols. In Japan, the National Institute for Environmental Studies (NIES) also has a LIDAR Monitoring Network, which is called AD-NET (Sugimoto and Huang, 2014) [24]. Hayashida *et al.*[8] observed the volcanic disturbances in the stratospheric aerosol layer over Tsukuba, Japan. Some Chinese institutions such as the Anhui Institute of Optics and Fine Mechanics [9],

Xi'an University of Technology [10], the Institute of Atmospheric Physics [11], and Lanzhou University [12] are also studying LIDAR systems to detect aerosols. Examples of studies are given by Yan *et al.* [13], who observed the boundary layer structure and aerosol properties over Xi'an, Jinhuan *et al.* [14], who analyzed the variation characteristics of the atmospheric aerosol optical depths and visibility in North China, and Huang *et al.* [15], who detected the dust aerosol vertical structure. For heavily industrialized cities such as Wuhan, Zhang *et al.* [16] analyzed the aerosol optical properties for a haze episode and Gong *et al.* [17,18] studied the aerosol optical properties and determined their size distribution. Nevertheless, research on the atmospheric boundary layer height (ABLH) and aerosol annual optical properties is relatively scarce. To fill this gap, the annual changes in aerosol optical characteristics and ABLH were studied for Wuhan.

Our study of the vertical structure and optical properties of the aerosol layer was performed using ground-based LIDAR. The LIDAR uses aerosols as tracers, which, through scattering, contribute to the backscatter intensity of the range-resolved measurement. Accordingly, the height and optical properties of the aerosol layer were determined from the LIDAR signal. We used two years of LIDAR data (January 2013–January 2015) to investigate the vertical structure and optical properties of the aerosol layer in Wuhan. The distribution of the ABLH was analyzed and the seasonal and monthly aerosol variation was studied. Finally, a typical haze event from 9–14 October 2014 was used to analyze the correlation between the concentration of air particles below 2.5 μm diameter ($PM_{2.5}$) and the aerosol optical depth (AOD).

2. STUDY SITE AND INSTRUMENTATION USED

2.1. Study Site Location

The LIDAR system was located at Wuhan University (114°21′E, 30°32′N), Guangbutun, in the business center of Wuhan, about 39 m away from the ground and surrounded by buildings [18]. Wuhan City is one of China's largest industrial and commercial cities, with serious impacts on the air quality from industrial emissions. Its population reached 10.02 million in 2011, with 5.7 million in the center where the atmospheric pollution is largest. This region has a subtropical monsoon climate, with abundant rainfall and high temperatures in the summer and abundant cloud cover, relatively high humidity, and low temperatures in winter. Because of its sustained extremely high temperatures in the summer, Wuhan is known as one of China's four furnaces.

2.2. LIDAR System Description

The light detection and ranging system included a laser emitting, an optical receiving, and a signal acquisition system. The laser emitting system consisted of three parts: a Nd: YAG (Neodymium-doped Yttrium Aluminium Garnet) laser, a laser beam expander, and a mirror (Figure 1). The emission wavelength of the Nd: YAG pulsed laser was 532 nm, with a laser pulse width, energy, and repetition frequency of 10 ns, 140 mJ, and 20 Hz, respectively. The laser beam was Gaussian. To further compress the divergence angle of the laser beam, a beam expander was used. The optical receiver system consisted of three parts: a telescope, a collimating lens, and narrowband interference filters. The return LIDAR signal backscattered by the atmosphere was collected by a 10-inch (254 mm) diameter Schmidt-Cassegrain telescope. A narrowband filter was used to suppress the sky background components.

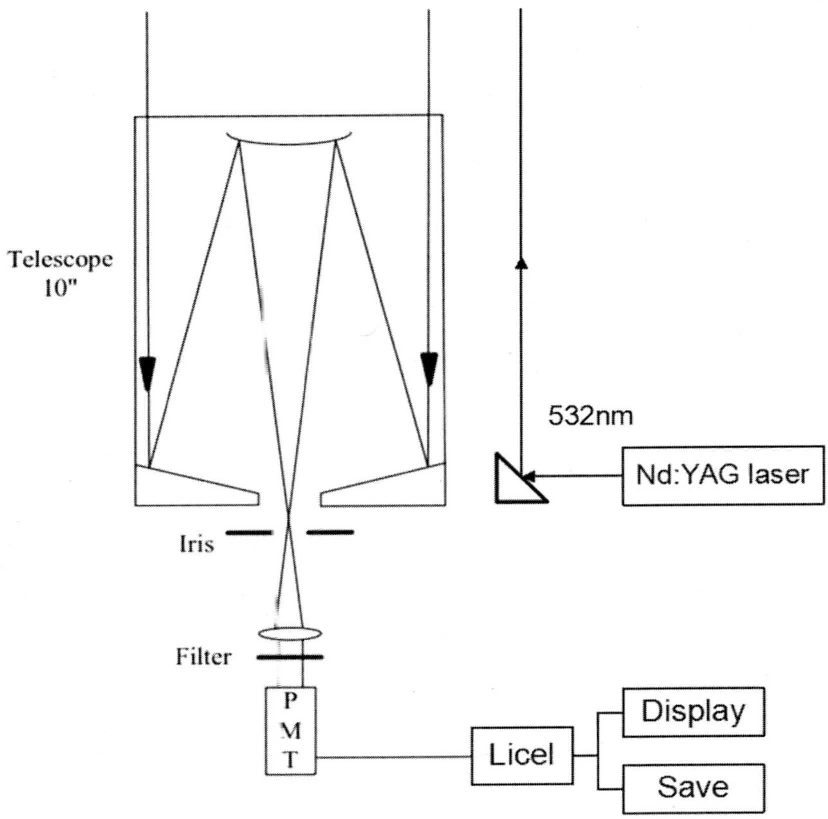

Figure 1. LIDAR image.

The collected backscattered signals were detected by a photomultiplier tube (PMT) with a high-speed preamplifier in an analog detection mode and recorded by a 16-bit data acquisition (ADQ) card at a 40 MHz sampling rate. The backscattered signals measured by the ADQ were stored in a personal computer, with vertical and temporal resolutions of 3.75 m and 60 s, respectively. More details can be seen in Table 1.

Table 1. Mie LIDAR specifications.

Transmitter	
Wavelength (nm)	532, Nd: YAG
Pulse width (ns)	10
Pulse repetition frequency (Hz)	20
Maximum pulse energy (mJ)	140
Laser beam divergence (mrad)	0.5
Receiver	
Optical design	Schmidt-Cassegrain
Diameter of telescope (mm)	254
Focal length (mm)	2500
Filter bandwidth (nm)	3
Field of view (mrad)	1
Spatial resolution (m)	3.75

3. DATA RETRIEVAL

The data processing consisted of two steps. First, the ideal profile fitting method to calculate the distribution of the boundary layer profile with the best stability for a good mixed situation to effectively obtain the ABLH by fitting the backscatter coefficient curve, was introduced [19]. Then, we obtained the distribution of the AOD and the extinction coefficient using the Fernald method [20,21,22,23], which regards the atmosphere as two components, for a more accurate calculation.

3.1. Ideal Profile Fitting Method

The ideal profile fitting method obtained the boundary layer parameters by fitting an atmosphere backscattering coefficient [19]. The atmospheric boundary layer becomes a mixed layer, characterized by approximate height invariant values for winds and scalar quantities under convective conditions. The mixed layer is capped by an entrainment zone, which provides a transition to the (often stable) lower troposphere. The mixed layer depth is of primary importance because it determines the volume of atmosphere through which surface-emitted pollutants can be diluted, and it is also a length scale governing the behavior of the largest scales of boundary layer turbulence [25]. Thus, we regard the mixed layer height as the ABLH top. The ideal profile was expressed by an error function:

$$X(r) = \frac{X_m + X_u}{2} - \frac{X_m - X_u}{2} erf\left(\frac{r - r_m}{s}\right) \tag{1}$$

Where r_m is the average depth of the boundary layer, X_m is the average of $X(r)$, X_u is the averaged $X(r)$ above the boundary layer, and s is related to the thickness of the entrainment layer. The above parameters were determined by least squares fitting the measured signal to the ideal profile. Finally, the ABLH was calculated.

3.2. Fernald Method

According to the Fernald method, the received laser pulse power $P(r)$ from distance r can be described as a function of the volume extinction coefficient $\alpha(r)$ and the backscatter coefficient $\beta(r)$ [21].

$$P(r) = CP_0 r^{-2}[\beta_m(r) + \beta_a(r)]\exp[-2\int_0^r [\alpha_m(r) + \alpha_a(r)]dr] + P_{nb} \tag{2}$$

Where C is a calibration constant including the losses in the transmitting and receiving optics and the effective receiver aperture, $P_0$is an output monitor pulse whose amplitude is proportional to the transmitted energy, $\beta_m(r)$ and $\beta_a(r)$ are the backscattering coefficients of the aerosols and molecules at slant range r, respectively, $\alpha_m(r)$ and $\alpha_a(r)$ are the extinction cross-sections of the aerosols and molecules at range r, respectively, and P_{nb} is the background noise. The molecular atmosphere scattering properties, $\beta_m(r)$ and $\alpha_m(r)$, were determined from the approximated appropriate standard atmospheres or the best available meteorological data, while the aerosol scattering properties, including $\beta_a(r)$ and $\alpha_a(r)$, had to be determined. According to the research [26], they obtained the vertical distribution of the LIDAR ratio by the Raman scattering principle on September 12 2009 at Wuhan. They found that the LIDAR ratio fluctuated around 50 sr in the range of 3 km. Considering that larger-sized aerosols are dropping and the atmosphere becomes clearer as the night is falling, they took the average of the LIDAR ratio (50 sr).The corresponding ratio for the aerosol scatterers was constant for the Wuhan area ($S_a = \alpha_a(r) / \beta_a(r) = 50$). The solution to Equation (2) for the aerosol backscattering coefficient then became:

$$\beta_a(r) = \frac{P(r)r^2 \exp[-2(S_a - S_m)\int_0^r \beta_m(r)dr]}{CP_0 - 2S_a\int_0^r P(r)r^2 \exp[-2(S_a - S_m)\int_0^r \beta_m(r')dr']dr} - \beta_m(r) \tag{3}$$

The atmospheric extinction coefficient is an important parameter to study the properties of the boundary layer, while the quality of the atmosphere can be characterized by the AOD [22]. The AOD is the integral of the extinction coefficient ($\alpha(r)$) along the optical path between R_1 and R_2:

$$AOD = \int_{R_1}^{R_2} \alpha(r)dr \tag{4}$$

In addition, there is an inherent problem with the Mie LIDAR, which is its overlap function. The overlap function of the Mie LIDAR has been calculated by a simple overlap factor calculation method based on laser intensity distribution [27].The relative error of the overlap factor correction is more than 20% when the range is below 200 m. Thus, we have removed the LIDAR signals from 0 to 200 m. Additionally, we used the least squares linear fitting to fit the range-corrected signal in the far range. Then we extrapolated the fitting function to get the range-corrected signals from 0 to 200 m. Finally, the range corrected signal was used to get the AOD inversion.

4. RESULTS AND DISCUSSION

We interpreted the ABLH and aerosol characteristics from January 2013–January 2015 based on 151 observations between 7 p.m. and 7 a.m. the next day. According to the date, the spatiotemporal distribution of the tropospheric aerosols was analyzed.

4.1. ABLH Characteristics

The statistics data represents the average height of the entire night atmospheric boundary layer. The frequency distribution histogram of the ABLH in Wuhan for 2013 and 2014 (Figure 2) shows an approximately normal distribution, with an ABLH of about 2 km in spring–summer and about 1.2 km in autumn–winter. A large fraction of the annual ABLH (33.1%) ranged 1.5–2 km, with an average of 1.66 km. In addition, it was mostly (70.8%) below 2 km.

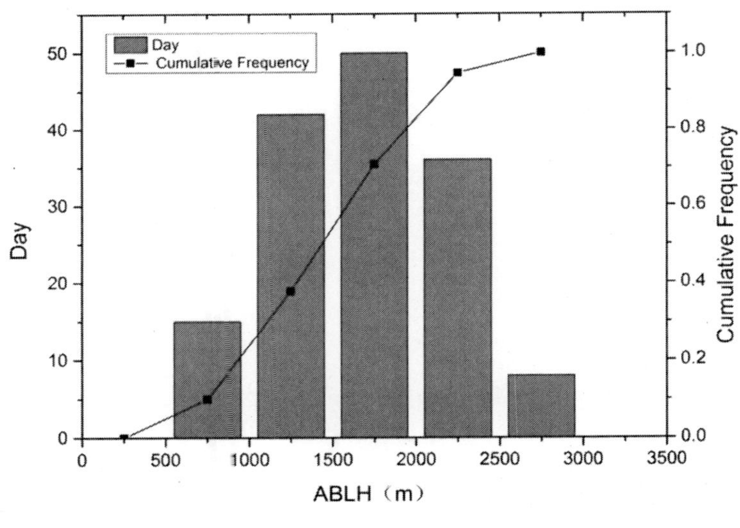

Figure 2. Cumulative probability distribution and probability distribution of the atmospheric boundary layer height (ABLH). The black line represents the cumulative frequency and the red histogram shows the number of occurrences in days.

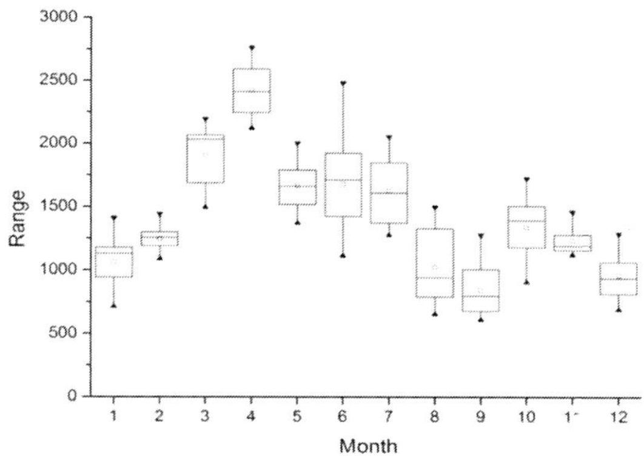

spring	Mar–May
summer	Jun–Aug
autumn	Sep–Nov
winter	Dec–Feb

Figure 3. Monthly statistics of the atmospheric boundary layer height (ABLH). The boxes represent the mean of the maximum and minimum for each month, the triangles represent the monthly maximum and minimum, and the short horizontal lines represent the monthly average.

Figure 3 shows the monthly and seasonal variation of the ABLH. The rainy season of Wuhan is in the summer. Half of month is raining in May, June, and July. Thus, the result is probably low in May, June, and July. The maximum value was about 2.76 km in April, the minimum was about 0.67 km in September, and the annual average was about 1.66 km. Spring and summer had higher values, with the highest in spring, compared to autumn and winter, with the lowest value in autumn. In Wuhan, in spring–summer and March–August, because of the strong solar radiation and high surface temperatures, the atmospheric convective motion is frequent and intense, enhancing the diffusion of aerosols.

4.2. AOD Characteristics

Figure 4 depicts the variations of the monthly AOD at 532 nm in January–December. The highest AOD (~0.79) appeared in October, while the smallest (~0.08) appeared in January. The annual mean was about 0.3, with higher values in spring and autumn and lower values in summer and winter. The higher AOD in spring was probably caused by dust particles in the free troposphere transported from the north, while the summer values seemed to result from hygroscopic aerosols and abundant moisture in the mixing layer and intense turbulence near the ground.

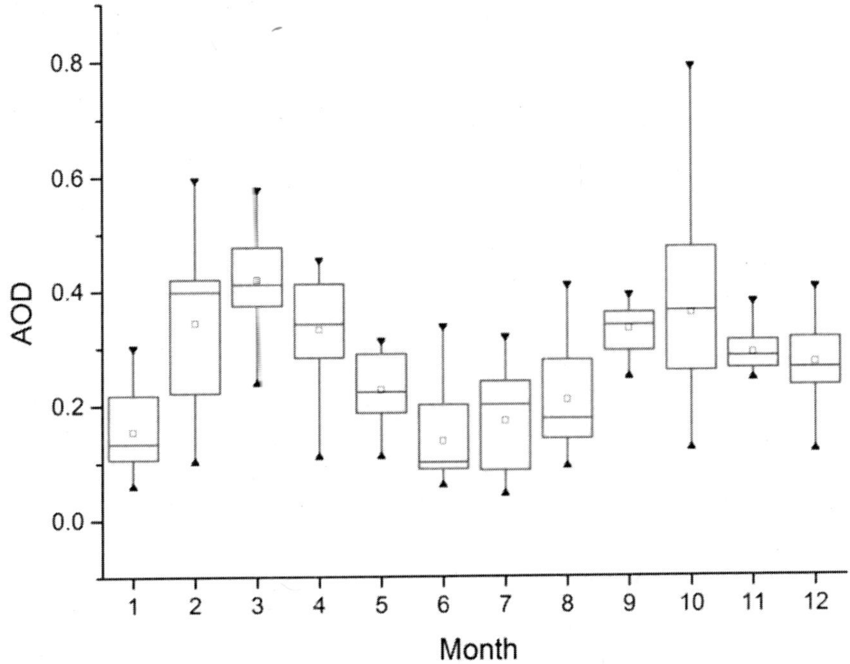

Figure 4. Monthly statistics of the aerosol optical depth (AOD) at 532 nm. The boxes represent the mean of the maximum and minimum for each month, the triangles represent the monthly maximum and minimum, and the short horizontal lines represent the monthly average.

4.3. Haze Event Analysis

A typical haze event on 9–14 October 2014 was analyzed to study the relationship between the $PM_{2.5}$ and the AOD. Like most haze events, it was characterized by a long duration, high concentrations, and significant urban pollution, and was not influenced by rain.

Figure 5 shows the extinction coefficient profiles during a typical haze event on 9–14 October 2014. We studied the changes in the aerosol extinction coefficient within 3 km. As seen in Figure 5a, the extinction coefficient was high on 9 October, ranging 0–1.5 km. The haze began on 9 October and reached its maximum on 11 October, when it was high in the 0–2.5 km range. The haze began to weaken after October 11 and, on 14October, the atmosphere returned to its normal state.

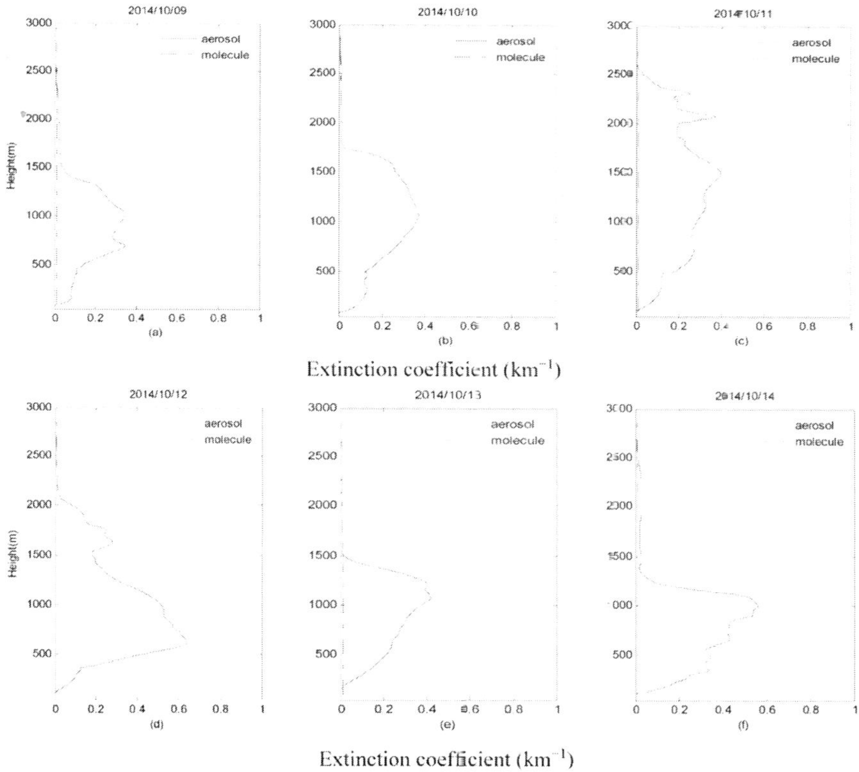

Figure 5. Extinction coefficient profiles for a typical haze event on 9–14 October 2014: (**a**) 9 October, (**b**) 10 October, (**c**) 11 October, (**d**) 12 October, (**e**) 13 October, and (**f**) 14 October. The solid lines represent the atmospheric aerosol extinction coefficient, while the dotted lines represent the atmospheric molecule extinction coefficient.

Figure 6 shows the night trends in AOD and $PM_{2.5}$ during the haze event. The $PM_{2.5}$ data was provided by the particulate matter detector from the Bureau of Meteorology of Wuhan. There is no significant relationship between AOD and $PM_{2.5}$ on 9 and 14 October due to the fact that there is no serious haze on those days. The trend in AOD and $PM_{2.5}$ was roughly the same, being relatively stable on 10–13 October. A sudden increase occurred on 11 October, with the beginning of the haze event, and peak values of AOD and $PM_{2.5}$ of 0.7 and 120, respectively, indicating serious pollution. This shows that there is a certain correlation between AOD and $PM_{2.5}$ during the haze. The haze began to weaken on 12 October and the air quality returned to normal on 13–14 October, consistent with the variation of the extinction coefficient.

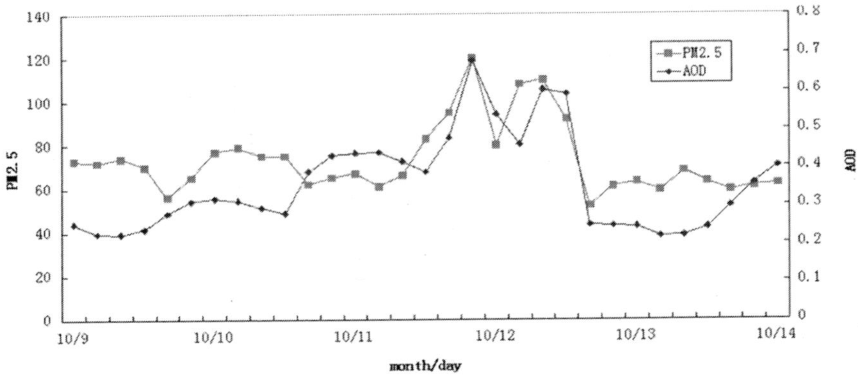

Figure 6. Variation in aerosol optical depth (AOD) and concentration of particles with a diameter below 2.5 μm ($PM_{2.5}$) on 9–14 October 2014. The red line shows the change trend of $PM_{2.5}$. The blue line shows the change trend of AOD.

Figure 7 shows the correlation between the $PM_{2.5}$ and the AOD in the haze period, which was 0.5165, indicating a linear relationship. The AOD is also indicative of the level of environmental pollution.

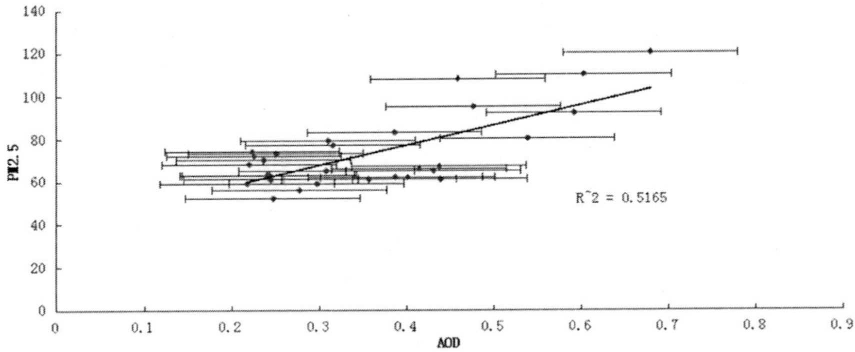

Figure 7. Correlation between the concentration of particles below 2.5 μm diameter ($PM_{2.5}$) and the aerosol optical depth (AOD). The blue points represent the date, the horizontal lines represent the positive and negative deviations, and the black line represents the linear regression.

5. CONCLUSIONS

To understand the influence of turbulence on the aerosol distribution, the nocturnal aerosol layer in the lower troposphere was studied over Guangbutun, Wuhan, using a ground-based LIDAR during January 2013 January 2015.

The optical properties of the aerosol layers were determined using the fluctuation of the LIDAR backscattered signals from the aerosols. The distribution trends of the night aerosol layer in Wuhan were higher in spring and summer (~2 km) and lower in autumn and winter (~1.2 km). The temperature distribution may result from the special characteristics of Wuhan, with higher values in spring–summer and lower values in autumn–winter. The atmospheric boundary layer height (ABLH) was mainly at an altitude of 1.5–2 km (~33.1% probability), with an annual average of 1.66 km. Additionally.it was mostly (70.8%) below 2 km. The AOD was higher in spring because of the anthropogenic aerosols from the building construction and manufacturing industries and stagnant weather, and it was lower in summer, possibly because of the large volume of precipitation and extremely high temperatures. The highest AOD (~0.79 at 532 nm) appeared in October, probably due to hazy weather, while the lowest appeared in January (~0.08 at 532 nm). The annual mean was 0.3. In regard to the relationship between AOD and $PM_{2.5}$, there is no significant relationship in normal weather conditions. However, in extreme weather conditions, the AOD and the $PM_{2.5}$ had a correlation coefficient of 0.5165, indicating a linear relationship in the troposphere. It shows that the AOD is also indicative of the level of environmental pollution. This research provides a good basis for further studies on atmospheric changes and urban pollutant generation since it is important to improve our understanding of the effects of aerosol on the climate and atmosphere.

ACKNOWLEDGMENTS

This study was financially supported by the Haze Program of the Wuhan Technological Bureau, Cheng guan project of Wuhan (2014070404010198),National Basic Research Program (Grant No. 2011CB707106) and the National Natural Science Foundation of China (NSFC) (Program No. 41127901No. 41401498). We would like to thank the Bureau of Meteorology of Wuhan for providing the $PM_{2.5}$ data. We also thank the editor for the help in improving the manuscript.

AUTHOR CONTRIBUTIONS

The study was completed with cooperation between all authors. Wei Gong and Liu Boming designed the research topic, Miao Zhang and Liu Boming conducted the experiment and wrote the paper, and Yingying Ma checked the experimental results. All authors agreed with the submission of the manuscript.

CONFLICTS OF INTEREST

The authors declare no conflicts of interest.

REFERENCES

1 Bösenberg, J.; Hoff, R.; Ansmann, A.; Müller, D.; Antuna, J.C.; Whiteman, D.; Freudenthaler, V. Plan for the implementation of the GAW aerosol LIDAR observation network GALION. *WMO TD n°* **2007**, *1443*, 178–178.

2 Haines, A. *Climate Change 2001: The Scientific Basis. Contribution of Working Group 1 to the Third Assessment report of the Intergovernmental Panel on Climate Change*; Houghton, J.T., Ding, Y., Griggs, D.J., Noguer, M., van der Winden, P.J., Dai, X., Eds.; Cambridge University Press: Cambridge, UK, 2001; p. 321.

3 Nakajima, T.; Yoon, S.; Ramanathan, V.; Shi, G.; Takemura, T.; Higurashi, A.; Takamura, T.; Aoki, K.; Sohn, B.; Kim, S.; *et al.* Overview of the atmospheric brown cloud east Asian regional experiment 2005 and a study of the aerosol direct radiative forcing in East Asia. *J. Geophys. Res.* **2007**, *112*.

4 Holben, B.N.; Eck, T.F.; Slutsker, I.; Tanre, D.; Buis, J.P.; Setzer, A.; Vermote, E.; Reagan, J.A.; Kaufman, Y.J.; Nakajima, T. AERONET—A federated instrument network and data archive for aerosol characterization. *Remote Sens. Environ.* **1998**, *66*, 1–16.

5 Matthias, V.; Freudenthaler, V.; Amodeo, A.; Balin, I.; Balis, D.; Bösenberg, J.; Chaikovsky, A.; Chourdakis, G.; Comeron, A.; Delaval, A.; *et al.* Aerosol LIDAR intercomparison in the framework of the EARLINET project. 1. Instruments. *Appl. Opt.* **2004**, *43*, 961–976.

6 Matthias, V.; Balis, D.; Bösenberg, J.; Eixmann, R.; Iarlori, M.; Komguem, L.; Mattis, I.; Papayannis, A.; Pappalardo, G.; Perrone,

M.R.; *et al.* Vertical aerosol distribution over Europe: Statistical analysis of Raman Lidar data from 10 European Aerosol Research Lidar Network (EARLINET) stations. *J. Geophys. Res. Atmos.* **2004**, *109*, 159–172.

7 Welton, E.J.; Campbell, J.R.; Spinhirne, J.D.; Scott, V.S., III. Global monitoring of clouds and aerosols using a network of micropulse Lidar systems. In Presented at the Second International Asia-Pacific Symposium on Remote Sensing of the Atmosphere, Environment, and Space, Sendai, Japan, 9–12 October 2000; pp. 151–158.

8 Hayashida, A.S.; Sasano, Y.; Iikura, Y. Volcanic disturbances in the stratospheric aerosol layer over Tsukuba, Japan, observed by the National Institute for Environmental Studies Lidar from 1982 through 1986. *J. Geophys. Res. Atmos.* **1991**, *96*, 15469–15478.

9 Wu, D.; Zhou, J.; Liu, D.; Wang, Z.; Zhong, Z.; Xie, C.; Qi, F.; Fan, A.; Wang, Y. 12-year LIDAR Observations of Tropospheric Aerosol over Hefei (31.9 N, 117.2 E), China. *J. Opt. Soc. Korea* **2011**, *15*, 90–95.

10 Gao, F.; Bergant, K.; Filipčič, A.; Forte, B.; Hua, D.-X.; Song, X.Q.; Stanič, S.; Veberič, D.; Zavrtanik, M. Observations of the atmospheric boundary layer across the land–sea transition zone using a scanning Mie Lidar. *J. Quant. Spectrosc. Radiat. Transfer* **2011**, *112*, 182–188.

11 Jinhuan, Q.; Siping, Z.; Qirong, H.; Qilin, X.; Liquan, Y.; Wenming, W.; Jidong, P.; Jinhui, S. Lidar Measurements of Cloud and Aerosol in the Upper Troposphere in Beijing. *Chin. J. Atmos. Sci.* **2003**, *27*, 1–7.

12 Huang, J.; Zhang, W.; Zou, J.; Bi, J.; Shi, J.; Wang, X.; Chang, Z.; Huang, Z.; Yang, S.; Zhang, B.; *et al.* An overview of the Semi-arid Climate and Environment Research Observatory over the Loess Plateau. *Adv. Atmos. Sci.* **2008**, *25*, 906–921.

13 Yan, Q.; Hua, D.; Wang, Y.; Li, S.; Gao, F.; Zhou, Z.; Wang, L.; Liu, C.; Zhang, S. Observations of the boundary layer structure and aerosol properties over Xi'an using an eye-safe Mie scattering Lidar. *J. Quant. Spectrosc. Radiat. Transfer* **2013**, *122*, 97–105.

14 Jinhuan, Q.; Liquan, Y. Variation characteristics of atmospheric aerosol optical depths and visibility in North China during 1980–1994. *Atmos. Environ.* **2000**, *34*, 603–609.

15 Huang, Z.; Huang, J.; Bi, J.; Wang, G.; Wang, W.; Fu, Q.; Li, Z.; Tsay, S.C.; Shi, J. Dust aerosol vertical structure measurements using three MPL lidars during 2008 China-US joint dust field experiment. *J. Geophys. Res. Atmos.* **2010**, *115*, D7.

16 Zhang, M.; Ma, Y.; Gong, W.; Zhu, Z. Aerosol Optical Properties of a Haze Episode in Wuhan Based on Ground-Based and Satellite Observations. *Atmosphere* **2014**, *5*, 699–719.

17 Gong, W.; Zhang, M.; Han, G.; Ma, X.; Zhu, Z. An Investigation of Aerosol Scattering and Absorption Properties in Wuhan, Central China. *Atmosphere* **2015**, *6*, 503–520.

18 Gong, W.; Zhang, S.; Ma, Y. Aerosol optical properties and determination of aerosol size distribution in Wuhan, China. *Atmosphere* **2014**, *5*, 81–91.

19 Wang, L.C.; Gong, W.; Ma, Y.Y.; Zhang, M. Modeling regional vegetation NPP variations and their relationships with climatic parameters in Wuhan, China. *Earth Interact.* **2013**, *17*, 1–20.

20 Steyn, D.G.; Baldi, M.; Hoff, R.M. The detection of mixed layer depth and entrainment zone thickness from lidar backscatter profiles. *J. Atmos. and Ocean. Tech.* **1999**, *16*, 953–959.

21 Fernald, F.G.; Herman, B.M.; Reagan, J.A. Determination of Aerosol Height Distributions by Lidar. *J. Appl. Meteorol.* **1972**, *11*, 482–489.

22 Fernald, F.G. Analysis of atmospheric lidar observations: Some comments. *Appl. Opt.* **1984**, *23*, 652–653.

23 Klett, J.D. Stable analytical inversion solution for processing LIDAR returns. *Appl. Opt.* **1981**, *20*, 211–220.

24 Sugimoto, N.; Huang, Z.W. Lidar methods for observing mineral dust. *J. Meteor. Res.* **2014**, *28*, 173–184.

25 Stull, R.B. *An Introduction to Boundary Layer Meteorology*; Kluwer Academic Publishers: Dordrecht, The Netherlands, 1988; p. 666.

26 Gong, W.; Zhang, J.; Mao, F.; Jun, L. Measurements for profiles of aerosol extinction coefficient, backscatter coefficient, and lidar ratio over Wuhan in China with Raman/Mie lidar. *Chin. Opt. Lett.* **2010**, *8*, 533–536.

27 Gong, W.; Mao, F.; Li, J. OFLID: Simple method of overlap factor calculation with laser intensity distribution for biaxial lidar. *Opt. Commun.* **2011**, *284*, 2966–2971.

CHAPTER 4

Simulating the Impacts of Marine Organic Emissions on Global Atmospheric Chemistry and Aerosols Using an Online-Coupled Meteorology and Chemistry Model

Brett Gantt[1,2], Timothy Glotfelty[1], Nicholas Meskhidze[1], Yang Zhang[1*]

[1]Department of Marine, Earth, and Atmospheric Sciences, North Carolina State University, Raleigh, USA

[2]Now at Office of Air Quality Planning and Standards, U.S. Environmental Protection Agency, Research Triangle Park, Raleigh, USA

ABSTRACT

To realistically simulate the impacts of marine isoprene and primary organic aerosols (POA) on atmospheric chemistry, a unified model framework with online emissions, comprehensive treatment of gas-phase chemistry, and advanced aerosol microphysics is required. In this work, the global-through-urban WRF/Chem model (GU-WRF/Chem) implemented with the online emissions of marine isoprene and size-resolved marine POA is applied to examine such impacts. The net effect of these emissions was increased surface concentrations of isoprene and organic aerosols and decreased surfaced concentrations of hydroxyl radical and ozone over most marine regions. With the inclusion of these emissions, GU-WRF/Chem better predicted the surface concentrations of isoprene and organic aerosols and the aerosol number size distribution when compared to measurements in clean marine conditions.

KEYWORDS

Marine Organic Emissions, Marine Organic Aerosol, Atmospheric Chemistry, Online-Coupled Model, GU-WRF/Chem

1. INTRODUCTION

The chemical composition and aerosols of the natural background atmosphere are important to model predictions of air quality and climate [1] [2]. Most models, however, have an incomplete emission inventory of natural marine aerosols and precursor gases beyond sea-salt aerosol and dimethyl sulfide. Two major emission types neglected from the default setting of many current models are biogenic volatile organic compounds (BVOCs) of marine origin and primary organic aerosol (POA) associated with sea spray. Isoprene is the most widely-re- ported BVOC species with a marine emission source [3] and has been shown to have a small impact on surface ozone (O_3) and secondary organic aerosol (SOA) formation in coastal regions [4] . Marine POA has been suggested to affect the surface aerosol mass concentration and cloud condensation nuclei (CCN) in remote marine regions [5] - [8] . However, large uncertainties in chemical composition [9] [10] and magnitude [11] [12] of sea spray aerosol have been shown to play an important role in determining the direct and indirect radiative forcing of marine aerosols [13] . Previous modeling efforts have included marine isoprene and POA emissions in several atmospheric chemistry [14] - [19] and online-coupled global climate models [6] [13] [20] [21] .

In this study, we use the online-coupled global-through-urban weather and forecasting model with chemistry (GU-WRF/Chem) to quantify the impact of online emissions of marine isoprene and POA on atmospheric chemistry and aerosol abundance in remote marine and coastal regions. GU-WRF/Chem is capable of simulating these impacts across a wide range of spatial scales while reducing uncertainties from the use of offline-coupled model systems with inconsistent model treatments [22] . Comprehensive treatments of gas-phase chemistry and aerosol microphysics, in conjunction with online emissions of marine BVOC and POA, are expected to improve model predictions of surface concentrations of BVOCs, O_3, and organic aerosols. The overall objective of this study is to estimate the impact of marine isoprene and POA emissions on remote marine atmospheric chemistry through comprehensive evaluation of model predictions against observations from marine regions.

2. MODEL DESCRIPTION AND SETUP

GU-WRF/Chem is developed upon the global WRF (GWRF) version 3.0 [23] -[26] and the mesoscale WRF/ Chem version 3.0 [27] , and described in detail by Zhang et al. [22] . Simulations in this study are conducted with the 2005 Carbon Bond chemical mechanism with global extension (CB05_GE) [28] . The Model of Aerosol Dynamics, Reaction, Ionization, and Dissolution (MADRID) aerosol microphysics module [29] [30] used in GU-WRF/Chem treats all major aerosol chemical and microphysical processes including inorganic aerosol thermodynamic equilibrium, SOA formation, nucleation, gas/particle mass transfer, condensation, and coagulation with eight size sections to represent the particle size distribution. MADRID was selected from among the three aerosol modules in this version of GU-WRF/Chem because it is the only one that treats isoprene as an SOA precursor. 32 aerosol types were considered in GU-WRF/Chem with anthropogenic and terrestrial emissions described in detail in Zhang et al. [22]. Model configurations not pertaining to chemistry were consistent with the baseline GWRF simulation described in Zhang et al. [26] . The model was reinitialized from the NCEP Final Global Data Assimilation System every 24 hours to allow meteorological feedbacks while constraining the simulation with re-initialized meteorological conditions based on observations.

Marine isoprene emissions implemented into GU-WRF/Chem were based on the parameterization developed by Gantt et al. [31] , which experimentally determined that marine isoprene production is a function of phytoplankton abundance, speciation, and photosynthetically active radiation. Model implementation of marine isoprene emissions of Gantt et al. [31] has been previously described in detail in Gantt et al. [4] , which uses the isoprene production rate for diatoms, model-predicted solar radiation, and Level 3 monthly average chlorophyll- a concentration ([Chl-a]) and diffuse attenuation coefficient derived from the Sea-viewing Wide Field-of- view Sensor (SeaWiFS) satellite to drive the emissions. Despite evidence for marine production of monoterpenes [32] , this emission source was not included in GU-WRF/Chem due to the very low production rates relative to marine isoprene emissions [33] . Marine isoprene emissions were calculated online within GU-WRF/Chem and added to the existing terrestrial isoprene emissions of Guenther et al. [34] .

Marine POA emissions implemented into GU-WRF/Chem were based on the parameterization developed by Gantt et al. [35] . These size-dependent aerosol emissions were based on the calculation of the organic mass fraction of sea spray aerosol (OM_{SSA}), which Gantt et al. [35] found to be positively related to phytoplankton abundance and aerosol diameter and negatively related to surface wind speed. Within GU-WRF/Chem, $OM_{SSA}$was calculated online using monthly-mean SeaWiFS-derived [Chl-a] and model-predicted surface wind speed and applied to the existing sea spray emission module of Gong et al. [36] and O'Dowd et al. [37] . Model implementation of these emissions performed here was similar to the method described in detail by Meskhidze et al. [6] . The size-dependent mass and number emissions of marine POA were added online to the existing terrestrial POA based on the sectional size bins of MADRID in GU-WRF/Chem, with the atmospheric evolution of marine/terrestrial POA and SOA is simulated explicitly.

Two sets of GU-WRF/Chem simulations were performed: 1) the Default simulation without marine organic emissions and 2) the MarOrg simulation with marine isoprene and POA emissions described above. Each set of simulations was conducted over a 12-month period for two years (2001 and 2009) with a one-month spin-up for each year at a spatial resolution of $4^\circ \times 5^\circ$ and 27 vertical layers (1st vertical layer height: 35 m) up to 50 mb. These years were chosen for direct evaluation with field campaigns [38] -[40] measuring marine isoprene emissions and surface concentrations of isoprene and organic aerosols in remote marine environments. Other than the marine isoprene and POA emissions, all other chemical and meteorological model configurations for the Default and MarOrg simulations were kept constant.

3. RESULTS AND DISCUSSIONS

3.1. Trace Gases

Figure 1 shows that marine isoprene emissions in the MarOrg simulation resulted in a change in surface iso- prene concentrations from the simulation with default treatments over marine regions. The strong summertime solar radiation and high [Chl-a] levels in northern Atlantic and Pacific Oceans contributed to the high emission rates in those regions, while moderate ocean biological activity throughout the year over much of the Southern Ocean also resulted in high marine isoprene emission rates. Over remote marine regions, marine isoprene emissions increased surface isoprene concentrations by 5 - 10 ppt (see Figure 1). This increase was within the range of most isoprene concentrations (<1 to 300 ppt, typically < 50 ppt) observed over remote marine regions [3] . The large emission rates occurring near coastal regions (and some inland lakes) were due to the combination of isolated regions of very high [Chl-a] and coarse (4° × 5°) grid resolution. The greatest increases in surface isoprene concentrations were predicted over the northern Atlantic/Pacific and Southern Oceans where marine isoprene emission rates are high. These areas also had low oxidant levels and therefore long lifetimes of isoprene emitted from marine sources and transported from terrestrial regions. Over terrestrial areas such as the rainforests of South America, Southeast Asia, and Africa, a mixture of positive and negative changes in the surface isoprene concentrations is evident. Such changes were likely due to the fact that terrestrial emissions were computed online and were sensitive to the slight variability in meteorology occurring due to model reinitializations. Although the changes in surface isoprene concentrations over terrestrial areas were of similar magnitude as those of marine regions, they were much smaller when calculated in terms of percentage changes (typically < 5%).

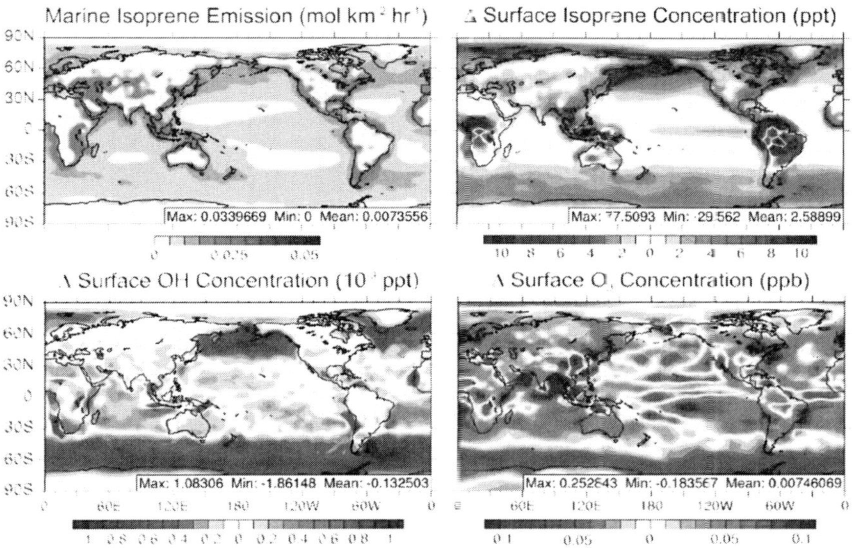

Figure 1. Multi-year (2001 and 2009) annual average marine isoprene emissions and differences in the predicted surface concentrations of isoprene, OH, and O_3 between the Default and MarOrg simulations.

Predictions of other atmospheric chemistry-relevant gases in remote marine regions were also affected by emissions of marine isoprene. The surface concentrations of hydroxyl radical (OH) decreased by 10^{-3} ppt or more in parts of the northern Pacific and Southern Oceans(see Figure 1) because of its reaction with the additional isoprene from marine sources. Small changes in surface O_3 concentration due to marine isoprene emissions(both increases and decreases of over 0.1 ppb) were also predicted; oceanic regions immediately downwind of anthropogenic pollution centers (China, U.S., etc.) experienced increased concentrations while remote marine regions typically have decreased concentrations (Figure 1). Decreases in predicted surface O_3 concentrations over most of the eastern North Pacific Ocean were consistent with Gantt et al. [16] , who found that the emission of isoprene and monoterpenes from marine sources caused a small (−0.1%) decrease in CMAQ-predicted O_3 mixing ratios off the Pacific coast of the U.S.

Although coincident measurements of isoprene emissions and concentrations over remote marine regions are very limited [3] , the May 2001 simulation period allowed for model evaluation of predicted isoprene emissions/ concentrations in the northwest Pacific Ocean [38] . Compared to the Matsunaga et al. [38] observations, the model overpredicted marine isoprene emissions (especially during the daytime) and underpredicted surface concentrations (Figure 2). Concurrent over predictions of emissions and under predictions of surface concentra- tions suggested that the simulated isoprene lifetime by GU-WRF/Chem over the marine boundary layer was too short, potentially due to the coarse grid resolution. The low temporal resolution of the observed emissions in Matsunaga et al. [38] made a more quantitative analysis of the diurnal cycle of emissions difficult, but the few observations available were inconsistent with the large increase in daytime emissions predicted by the model. The very low surface isoprene concentrations predicted by the Default simulation do not support the suggestion from Matsunaga et al. [38] that terrestrial isoprene sources influence the surface concentrations in this region; although we cannot definitely rule out the possibility of terrestrial isoprene sources affecting surface concentrations because of the use of a coarse model grid resolution in our simulations, proximity to the Japanese coast (~500 km) and the short isoprene lifetime predicted by GU-WRF/Chem in the marine boundary layer. The chemical mechanism used in GU-WRF/Chem, CB05-GE, is part of series developed for urban photochemistry [41] that may have difficulty representing remote marine conditions despite recent updates [38] . Nonetheless, according to Figure 2, the inclusion of marine isoprene emissions in the MarOrg simulation brought the predicted concentrations closer to observations.

3.2. Aerosols

Globally, the implementation of marine isoprene and POA emissions resulted in increased surface $PM_{2.5}$ organic aerosol concentrations (> 0.5 $\mu g \cdot m^{-3}$) over biologically-active remote marine regions in the northern Atlantic/Pacific and Southern Oceans (Figure 3). While the predicted increases in organic aerosol concentration occurred year-round in the Southern Ocean due to the persistent strong winds and moderate [Chl-a], increases in northern Atlantic and Pacific were the highest in the spring and lowest in the winter months. Surface organic aerosol concentrations in tropical oceanic regions were largely unchanged between the Default and Mar Org simulations due to low biological activity and sea spray aerosol emissions relative to mid- and high-latitude oceanic regions. Comparison of the predicted concentrations with observations in clean marine conditions com- piled from the years 1973 to 2009 [9] indicated a latitudinal model bias, with over predictions in mid- and high- latitude regions and underpredictions in the tropics. Although only a few sites have measurements throughout the year (e.g., Mace Head, Amsterdam Island), comparison with model predictions inFigure 3 suggests that the MarOrg simulation did not replicate the observed seasonal cycle of concentrations. This result is consistent with Gantt et al. [42] who found that none of the existing sea spray aerosol-based marine POA emission parameterizations are able to replicate the observed seasonality of organic aerosol concentrations.

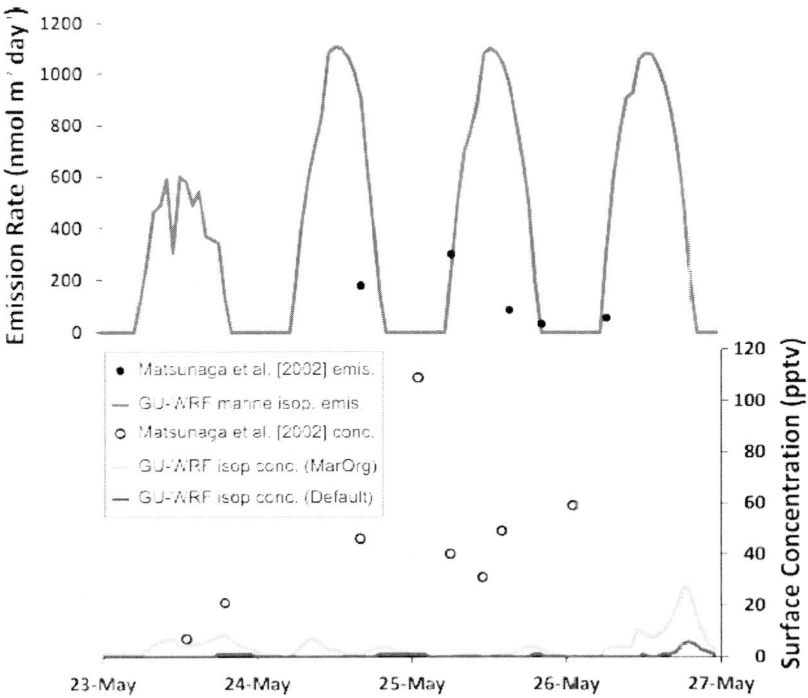

Figure 2. Time series comparison of observed [38] and modeled isoprene emissions (top) and surface concentrations (bottom) from the northwest Pacific Ocean in May 2001.

Comparison of GU-WRF/Chem and aerosol mass spectrometer observations from Mace Head, Ireland [39] for the year 2009 revealed that the inclusion of marine isoprene and POA emissions in the MarOrg simulation improved the prediction of surface submicron organic aerosol mass concentrations (see Figure 4). Without marine isoprene and POA emissions, the Default simulation frequently predicted surface organic aerosol concentrations at Mace Head < 0.025 $\mu g \cdot m^{-3}$ during clean marine conditions (defined as periods having aerosol number concentration < 700 cm^{-3}, onshore wind direction, and black carbon concentrations < 0.05 $\mu g \cdot m^{-3}$) while the observed concentrations typically ranged between 0.05 and1 $\mu g \cdot m^{-3}$ [39] . Organic aerosol concentrations rarely fell below 0.05 $\mu g \cdot m^{-3}$ in the MarOrg simulation. Model-measurement discrepancies persisted in the MarOrg simulation, especially during the marine organic aerosol plume period (16 August 2009) observed by Ovadnevaite et al. [40] . During this plume event, the MarOrg simulation predicted organic aerosol concentrations (0.2 $\mu g \cdot m^{-3}$) that were an order of magnitude lower than the observed maximum concentrations (3.8 $\mu g \cdot m^{-3}$). The change of annual average submicron organic (secondary and primary) aerosol concentrations at Mace Head between the Default and MarOrg simulations was 0.15 $\mu g \cdot m^{-3}$, a value similar to the observed annual concentrations (0.25 $\mu g \cdot m^{-3}$) at Mace Head during clean marine conditions.

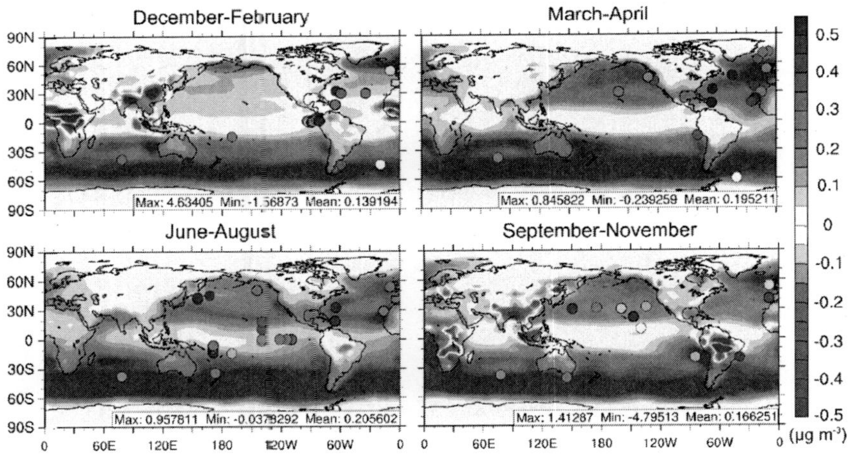

Figure 3. Multi-year (2001 and 2009) seasonally-averaged changes in the surface $PM_{2.5}$ organic aerosol concentrations between the Default and MarOrg simulations superimposed with observed surface organic aerosol concentrations in clean marine conditions compiled by Gantt and Meskhidze [9] .

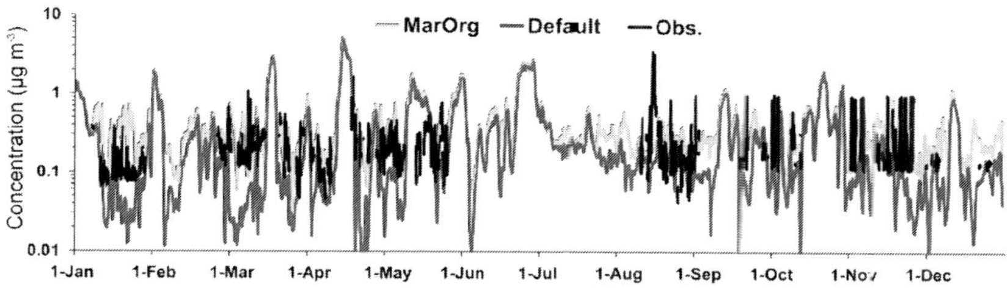

Figure 4. Time series comparison of modeled organic aerosol mass concentrations by GU-WRF/Chem with those observed at Mace Head, Ireland for the year 2009.

During the marine organic aerosol plume (16 August 2009) observed by Ovadnevaite et al. [40] (see Figure 5), the MarOrg simulation had a slightly higher aerosol number concentration than the Default simulation in the size range over which both simulations had the largest underpredictions (0.05 to 0.5 μm in diameter) relative to observations. Much of this difference is attributable to the increased organic aerosol number concentrations for diameters > 0.05 μm. In other clean marine periods where the MarOrg and Default simulations predicted substantially different submicron organic aerosol mass concentrations, the increase in the aerosol number concentration for the MarOrg simulation in the 0.05 to 0.5 μm diametersize range is more evident. These model results are consistent with the observations of multiple marine organic aerosol plumes from Ovadnevaite et al. [40] , who found that periods with the highest mass fraction of organics in submicron aerosols also had aerosol number size distributions peaking between 0.05 and 0.2 μm in diameter. It's clear, however, that the MarOrg simulation did not accurately simulate this marine organic aerosol plume both in terms of magnitude and size distribution.

4. CONCLUSION

In this study, we implemented marine isoprene and size-resolved marine POA emissions into GU-WRF/Chem, an online-coupled meteorology and chemistry model. The resulting increases in the surface concentrations of isoprene and organic aerosols over remote marine regions led to small decreases in the surface concentration of OH and O_3 and brought the model predictions into better agreement with isoprene emissions and surface concentrations in the northwest Pacific Ocean. Predictions of the aerosol number size distribution during an observed marine organic aerosol plume were also improved when these emissions were included. Although the inclusion of these emission sources did not bring the model into complete agreement with observed concentrations of isoprene and organic aerosols in remote marine regions, the improved predictions suggested that these emissions are important processes affecting atmospheric chemistry in these areas. As many coastal cities experience periods when clean marine conditions impact the background concentrations of pollutants, the inclusion of these marine emissions would benefit future air quality studies. The improved representation of aerosol number size distributions in clean marine conditions may also benefit the prediction of cloud condensation nuclei concentrations and associated cloud properties for future climate studies.

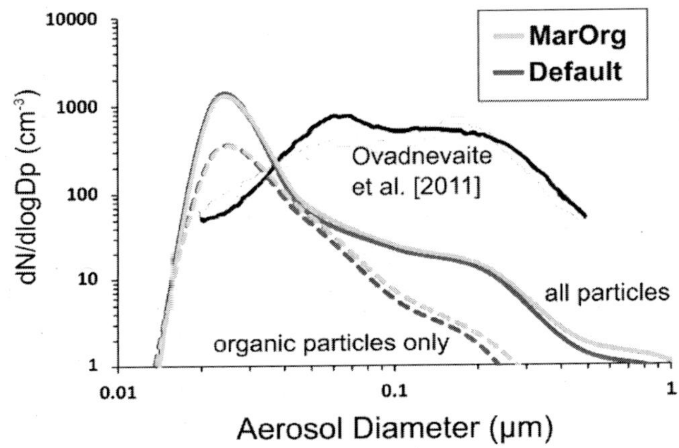

Figure 5. Observed [39] and predicted aerosol size distribution for the Default and MarOrg simulations on 16 August 2009. For reference, the organic aerosol size distribution for both simulations is given by the dotted lines. Note that the sectional size distribution predicted by MADRID in GU-WRF/Chem has been mapped to a modal size distribution through smoothing to better compare with observations.

ACKNOWLEDGEMENTS

This project was sponsored by the U.S. Department of Energy grant #DE-FG02-08ER64508, the EPA STAR grant R83337601, and the National Science Foundation EaSM program (AGS-1049200). We also thank Jurgita Ovadnevaite, Darius Ceburnis, and Colin O'Dowd for providing the Mace Head data.

REFERENCES

1 Lohmann, U., Feichter, J., Penner, J. and Leaitch, R. (2000) Indirect Effect of Sulfate and Carbonaceous Aerosols: A Mechanistic Treatment. Journal of Geophysical Research, 105, 12193-12206.

2 Fiore, A.M., Jacob, D.J., Bey, I., et al. (2002) Background Ozone over the United States in Summer: Origin, Trend, and Contribution to Pollution Episodes. Journal of Geophysical Research, 107, 2156-2202.

3 Shaw, S.L., Gantt, B. and Meskhidze, N. (2010) Production and Emissions of Marine Isoprene and Monoterpenes: A Review. Advances in Meteorology, 2010, Article ID: 408696.

4 Gantt, B., Meskhidze, N., Zhang, Y. and Xu, J. (2010) The Effect of Marine Isoprene Emissions on Secondary Organic Aerosol and Ozone Formation in the Coastal United States. Atmospheric Environment, 44, 115-121.

5 O'Dowd, C.D., Facchini, M.C., Cavalli, F., et al. (2004) Biogenically Driven Organic Contribution to Marine Aerosol. Nature, 431, 676-680.

6 Meskhidze, N., Xu, J., Gantt, B., et al. (2011) Global Distribution and Climate Forcing of Marine Organic Aerosol: 1. Model Improvements and Evaluation. Atmospheric Chemistry and Physics, 11, 11689-11705.

7 Collins, D.B., Ault, A.P., Moffet, R.C., et al. (2013) Impact of Marine Biogeochemistry on the Chemical Mixing State and Cloud Forming Ability of Nascent Sea Spray Aerosol. Journal of Geophysical Research, 118, 8553-8565.

8 Hu, Q.-H., Xie, Z.-Q., Wang, X.-M., et al. (2013) Secondary Organic Aerosols over Oceans via Oxidation of Isoprene and Monoterpenes from Arctic to Antarctic. Scientific Reports, 3, Article Number: 2280.

9 Gantt, B. and Meskhidze, N. (2013) The Physical and Chemical Characteristics of Marine Primary Organic Aerosol: A Review. Atmospheric Chemistry and Physics, 13, 3979-3996.

10 Prather, K.A., Bertram, T.H., Grassian, V.H., et al. (2013) Bringing the Ocean into the Laboratory to Probe the Chemical Complexity of Sea Spray Aerosol. Proceedings of the National Academy of Sciences of the United States of America, 110, 7550-7555.

11 Lewis, E.R. and Schwartz, S.E. (2004) Sea Salt Aerosol Production: Mechanisms, Methods, Measurements and Models —A Critical Review. Geophysical Monograph Series, 152, 413.

12 de Leeuw, G., Andreas, E.L., Anguelova, M.D., Fairall, C.W., Lewis, E.R., O'Dowd, C., et al. (2011) Production Flux of Sea Spray Aerosol. Reviews of Geophysics, 49, Article ID: RG2001.

13 Tsigaridis, K., Koch, D.M. and Menon, S. (2013) Uncertainties and Importance of Sea-Spray Composition on Aerosol Direct and Indirect Effects. Journal of Geophysical Research, 118, 220-235.

14 O'Dowd, C.D., Langmann, B., Varghese, S., Scannell, C., Ceburnis, D. and Facchini, M.C. (2008) A Combined Organic-Inorganic Sea-Spray Source Function. Geophysical Research Letters, 35, Article ID: L01801.

15 Arnold, S.R., Spracklen, D.V., Williams, J., Yassaa, N., Sciare, J., Bonsang, B., et al. (2009) Evaluation of the Global Oceanic Isoprene Source and Its Impacts on Marine Organic Carbon Aerosol. Atmospheric Chemistry and Physics, 9, 1253-1262.

16 Gantt, B., Meskhidze, N. and Carlton, A.G. (2010) The Contribution of Marine Organics to the Air Quality of the Western United States. Atmospheric Chemistry and Physics, 10, 7415-7423.

17 Anttila, T., Langmann, B., Varghese, S. and O'Dowd, C. (2010) Contribution of Isoprene Oxidation Products to Marine Aerosol over the North-East Atlantic. Advances in Meteorology, 2010, Article ID: 482603.

18 Myriokefalitakis, S., Vignati, E., Tsigaridis, K., Papadimas, C., Sciare, J., Mihalopoulos, N., et al. (2010) Global Modeling of the Oceanic Source of Organic Aerosols. Advances in Meteorology, 2010, Article ID: 939171.

19 Vignati, E., Facchini, M.C., Rinaldi, M., Scannell, C., Ceburnis, D., Sciare, J., et al. (2010) Global Scale Emission and Distribution of Sea-Spray Aerosol: Sea-Salt and Organic Enrichment. Atmospheric Environment, 44, 670-677.

20 Long, M.S., Keene, W.C., Kieber, D.J., Erickson, D.J. and Maring, H. (2011) A Sea-State Based Source Function for Size- and Composition-Resolved Marine Aerosol Production. Atmospheric Chemistry and Physics, 11, 1203-1216.

21 Westervelt, D.M., Moore, R.H., Nenes, A. and Adams, P.J. (2012) Effect of Primary Organic Sea Spray Emissions on Cloud Condensation Nuclei Concentrations. Atmospheric Chemistry and Physics, 12, 89-101.

22 Zhang, Y., Karamchandani, P., Glotfelty, T., Streets, D.G., Grell, G., Nenes, A., et al. (2012) Development and Initial Application of the Global-through-Urban Weather Research and Forecasting Model with Chemistry (GU-WRF/Chem). Journal of Geophysical Research, 117, Article ID: D20206.

23 Richardson, M.I., Toigo, A.D. and Newman, C.E. (2005) Non-Conformal Projection, Global, and Planetary Versions of WRF. Proceedings of the 2005 Joint WRF/MM5 User's Workshop, Mesoscale and Microscale Meteorology Division, National Center for Atmospheric Research, Boulder, 27-30 June 2005, Paper #7.1.

24 Richardson, M.I., Toigo, A.D. and Newman, C.E. (2007) Planet WRF: A General Purpose, Local to Global Numerical Model for Planetary Atmosphere and Climate Dynamics. Journal of Geophysical Research, 112, Article ID: E09001.

25 Skamarock, W.C., Klemp, J.B., Dudhia, J., et al. (2008) A Description of the Advanced Research WRF Version 3, Tech. Note, NCAR/TN-475+STR. National Center for Atmospheric Research, Boulder.

26 Zhang, Y., Hemperly, J., Meskhidze, N. and Skamarock, W.C. (2012) The Global Weather Research and Forecasting (GWRF) Model: Model Evaluation, Sensitivity Study, and Future Year Simulation. Atmospheric and Climate Sciences, 2, 231-253.

27 Grell, G.A., Peckham, S.E., Schmitz, R., McKeen, S.A., Frost, G., Skamarock, W.C. and Eder, B. (2005) Fully Coupled "Online" Chemistry within the WRF Model. Atmospheric Environment, 39, 6957-6975.

28 Karamchandani, P., Zhang, Y., Chen, S.Y. and Balmori-Bronson, R. (2012) Development and Testing of an Extended Chemical Mechanism for Global-through-Urban Applications. Atmospheric Pollution Research, 3, 1-24.

29 Zhang, Y., Pun, B., Vijayaraghavan, K., Wu, S.Y., Seigneur, C., Pandis, S.N., et al. (2004) Development and Application of the Model of Aerosol Dynamics, Reaction, Ionization and Dissolution (MADRID). Journal of Geophysical Research, 109, Article ID: D01202.

30 Zhang, Y., Pan, Y., Wang, K., Fast, J.D. and Grell, G.A. (2010) WRF/Chem-MADRID: Incorporation of an Aerosol Module into WRF/Chem and Its Initial Application to the TexAQS2000 Episode. Journal of Geophysical Research, 115, Article ID: D18202.

31 Gantt, B., Meskhidze, N. and Kamykowski, D. (2009) A New Physically-Based Quantification of Marine Isoprene and Primary Organic Aerosol Emissions. Atmospheric Chemistry and Physics, 9, 4915-4927.

32 Yassaa, N., Peeken, I., Zollner, E., Bluhm, K., Arnold, S., Spracklen, D. and Williams, J. (2008) Evidence for Marine Production of Monoterpenes. Environmental Chemistry, 5, 391-401.

33 Meskhidze, N., Sabolis, A., Reed, R. and Kamykowski, D. (2015) Quantifying Environmental Stress-Induced Emissions of Algal Isoprene and

Monoterpenes Using Laboratory Measurements. Biogeosciences, 12, 637-651.

34 Guenther, A., Zimmerman, P., Harley, P., Monson, R. and Fall, R. (1993) Isoprene and Monoterpene Emission Rate Variability: Model Evaluation and Sensitivity Analysis. Journal of Geophysical Research, 98, 12609-12617.

35 Gantt, B., Meskhidze, N., Facchini, M.C., Rinaldi, M., Ceburnis, D. and O'Dowd, C.D. (2011) Wind Speed Dependent Size-Resolved Parameterization for the Organic Mass Fraction of Sea Spray Aerosol. Atmospheric Chemistry and Physics, 11, 8777-8790.

36 Gong, S., Barrie, L.A. and Blanchet, J.P. (1997) Modeling Sea Salt Aerosols in the Atmosphere: 1. Model Development. Journal of Geophysical Research, 102, 3805-3818.

37 O'Dowd, C.D., Michael, H.S., Consterdine, I.E. and Lowe, J.A. (1997) Marine Aerosol, Sea-Salt, and the Marine Sulphur Cycle: A Short Review. Atmospheric Environment, 31, 73-80.

38 Matsunaga, S., Mochida, M., Saito, T. and Kawamura, K. (2002) In Situ Measurement of Isoprene in the Marine Air and Surface Seawater from the Western North Pacific. Atmospheric Environment, 36, 6051-6057.

39 Ovadnevaite, J., O'Dowd, C., Dall'Osto, M., Ceburnis, D., Worsnop, D.R. and Berresheim, H. (2011) Detecting High Contributions of Primary Organic Matter to Marine Aerosol: A Case Study. Geophysical Research Letters, 38, Article ID: L02807.

40 Ovadnevaite, J., Ceburnis, D., Martucci, G., Bialek, J., Monahan, C., Rinaldi, M., et al. (2011) Primary Marine Organic Aerosol: A Dichotomy of Low Hygroscopicity and High CCN Activity. Geophysical Research Letters, 38, Article ID: L21806.

41 Gery, M.W., Whitten, G.Z., Killus, J.P. and Dodge, M.C. (1989) A Photochemical Kinetics Mechanism for Urban and Regional Scale Computer Modeling. Journal of Geophysical Research, 94, 12925-12956.

42 Gantt, B., Johnson, M.S., Meskhidze, N., Sciare, J., Ovadnevaite, J., Ceburnis, D. and O'Dowd, C.D. (2012) Model Evaluation of Marine Primary Organic Aerosol Emission Schemes. Atmospheric Chemistry and Physics, 12, 8553-8566.

CHAPTER 5

High Contribution of Sea Salt Aerosols on Atmospheric Particles Measured at an Urban Tropical Location in Reunion Island

Chatrapatty Bhugwant[1*], Miloud Bessafi[2], Olivier Favez[3], Laura Chiappini[3], Bruno Sieja[1], Eva Leoz-Garziandia[3]

[1]Observatoire Réunionais de l'Air (ORA), Réunion Island, France

[2]LE2P, University of Réunion Island, Réunion Island, France

[3]Institut National de l'Environnement Industriel et des Risques (INERIS), Verneuil en Halatte, France.

ABSTRACT

PM10 was monitored during 2008-2011 period at LUT and BON, two urban tropical stations close to each other at Saint-Pierre city, in Reunion Island (south-western Indian Ocean). During the study period, notable PM10 concentrations are observed at BON close to the coasts. At LUT, a more inland site, the daily PM10 concentration range between 13 and 70 µg/m^3. Importantly, the limit value for the protection of the human health is systematically exceeded at BON while it is never exceeded at LUT. Also, the quality objective (QO: 30 μg/m^3, on annual average) and the limit value for the human health protection (LV: 40 μg/m^3, on annual average) are exceeded at BON each year during 2008-2011, while at LUT no regulatory values are exceeded. Nitrogen dioxide, an atmospheric tracer of anthropogenic activities was also monitored at LUT and BON. The mean diurnal NO_2 variation is of the same level and order of magnitude and exhibits a similar pattern at both stations suggesting that distinct sources influence the PM10 at LUT and at BON. Chemical analysis was performed on daily filters sampled in September-November 2011 at the two stations to determine the sea salt contribution on PM10 across Saint-Pierre city. It showed that the sea salt contribution to the PM10 is 55% at BON in September 2011. The sea salt particles are therefore the main cause for the exceedances of the regulatory values of PM10 recorded at BON. The results importantly suggest that the notable PM10 concentrations measured at this urban marine site might have some but minor impact on human health.

KEYWORDS

PM10; nitrogen dioxide; anthropogenic; sea salt aerosols; chemical characterization; regulatory values; human health; lung cancer; urban; tropical

1. INTRODUCTION

The study of particulate matter is of increasing interest to the scientific community (epidemiologists, researchers…) and to decision makers, in particular to determine the impact of this pollutant on human health [1-3]. It is thus important to conduct the monitoring of this pollutant especially in densely inhabited regions, in order to take adequate measures (e.g. prevent the surrounding population, in particular sensible persons such as children and old ones) if regulatory values are exceeded and also to study its long-term trend and effects. In this sense, since a few decades, decision makers have established regulatory values for the PM10 concentration levels, following the severity of the impact of this pollutant on human health and on the environment [4-6]. European directives, have set air quality standards (limit values, quality objectives) to be achieved for more than a dozen pollutants, among which the PM10 [7].

The European regulations for PM10 (e.g. Decree of 15th February 2002) is based on the following:

- Quality Objective (QO): 30 μg/m^3 on annual average;

- Limit values (LV): 90.4% of the daily average must be below 50 μg/m^3 (i.e., 35 days of exceedances allowed per calendar year) ;
- Limit values (LV): 40 μg/m^3 on annual average.

Importantly, air quality organizations which provide air quality index daily to the population are confronted with the contribution of natural aerosols on particles measurements, especially in marine urban regions, which may in part hamper the air quality index.

Anthropogenic activities such as traffic circulation emit important amounts of particles and gases [8]. Several epidemiological studies indicate that lung cancer and other health effects are possibly associated with atmospheric particles [2,9-14]. In this sense, in June 2012, the International Agency for Research on Cancer (IARC) classified diesel engine exhaust as carcinogenic to humans (Group 1), based on sufficient evidence that this exposure is associated with an increased risk for lung cancer [3].

Natural sources are also a major cause for atmospheric particles [15-18]. In particular, marine aerosols contribute significantly to the global aerosols load and consequently influence the Earth's radiative budget [19]. Some studies suggest that on coastal regions, sea salt aerosols may notably contribute to the annual mean particulate mass such as PM10 [20-22]. They emit a large amount of halogens, which are highly reactive and thus have a notable impact on the chemistry of hydrocarbons and/or ozone in the atmosphere [23]. Furthermore, sea salt plays an important role in a number of physical and chemical atmospheric processes [24,25]. For example, the reaction of sea salt particles with nitric acid forms sodium nitrate, which may be integrated in the particulate phase of the nitrogen budget [26]. The halogens released by the reaction of acidic gases with sea salt contribute, via heterogeneous reactions, to ozone destruction [27,28]. Sea salt may also indirectly impact vulnerable ecosystems via acid deposition [29,30]. Besides, sea salt aerosols contribute to corrosion of materials (e.g. metals) in coastal regions which cost billions for their rehabilitation [31]. Sea salt aerosols also contribute to cloud condensation nuclei (CCN), especially in marine regions [32,33]. Consequently, sea salt aerosols are essential components of atmospheric constituents at local, regional and global scales [34,35].

A number of studies have been dedicated to the modeling of atmospheric particles in general and sea salt particles in particular [36-38]. Many studies of marine aerosols and their role in the climate system were performed on the global scale [34,39-42]. For air quality assessment on local-to-regional scales, it is important to determine the gradients in sea salt levels and a number of regional models include a large description of sea salt [43-46]. The validation of these models is notably concerned by the limited number of available measurements. However, up to now, few atmospheric measurements have been undertaken with links to mixed sources (e.g. urban and marine aerosols) in particular the southern hemisphere [47-49].

The objective of this study is to quantify the sea salt contribution on the PM10 particles measured at two marine urban locations close to each other. Continuous PM10 measurements and chemical analysis of particles sampled daily on filters to determine the major ions (specific tracers of sea salt aerosols) was undertaken, to establish the distribution of sea salt aerosols on PM10 measured across Saint-Pierre city. These data were analyzed in conjunction with meteorological parameters in order to confirm the air masses origin. Finally, an assessment of the sea salt contribution based on the retrieval of the sea salt contribution on the PM10 measurements is proposed for the recalculation of the air quality index more realistic of anthropogenic activities.

2. EXPERIMENTAL PROCEDURES—MEASUREMENT LOCATIONS

Réunion Island is a French department located in the South-Western Indian Ocean (21 S; 55 E). This tiny (area: ~2500 km^2) and mountainous island (highest point: ~3075 m asl (above sea level)) holds two agglomerations with ~100,000 inhabitants: Saint-Denis, located to the North and Saint-Pierre located to the South of it (see **Figure 1**). **Figure 1**(a) is a zoom on Saint-Pierre city illustrating the geographical context of the measurement locations, in particular with respect to the coasts.

Saint-Pierre city is exempt of light/heavy industries susceptible emitting any atmospheric pollution and a significant proportion (~49%) of private cars as well as quite all trucks and buses operate with diesel at Reunion Island [50].

Figure 1(b) shows the configuration of Bons Enfants (BON hereafter) station, highlighting in particular:

- Its position, located about 200 m from the south coast of Reunion Island;
- Its proximity (about 1.5 km) with another urban station Luther King (LUT hereafter), the latter not showing any limit value of PM10 exceeded since 2008 (see Section 4.1.4 here after);
- The two measurement stations are located in the same urban environment, with light traffic circulation and at more than 0.5 to 1 km away from heavy highways. Hence, they are deemed representative of the urban background pollution level linked to anthropogenic activities (mainly traffic circulation);

- The presence of shallow water between the coast and the coral reefs which contributes to an efficient daily sea spray via wind flows and a potential sea salt source over this area. This is not the case for LUT which is downstream the coast too but where there is deep sea water with quite no coral reef and thus much lower sea spray emissions than at BON.

3. EXPERIMENTAL SET-UPS AND MEASUREMENTS

The PM10 and NO_2 (nitrogen dioxide) concentrations have been recorded continuously at LUT and BON since August 2007.

The PM10 measurements are subsequently compared with meteorological data. In September 2011, atmospheric samples were also collected on filter on a daily basis for subsequent laboratory analysis in order to characterize the chemical composition of the PM10 particles.

3.1. Atmospheric Sampling Strategy

Figure 2 illustrates BON (upper left) and LUT (upper right) stations implemented at Saint-Pierre since mid- 2007. It may be seen that these two stations have the same characteristics and a similar environment.

3.1.1. Particles (PM10) Continuous Measurement

The PM10 concentration has been measured at LUT and BON using Tapered Element Oscillating Microbalances (TEOM, 1400ab models, commercialized by Thermo Environmental Instruments) equipped with Sample Equilibration and Filter Dynamic Measurements Systems (8500c models). The latter two devices (respectively referred to as SES and FDMS) allow limiting and taking into account Semi-Volatile Matter (SVM) losses inside the instrument. Indeed, the SES allows the reduction of heating at the inlet from 50 to 30°C, and maintaining sample RH (relative humidity) below 25%, by means of a Nafion dryer. On the other hand, volatilization of SVM at 30°C is resolved by the use of the FDMS system, which provides a full determination of volatile mass through a self-referencing gas conditioning scheme. PM measurements obtained by a TEOM equipped with both SES and FDMS systems have shown to compare very well with other real-time automatic analyzers accounting SVM [51-53]. The resulting measurement device is referred to as TEOM-FDMS hereafter. It allows continuous monitoring of the PM10 concentrations at a 15-mins time-base at the two Saint Pierre measurement locations.

Figure 1. Map of Réunion Island. (a) (Source: Google Earth, 2011) and a zoom on Saint-Pierre agglomeration; (b) With location of Bons Enfants (BON) and Luther King (LUT) stations (Source: ©IGN-®authorization N° 9876).

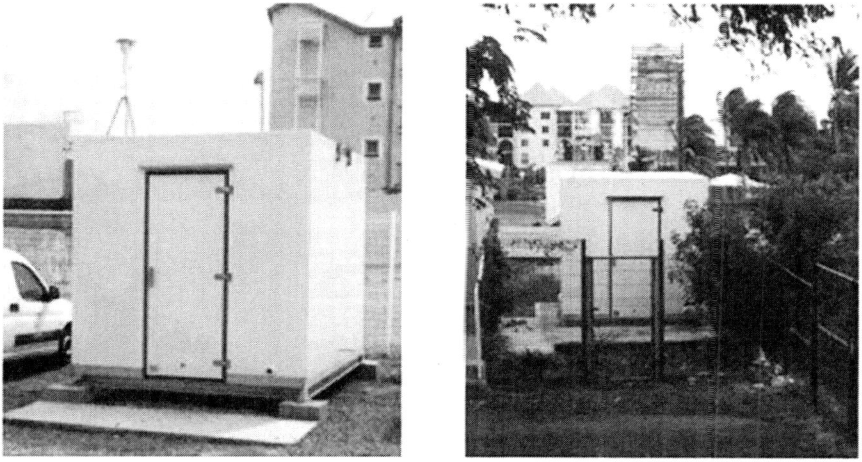

Figure 2. Photography of the BON (upper left) and LUT (upper right) measurement stations.

3.1.2. NO_2 Continuous Measurement

In order to determine the origin and main causes of the PM10 concentration variability recorded at LUT and BON, nitrogen dioxide (NO_2) concentration, an atmospheric tracer of anthropogenic origin (mainly traffic circulation, in absence of any other source such as heavy industries and/or volcanic eruption) [49], was also monitored in parallel with the PM10 measurements by a AC31M analyzer commercialized by Environnement SA.

The principle of the NOx measurement, by the chemiluminescence's technique is described in Bhugwant and Hoareau (2003) [54].

3.1.3. Determination of Aerosols Composition from Filter Samples

PM10 samples were collected daily on filters at LUT and BON in parallel with continuous atmospheric (PM10 and NO_2) measurements during experiments carried out over a half-year, from July 2011 to December 2011.

The PM10 sampling was performed continuously and simultaneously at both LUT and BON sites by ORA using Partisol samplers type 2025 provided by the LCSQA (Laboratoire Central de Surveillance de la qualité de l'Air). The sampling was performed for two weeks sequences interspersed with a day off.

The filters used are of Tissuquartz Qat-Up 2500 type (diameter: 47 mm) commercialized by Pall, chosen for their low sodium and other mineral species levels (except silica). Filters were initially pre-fired at 500°C during 2 hours to minimize filter blanks. Sampling was performed at a constant air-flow rate of 16 l/hr, and for a duration of 24-hour per filter. They were then stored at a temperature below 20°C, including during transport (in cooler).

Chemical analyzes of selected filters sampled at LUT and BON were performed by LCSQA/INERIS. The analyses were performed in accordance with the CEN/TR 16269 and 16243 technical reports for the quantification of anions/cations (Cl^-, NO_3^-, SO_4^{2-}, Na^+, NH_4^+, K^+, Mg^{2+} and Ca^{2+}) respectively using ion chromatography, and carbon fractions (elemental carbon: EC and organic carbon: OC) using a thermal-optical protocol [55] implemented on a Sunset Lab. Instrument [56].

3.2. Complementary Data

3.2.1 Meteorological Data

The local meteorological parameters (winds speed and direction, etc.) are measured from the instruments onboard a tower at 10 m above the ground level in meteorological stations located close to LUT and BON locations. These instruments are part of the French Meteorological Service [57]. The meteorological data was collected from January 2008 to December 2011 at Pierrefonds (PIE), a station located in the vicinity of Saint Pierre city. This meteorological station is close to LUT and BON (distance: ~3.4 and 4.9 km respectively, and located to the north-west from them) and representative of the meteorological conditions prevailing over Saint-Pierre.

The meteorological data measured continuously with a 1/2-hour time-base were subsequently brought to a 1- hour basis in order to be compared with the atmospheric (PM10 and NO_2) data.

4. RESULTS AND DISCUSSION

The analysis of the atmospheric and meteorological datasets is first focused on their mean diurnal variation from one year to another, in order to study their trend and levels and also to look for possible links between them. The datasets are subsequently analyzed on a daily basis, in order to look closely at their variability and also to compare with the PM10 regulatory values. The daily PM10 variability at LUT and BON are then compared with the sea salt (cations and anions, in particular Na^+ and Cl^-) concentrations analyzed on the daily filter samples.

4.1. Diurnal Variation during the Study Period

4.1.1. Mean Diurnal PM10 Concentration Variation

The mean diurnal variation of PM10 concentration has been calculated at LUT and BON from 2008 to 2011, period during which data are available for a whole year (see **Figure 3**).

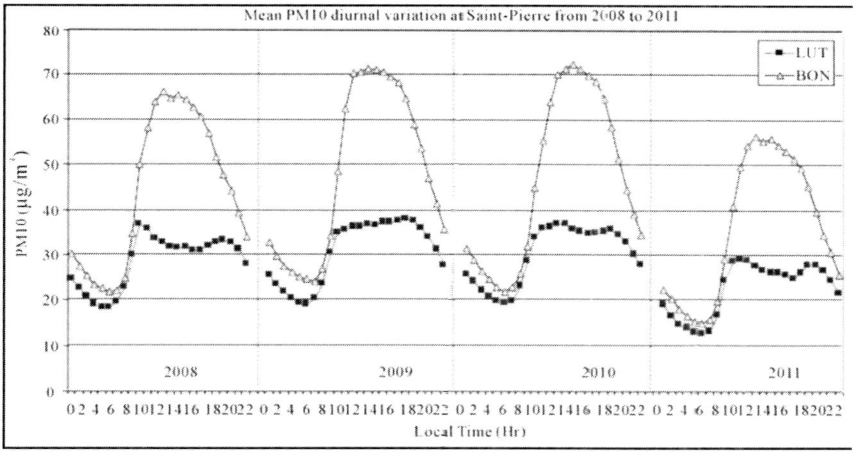

Figure 3. Mean diurnal PM10 concentration calculated annually at LUT and BON stations from 2008 to 2011.

The PM10 diurnal variation exhibits a particular pattern at BON and at LUT, with maximum levels at daytime (during significant anthropogenic activities, mainly traffic circulation) and minimum levels at night-time (absence of anthropogenic activities). However, no peak, corresponding to the morning and evening traffic circulation peaks is noticed except for LUT in 2008, 2010 and 2011, where such a tendency is slightly noticed, suggesting some influence of traffic circulation on the PM10 concentration at this site. Some studies show that PM10 measurements in urban locations mainly influenced by anthropogenic activities (in absence of industrial and natural/marine sources) exhibit a good correlation with traffic density [47,58]. Importantly, the PM10 levels are systematically and notably higher at BON than at LUT, especially during daytime (e.g. at 2:00 p.m., mean PM10 at BON: ~65 $\mu g/m^3$ and at LUT: ~35 $\mu g/m^3$).

Moreover, the mean diurnal evolution exhibits a notable variation from one year to another. Indeed, higher PM10 levels are observed in 2009 and 2010 (mean: 48 and 47 $\mu g/m^3$ at BON and 31 and 30 $\mu g/m^3$ at LUT respectively) while moderate levels are noticed in 2008 (mean: 44 $\mu g/m^3$ at BON and 28 $\mu g/m^3$ at LUT respectively) and low levels are measured in 2011 (mean: 36 $\mu g/m^3$ at BON and 21 $\mu g/m^3$ at LUT respectively). Nev ertheless, as the two locations are representative of the urban pollution over Saint Pierre city and close to each other, they are expected to record the same levels, tends and order of magnitude of PM10 concentrations, which is not the case. This variability may in part be due to the influence of meteorological conditions and other particles sources. Importantly, there is not any industry emitting atmospheric pollutions at Saint Pierre, especially in the vicinity of LUT and BON and both stations are quite equally distant from light/heavy highways (see **Figure 1**(b)).

4.1.2. Mean Diurnal NO_2 Concentration Variation

It is essential to look for the influence of the anthropogenic activities (i.e., mostly traffic circulation) at SaintPierre city in order to establish links with the PM10 diurnal variability. In this sense, the mean diurnal variation of NO_2 concentration measured at LUT and BON has been calculated each year from 2008 to 2011 (see **Figure 4**).

Firstly, the diurnal NO_2 concentration variation exhibits the same level and trend at the two sites. It exhibits systematically a bimodal pattern, with a maximum peak at 7:00 a.m. and a secondary peak at 8:00 p.m. at both LUT and BON, attributable to morning and evening traffic peaks [48,58,59]. The hourly NO_2 concentration is of the same order of magnitude at both sites but is in general slightly higher at BON than at LUT suggesting that the former site is a little more influenced by atmospheric pollutants induced by the surrounding traffic circulation. It may also be noticed that the mean diurnal evolution exhibits an annual variation, with however a difference with the PM10 diurnal profile (see **Figure 3**).

As observed for the PM10, high NO_2 levels are also observed for 2010 (mean: 17 $\mu g/m^3$ at BON and 14 $\mu g/m^3$ at LUT respectively) while moderate levels are noticed for 2008, 2009 and 2011 (mean: 13, 14 and 13 $\mu g/m^3$ at BON and 13, 11 and 12 $\mu g/m^3$ at LUT respectively). The difference on the PM10 variability observed at the two sites is not noticeable on the NO_2 concentration variation. The results imply that the anthropogenic activities contribution on the PM10 is quite similar and of the same order at both LUT and BON. Therefore, the potential factors which might explain the differences on the PM10 observed at LUT and BON, mainly during daytime, might be the conjugate effect of other (natural sources such as marine aerosols) contribution and dynamical processes. Indeed, as described earlier, BON is closer to the sea side (~200 m away from it) than LUT (~500 m away from it). Moreover, coral reefs are located directly downwind to the south of BON while they are quite far from LUT (see **Figure 1**(b)).

Due to the geographic configuration of the island (important relief) and the existence of coral reefs especially downwind BON station, the sea waves break constantly over the coral reefs and continuously generate marine aerosols via sea spray. This process might explain the higher PM10 levels systematically observed at BON than at LUT, especially during daytime when sea breezes prevail.

4.1.3. Mean Diurnal Winds (Direction and Speed) Variation

We tried to gain more information about the origin of air samples attaining Saint-Pierre agglomeration during the study period. We therefore analyzed meteorological data collected at PIE meteorological station. However, due to technical problems during the study period (from 2008 to 2011), only meteorological data measured in 2008 and 2009 are available. The mean winds (direction and speed) diurnal variation were calculated from winds data measured at PIE in 2008 and 2009 (see **Figure 5**).

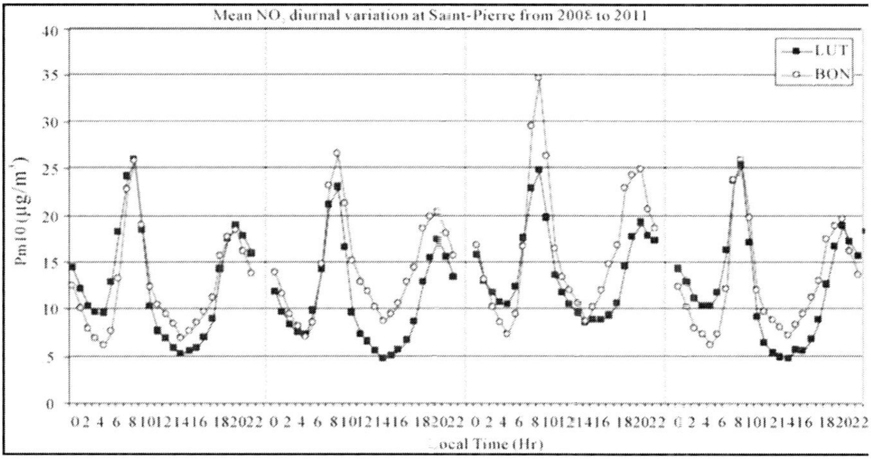

Figure 4. Mean diurnal NO_2 concentration calculated annually at LUT and BON stations from 2008 to 2011.

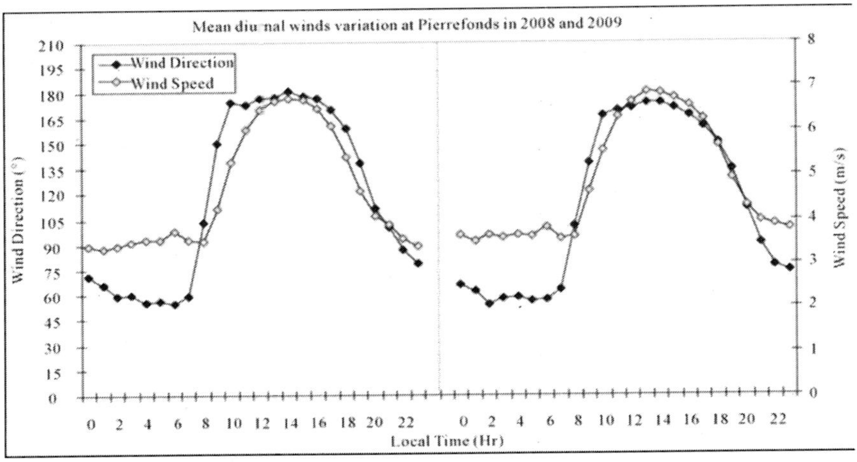

Figure 5. Annual winds diurnal variation calculated at Pierrefonds (PIE) meteorological station in 2008 and 2009.

The analysis being focused on the diurnal variation for each year, the change from one year to another on the diurnal pattern may only be on the winds amplitude of variation. The two years data available provide sufficient information about the daily variation and may be extrapolated to the 2010-2011 period. The mean winds (direction and speed) diurnal variation shows similar trends, patterns and levels in 2008 and 2009. High wind speeds (~5.7 m/s) are noticed at daytime (9:00 a.m. to 8:00 p.m.) while at night-time (from 9:00 p.m. to 8:00 a.m.) they are low (~3.6 m/s). This difference is mainly due to the conjugate effect of prevailing easterlies and the land/sea breeze processes.

During night-time, Saint-Pierre city is under the inland airflow (~70°), with a north-east origin of air masses. During daytime however, it is mainly of marine origin and thus under sea-spray flow, as the wind is in a south-east direction (~150°). In addition to the large scale trade winds prevailing over the region, the daily abrupt change in the morning and afternoon airflow is typical of a land/sea-sea/land breezes process and is experienced by day-to-day variation at Saint-Pierre throughout the year [57]. Thus, a conjunction effect of winds (breezes and easterlies) and human activities occurs simultaneously at Saint-Pierre at daytime. The analysis points out that at night-time when the anthropogenic activities are quite negligible and the air masses prevailing over Saint-Pierre originate on average from inland (north-east), the PM10 and NO_2 concentration levels are lowest and correspond to the background atmospheric pollution level over Saint-Pierre city. Conversely, at daytime when the anthropogenic activities are notable and the air masses prevailing over Saint-Pierre originate from south-east to south (mainly marine), the PM10 and NO_2concentration levels are highest and variable. The results suggest that at daytime, while the NO_2concentration variation is mainly due to local anthropogenic activities, the PM10 concentration is influenced by local anthropogenic activities and sea spray flow and thus marine sources as well, mostly at BON, as the latter site is closest to the coast than LUT.

4.1.4. Mean Monthly and Daily PM10 Concentration Variation

The monthly mean PM10 concentration calculated for 2008-2011 (not shown) indicates that during austral winter (August to October), the PM10 concentration is highest, with a mean value of 30 $\mu g/m^3$ and 47 $\mu g/m^3$ at LUT and BON respectively. Then, during austral summer (January to March), moderate PM10 concentration levels are observed (monthly mean: 28 $\mu g/m^3$ and 43 $\mu g/m^3$at LUT and BON respectively). However, from April to July then from November to December, minimum PM10 values are measured at both sites (monthly mean: 27 and 41 $\mu g/m^3$ at LUT and BON respectively). This PM10 seasonal variability may in part be explained by whitecaps and bursts, injecting seawater drops into the atmosphere, in particular when strong easterlies prevail in this region [42]. Also, the monthly PM10 evolution is corroborated with the swell seasonal variation, which exhibits high wave levels, in general in February and in October over the South-western Indian Ocean region [49,60]. Besides, during austral summer, the trade winds are stronger and thus the sea spray production due to wind shear on the sea surface is higher [57].

The mean daily PM10 concentration calculated at LUT and BON from 2008 to 2011 was also analyzed (not shown).We notice that the daily PM10 concentration exhibits distinct trend and level at both sites, from one year to another.

Indeed, during 2008, highest daily PM10 concentrations are observed from January to March then from September to November, while they are lowest from May to August at both sites. In 2009 however, while the daily PM10 concentration is high at BON from January to March, it exhibits moderate levels at LUT. Higher daily PM10 levels are observed both at BON and at LUT from April to September. Then, From October to December, lowest daily PM10 levels are observed at LUT while they remain high at BON. In 2010, the daily PM10 concentration is highest at BON from January to May while it is at moderate levels at LUT during this period. From June to October, the daily PM10 concentration is at moderate levels at BON while it is highest at LUT. Then, from November to December the daily PM10 levels are low both at BON and at LUT. In 2011, from January to March the daily PM10 concentration is highest at LUT while it exhibits moderate levels at

BON. Then, from May to December, no clear trend is observed at both sites, with systematically higher PM10 levels at BON than at LUT.

The differences in the daily PM10 concentration seasonal and annual variation at BON and LUT may be explained by the conjugate effect of changes in meteorological conditions (winds, insolation, rainfall etc.), in source effects (anthropogenic activities and marine sources) and also in regional events such as swell and sea waves activities.

Importantly, the quality objective (30 μg/m³, on annual average) as well as the limit value for the protection of the human health (Centile 90.4 = 50 μg/m³, on daily average not to be exceeded more than 35 days/year and the limit value (LV = 40 μg/m³, on annual average) are exceeded each year at BON, when the daily PM10 datasets are analyzed on a civil year.

The number of days exceeding the limit value and the annual average for the PM10 concentration at BON and LUT from 2008 to 2011 were calculated (see **Table 1**). It may be seen that the quality objective (QO = 30 μg/m³, on annual average) as well as the limit value for the protection of the human health (LV = 40 μg/m³, on annual average) are systematically exceeded every year at BON (except for the VL which is not exceeded in 2011), while at LUT only the QO is exceeded in 2010.

Also, the centile 90,4 (i.e., the number of days exceeding the limit value of 50 μg/m³, not to be exceeded more than 35 days/year) is systematically exceeded at BON each year, while at LUT the number of days exceeding the limit value of are quite lower than 35 days.

The results point out that the origin and main causes of the PM10 concentration at BON is quite different than at LUT, although both urban sites are close to each other.

4.2. Determination of the Chemical Composition of the PM10

The knowledge of the particles composition measured in Saint-Pierre is a key to confirm and quantify the contribution of sea salt in the PM10 concentration exceedances. In a preliminary study undertaken in late 2008, it has been suggested that one of the main causes of the PM10 concentration variability at BON may be an additional contribution of sea salt aerosols [60]. Based on in-situ observations, such as large deposits of sea salt crystals inside the samples head (see **Figure 6**), piecemeal analysis of the chemical composition of filters collected on the automatic PM10 (TEOM-FDMS) analyzers, and analysis of the waves and swell data with the PM10 concentrations in late 2008, the probable role played by the sea salts in the occurrence of exceedances of limit values on PM was highlighted [60].

Table 1. Number of days exceeding the limit value and the annual average calculated for the PM10 concentration measured at BON and LUT from 2008 to 2011.

Year	Number of days exceeding the limit value of 50 µg/m^3		Annual average (µg/m^3)	
	BON	*LUT*	*BON*	*LUT*
2008	81	2	44	28
2009	124	9	48	30
2010	85	14	46	31
2011	41	2	34	23

In this sense, since July 2011, particles sampled daily on filters in parallel with continuous PM10 concentration measurements at BON and LUT were analyzed in laboratory by INERIS. The analyses were performed for 39 days of filter samples during which the daily PM10 concentration mostly exceeded the limit value of 50 µg/m^3 between September 2011 and November 2011.

The aim of this chemical characterization was to determine the fraction of daily sea salt particles measured in the PM10 measurements at LUT and BON stations.

The sea salt concentrations are determined following the recommendations of the European guide for estimating the contributions from natural sources, as follows:

$$[\text{Sea Salt}] = \left(\left[\text{Na}^{+}\right] + \left[\text{Cl}^{-}\right]\right) \times 1168 \qquad (1)$$

and after verification of the correct correlation and compliance reports characteristics of sea droplets between these two species and between Na^+ and Mg^{2+}, magnesium being also a good tracer of sea salt.

The sea salt sulfate (SO_4^{2-}) concentrations contribution is determined using the SO_4^{2-}/Na^+ ratio of seawater (~0.25), allowing then to determine the concentrations of non sea salt sulfate (nssSO_4^{2-}). Finally, particulate organic matter (OM) concentrations are obtained using a conversion factor of 1.6 * OC à OM, [61,62].

Figure 6. Photography showing the deposition of sea salt crystals inside the PM10 sampling head at BON.

The daily PM10 concentrations recorded at BON and LUT stations have been calculated from August 2011 to mid-December 2011, period during which the particles sampled on filters were available for chemical analysis (see **Figure 7**).

We have focused more specifically the chemical analysis on the September 2011 results, during which there is a marked episode of daily exceeded PM10 values (between 05/09/2011 and 10/09/2011). During this month, there are 12 exceedances of the daily limit value threshold on the PM10 at BON. It may be noted that during this period, LUT has not exceeded any daily threshold value.

4.2.1. Sea Salt Contribution on the PM10 Concentrations in September 2011

The average chemical composition was performed for September 2011 (from 1st to 13th then from 15th to 27/09) from daily filters sampled at LUT and BON stations (see **Figure 8**). It may be noted that the sea salt appears clearly as the main fraction of PM10 sampled on each site (~50% at LUT and ~55% at BON).

We also note the large proportion of chemical species other than the cat-ions/anions and carbonaceous species, which can not be explained only by other components such as heavy metals (not shown), the latter being a minor/negligible contribution. This part is reasonably, primarily, attributable to the presence of liquid water on marine aerosols. Indeed, sea salts are known to be highly hydrophilic and are in fact present in the atmosphere as saline suspension, in particular over coastal regions [16]. The validity of this assumption is confirmed by the quite good correlation established between the sea salt concentration and the difference between PM10 and the sum of the other chemical species {EC + OM + NH_4^+ + NO_3^- + nss SO_4^{2-}} (see **Figure 9**).

The value of the slope obtained for the linear regression suggests that the sea salt contribution to the PM10 concentrations recorded at LUT and BON stations is about 1.34 times larger than the suggested estimate of the sea salt concentrations in the non-hydrated solid form.

This hypothesis implies that a mass of water vapor (denoted ssH_2O hereafter) corresponding to 1/3 of the mass of dry sea salt is also measured by the TEOMFDMS.

It is in good agreement with the modeling results of hygroscopic properties of NaCl crystals placed at a relative humidity of 10% corresponding to the relative humidity in the dryer outlet of the FDMS modules [63]. It may also be noted that the intercept (~3.7 μg/m^3), obtained for the correlation presented in **Figure 9** might be primarily attributable to the presence of terrigenous particles in the PM10 as well as the instrumental TEOMFDMS blank.

As shown in **Figure 9**, the results allow a quite exhaustive "daily chemical closure" of PM10 concentrations measured on the two study sites during September 2011. It therefore appears that the sea salt contribution is maximal on days corresponding to exceedances of the 50 μg/m^3 daily threshold at BON, while this contribution is quite less at LUT, and therefore the notable differences observed between both sites can mainly be attributed to sea salt aerosols.

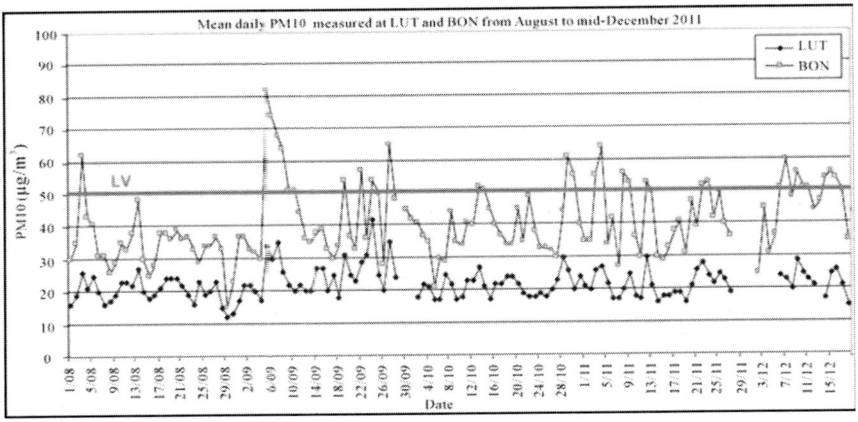

Figure 7. Mean daily PM10 concentrations measured at BON and LUT stations from August 2011 to mid-December 2011.

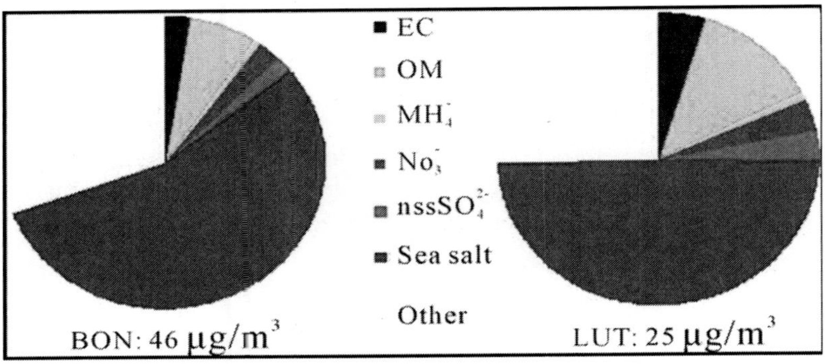

Figure 8. Average chemical composition of PM10 sampled daily on filters at LUT and BON in September 2011.

The daily concentrations of major chemical species which constitute the PM10 fraction at LUT and BON stations during September 2011 were calculated (see **Figure 10**).

The results presented in **Figure 10** suggest not only that the daily threshold exceedances recorded at BON station are mainly due to sea salt aerosols but also that almost all other chemical species studied present relatively homogeneous concentrations at both LUT and BON sites.

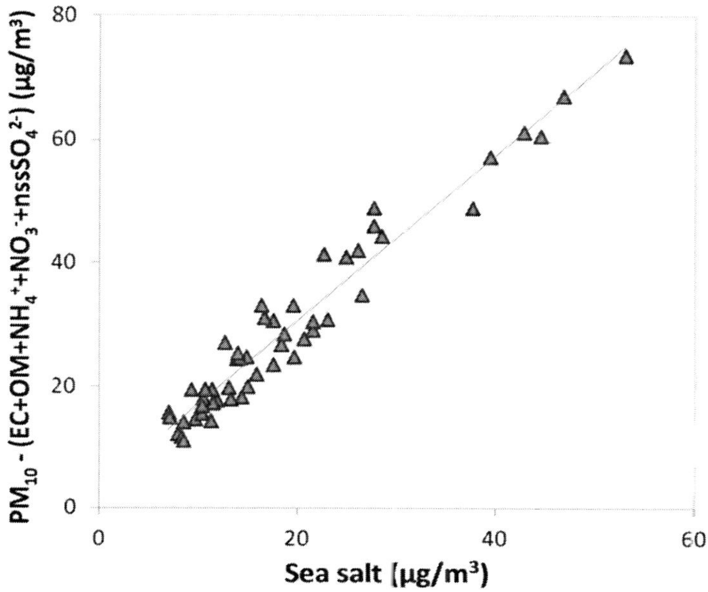

Figure 9. A proportional relationship obtained between the sea salt (non-hydrated) concentrations and an additional amount (other than EC, OM, NH_4^+, NO_3^+ and nss SO_4^{2-}).

This result of this analysis (see **Figure 11**(a)) indicating a very good correlation between the sum of {EC + OM + NH_4^+ + NO_3^+ + nssSO_4^{2-}} on the two stations for September 2011. It is also noteworthy that the difference of PM10 concentrations between these two stations can largely be attributed to the difference in concentrations of sea salt hydrates attaining each site (see **Figure 11**(b)).

We can therefore reasonably consider that there is an almost linear relationship between the difference of PM10 concentration measured at the two stations and the sea salt contribution at BON station. This relationship is illustrated by **Figure 12**, suggesting the following equation:

$$[\text{Sea salt}]_{\text{BON}} \approx 0.9 \times (\text{PM10}_{\text{BON}} - \text{PM10}_{\text{LUT}}) + 7 \quad \ldots \quad (2)$$

As mentioned earlier, the intercept of this equation is partially related to the presence of terrigenous particles and values of the instrumental PM10 analyzers blank used.

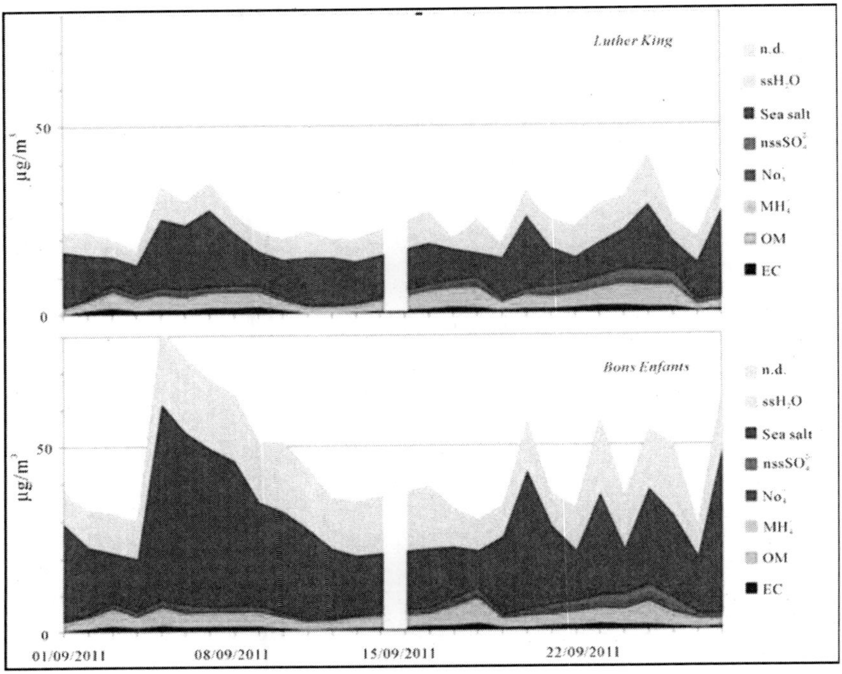

Figure 10. Daily concentration of major chemical species which form the PM10 fraction at LUT and BON stations during September 2011.

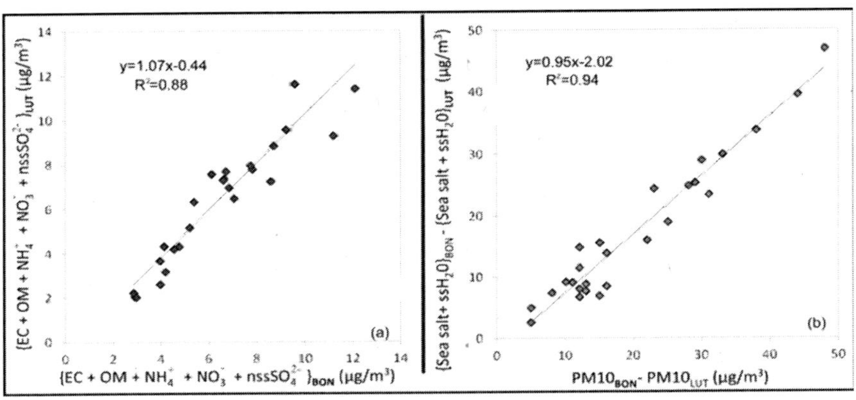

Figure 11. Homogeneity of particle concentrations mainly of anthropogenic origin observed at BON and LUT (a) and the major role played by sea salts on the difference in PM10 concentrations measured at the two stations (b).

However, its temporal variability and the uncertainty estimation it induces (possible risk of overestimation) can be considered as an offset, in the case of daily exceedances related to sea salt, by not taking into account the liquid water (ssH_2O) proposed in equation (2).

4.2.2. Differences between PM10 Concentrations Measured at Both Sites

Due to technical problems (related to the malfunction of the samplers as a result of high ambient relative humidity) and analytical costs, it was not possible to analyze in such detail all daily PM10 exceedances recorded in late 2011. Nevertheless, based on the results described in Section 4.2.1, it is possible to specify that all the exceedances observed in October and November 2011, can be mainly attributed to sea salt contribution (see **Table 2**).

The equation (1) proposed previously was applied to each daily threshold exceedances that could be investigated during late 2011. This application was performed in order to "re-calculate" the sea salt contribution at BON station from differences of PM10 concentrations between BON and LUT during days on which the daily exceedances were measured at BON. The data obtained were then compared to the sea salt concentrations estimated from chemical analyses during this period.

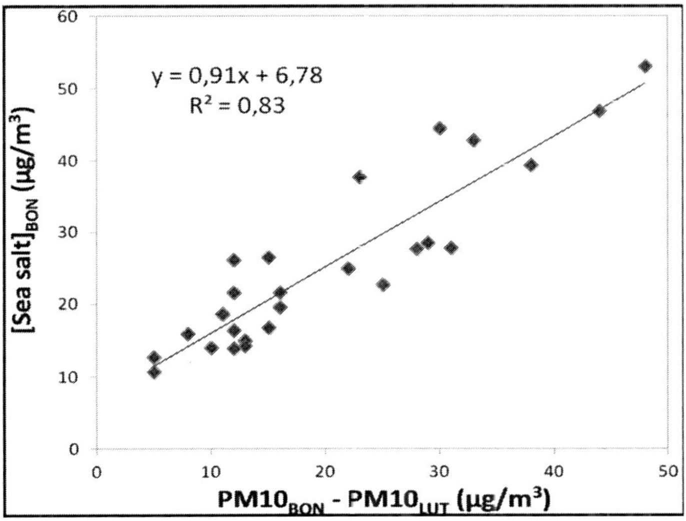

Figure 12. Correlation between the sea salt concentrations measured at BON and difference in PM10 concentrations recorded between LUT and BON stations.

Table 2. Subtraction cf daily sea salt contribution to PM10 concentrations >50 μg/m³ at BON in September, October and November 2011.

Date	PM10 (μg/m³)	Sea salts	PM10-sea salts	Sea salts + ssH_2O	PM10-(sea salts + ssH_2O)
05/09/2011	82	53	29	71	11
06/09/2011	74	47	27	63	11
07/09/2011	68	43	25	57	11
08/09/2011	65	39	26	53	12
09/09/2011	51	28	23	38	13
10/09/2011	51	28	23	37	14
19/09/2011	54	38	16	50	4
22/09/2011	57	28	29	37	20
24/09/2011	54	26	28	35	19
25/09/2011	51	23	28	30	21
27/09/2011	65	44	21	60	5
13/10/2011	52	35	17	46	6
14/10/2011	51	30	21	40	11
29/10/2011	61	N.A.	N.A.	N.A.	N.A.
30/10/2011	55	N.A.	N.A.	N.A.	N.A.
03/11/2011	55	27	28	36	19
04/11/2011	64	37	27	50	14
08/11/2011	57	23	34	31	26
09/11/2011	53	24	29	32	21
12/11/2011	53	N.A.	N.A.	N.A.	N.A.
22/11/2011	52	24	28	32	20
23/11/2011	53	19	34	25	28

N.A.: Data not available.

The results of this analysis (see **Figure 13**) show that there is good agreement between sea salt concentrations measured at BON and those estimated from differences in PM10 concentrations at the two stations.

These results suggest an overall uncertainty <50% for the empirical methodology. It may be emphasized that the latter estimate does not account for the influence of the amount of water associated with the presence of sea salts on the total PM10 concentrations. Thus, a possible overestimation due to the application of this methodology is partially offset by the non-inclusion of water concentrations.

Finally, as a test, this methodology was applied to all close to each other. The mean diurnal winds data measured at Pierrefonds a close meteorological station showed that at daytime the air samples with high wind speeds are mainly of south-eastern to southern (mainly marine) origin. At night-time, the air samples with moderate wind speeds are mainly of north-eastern to eastern (urban/terrigenous) origin. The proximity of coral reefs with the BON station and the conjugate effect of winds points out the notable influence of marine aerosols on the PM10 concentration variability observed at BON than at LUT.

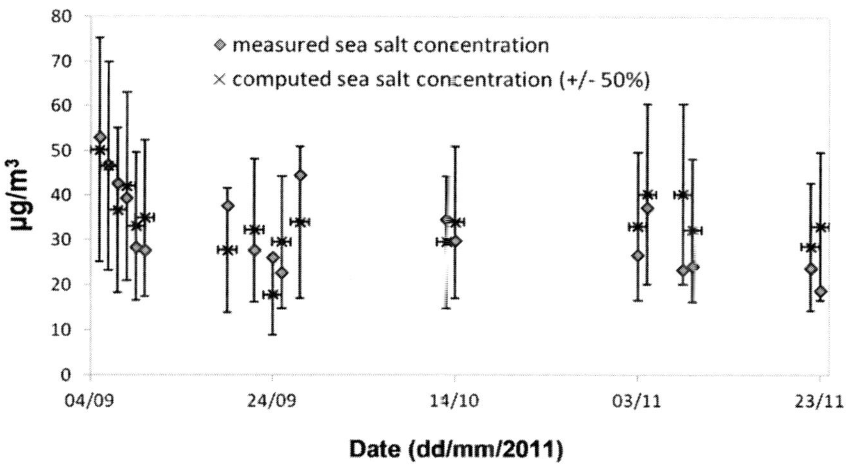

Figure 13. Comparison between sea salt concentrations measured and "re-calculated" at BON for which the daily PM10 >50 µg/m3 during September-November 2011 period.

5. CONCLUSIONS

The daily PM10 concentration profiles calculated at both sites also show that their trend and pattern are different. The PM10 concentration recorded at LUT and BON was compared with the regulatory values (daily and annual average). Importantly, the number of days exceeding the limit value for the protection of the human health (centile 90.4, i.e., the mean daily value of 50 μg/m^3 not to be exceeded more than 35 days/year) is systematically exceeded at BON while at LUT it is never exceeded. Also, the quality objective (QO: 30 μg/m^3 on annual average), the limit value for the human health protection (LV: 40 μg/m^3 on annual average) and the centile 90,4 are exceeded at BON each year during 2008-2011 period, while at LUT no other regulatory values are exceeded except for the QO value which is exceeded in 2010.

The main source of the PM10 was determined by the analysis of the composition of the sampled air. In September 2011, particles sampled daily on filters in parallel with continuous PM10 concentration measurements at BON and LUT were analyzed in laboratory in order to determine the presence of major ions (tracers of marine origin) and carbonaceous (elemental and organic) species (tracers of anthropogenic origin).

The analyses were performed for 39 days of filter samples during which the daily PM10 concentration mostly exceeded the limit value of 50 μg/m^3, between September and November 2011. The analysis results showed that the sea salt appears clearly as the main fraction of PM10 sampled on each site (~50% at LUT and ~55% at BON).

The results suggest not only that the daily threshold exceedances recorded on the BON station are mainly due to sea salts but also that all other chemical species studied present relatively homogeneous concentrations at both LUT and BON sites

A calculation based on the sampled filter analysis results was applied to the entire PM10 data set. This methodology was applied to all exceedances of 50μg/m^3 recorded at BON station and compared with the PM10 recorded at LUT station over the 2008-2011 period. The results showed that over this period, the daily PM10 concentration measured at BON station is quite below the limit value of 35 days/year for the threshold of 50μg/m^3, after subtraction of the sea salt contribution. It may therefore be assessed that the notable PM10 levels recorded at BON have minor impact on human health at this urban marine site.

Finally, this preliminary study points out the necessity of in-situ measurements to assess the origin and main causes of PM10 concentration variability at two close urban measurement stations, in response with local-toregional scale processes (e.g. sea salt emissions and anthropogenic activities), under 'particular' geographical (notable relief) and meteorological conditions (land/sea breezes, recirculation, etc.). It constitutes an adequate device for decision makers to take necessary steps, especially to inform the surrounding population, in order to give an appropriate air quality index to the population. It may also avoid giving misleading information pertaining to the potential health and environmental impacts of PM10, when QO and LV for human health are exceeded. This work can be further extended by conducting epidemiological studies to assess the health impact associated with atmospheric particles in urban tropical regions. However, it requires detailed investigation of other species (thorough chemical analysis of filter samples) in parallel with continuous monitoring of other atmospheric pollutants (ex. heavy metals and SO_2) for at least a civil year, especially "during" and "off" significant particular local/regional (e.g. swell) events. Also, this experiment must be undertaken at other more inhabited marine locations in order to confirm these results.

6. ACKNOWLEDGEMENTS

We acknowledge the French "Ministry of Environment" (MEDDE), ADEME, CINOR, TCO and CIVIS, for their financial support to the ORA air quality network at Réunion Island.

We gratefully acknowledge the ORA staff for technical support and for the atmospheric data sampling.

INERIS/LCSQA is acknowledged for the analysis of the particles sampled on filters in the frame of the CARA program.

We also gratefully acknowledge all the technical teams who have contributed to the sampling, the filter management and transportation of equipments and the implementation and validation of chemical analyses, especially C. Guadagno and G. Peris from ORA, and S. Verlhac, R. Aujay, V. Minguet, Y. Chatellier and N. A. Papin from INERIS/LCSQA.

Thanks also to Météo-France for providing the meteorological data.

REFERENCES

1 M. S. O'Neill, D. Loomis, V. M. Torres-Meza, A. Retama and D. Gold, "Estimating Particle Exposure in the Mexico City Metropolitan Area," Journal of Exposure Analysis and Environmental Epidemiology, Vol. 12, No. 2, 2002, pp. 145-156.

2 M. D. Attfield, P. L. Schleiff, J. H. Lubin, A. Blair, P. A. Stewart, R. Vermeulen, J. B. Coble and D. T. Silverman, "The Diesel Exhaust in Miners Study: A Cohort Mortality Study with Emphasis on Lung Cancer," Journal of the National Cancer Institute, Vol. 104, No. 11, 2011, pp. 869-883.

3 IARC (International Agency for Research on Cancer), "Diesel Engine Exhaust Carcinogenic," Press Release No 213, 2012.

4 C. Dir, "Council Directive 1999/30/EC of 22nd April 1999 Relating to Limit Values for Sulphur Dioxide, Nitrogen Dioxide and Oxides of Nitrogen, Particulate Matter and Lead in Ambient Air," Official Journal of the European Union, L163/41, 1999.

5 WHO, "Health Aspects of Air Pollution with Particulate Matter, Ozone and Nitrogen Dioxide," Report on a WHO Working Group, 2003.

6 WHO, "Air Quality Guidelines for Particulate Matter, Ozone, Nitrogen Dioxide and Sulphur Dioxide, Global Update 2005," Summary of risk Assessment, Geneva, 2006.

7 E. Dir, "European Directive 2008/50/EC of the European Parliament and the Council of 21st May 2008 on Ambient Air Quality and Cleaner Air for Europe," Official Journal of the European Union, L152/1, 2008.

8 IPCC (Intergovernmental Panel on Climate Change), S. Solomon, D. Qin, M. Manning, Z. Chen, M. Marquis, K. B. Averyt, M. Tignor and H. L. Miller, "Contribution of Working Group I to the Fourth Assessment Report of the Intergovernmental Panel on Climate Change," Cambridge University Press, Cambridge, 2007.

9 R. Bhatia, P. Lopipero and A. H. Smith, "Diesel Exhaust Exposure and Lung Cancer," Epidemiology, Vol. 9, No. 1, 1998, pp. 84-91.

10 EPA, "Fourth External Review Draft of Air Quality Criteria for Particulate Matter," 2003.

11 U. Gehring, J. Heinrich, U. Krämer, V. Grote, M. Hochadel, D. Sugiri, M. Kraft, K. Rauchfuss, H. G. Eberwein and H.-E. Wichmann, "Long-Term Exposure to Ambient Air Pollution and Cardiopulmonary Mortality in Women," Epidemiology, Vol. 17, No. 5, 2006, pp. 545-551.

12 L. Carbajal-Arroyo, A. Barraza-Villarreal, R. DurandPardo, H. Moreno-Macias, R. Espinoza-Lain, P. ChiarellaOrtigosa and I. Romieu, "Impact of Traffic Flow on the Asthma Prevalence among School Children in Lima, Peru," Journal of Asthma, Vol. 44, No. 3, 2007, pp. 197- 202.

13 C. A. Pope, R. T. Burnett, M. J. Thun, E. E. Calle, D. Krewski, K. Ito and G. D. Thurston, "Lung Cancer, Cardiopulmonary Mortality, and Long-Term Exposure to Fine Particulate Air Pollution," Journal of the American Medical Association, Vol. 288, No. 9, 2002, pp. 1132- 1141.

14 C. A. Pope, M. Ezzoti and D. W. Dockery, "Fine Particulate Air Pollution and Life Expectancy in the United States," The New England Journal of Medicine, Vol. 360, 2009, pp. 376-386.

15 M. O. Andreae and P. J. Crutzen, "Atmospheric Aerosols: Biogeochemical Sources and Role in Atmospheric Chemistry," Science, Vol. 276, No. 5315, 1997, pp. 1052- 1058.

16 C. D. O'Dowd, M. H. Smith, I. E. Consterdine and J. A. Lowe, "Marine aerosol, Sea-Salt, and the Marine Sulphur Cycle: A Short Review," Atmospheric Environment, Vol. 31, No. 1, 1997, pp. 73-80.

17 B. J. Finlayson-Pitts, "The Tropospheric Chemistry of Sea-Salt: A Molecular-Level View of the Chemistry of NaCl and NaBr," Chemical Reviews, Vol. 103, No. 12, 2003, pp. 4801-4822.

18 E. R. Lewis and S. E. Schwartz, "Sea Salt Aerosol Production: Mechanisms, Methods, Measurements, and Models—A Critical Review," Geophysical Monograph Series, Vol. 152, American Geophysical Union, Washington DC, 2004.

19 C. D. O'Dowd and G. De Leeuw, "Marine Aerosol Production: A Review of the Current Knowledge," Philosophical Transactions of the Royal Society A, Vol. 365, No. 1856, 2007, pp. 1753-1774.

20 N. W. Tindale and P. P. Pease, "Aerosols over the Arabian Sea: Atmospheric Transport Pathways and Concentrations of Dust and Sea Salt," Deep-Sea Research Part II, Vol. 46, No. 8-9, 1999, pp. 1577-1595.

21 J. Putaud, F. Raes, R. VanDingenen, E. Bruggemann, M. Facchini, S. Decesari, S. Fuzzi, R. Gehrig, C. Hueglin, P. Laj, G. Lorbeer, W. Maenhaut, N. Mihalopoulos, K. Mueller, X. Querol, S. Rodriguez, J. Schneider, G. Spindler, H. Ten Brink, K. Torseth and A. Wiedensohler, "European Aerosol Phenomenology-2: Chemical Characteristics of Particulate Matter at Kerbside, Urban, Rural and Background Sites in Europe," Atmospheric Environment, Vol. 38, No. 16, 2004, pp. 2579-2595.

22 M. M. Manders, M. Schaap, X. Querol, M. F. M. A. Albert, J. Vercauteren, T. A. J. Kuhlbusch and R. Hoogerbrugge, "Sea Salt Concentrations across the European Continent," Atmospheric Environment, Vol. 44, No. 20, 2010, pp. 2434-2442.

23 M. Hess, U. K. Krieger, C. Marcolli, T. Huthwelker, M. Ammann, W. A. Lanford and Th. Peter, "Bromine Enrichment in the Near-Surface Region of Br-Doped NaCl Single Crystals Diagnosed by Rutherford

Backscattering Spectrometry," The Journal of Physical Chemistry A, Vol. 111, No. 20, 2007, pp. 4312-4321.

24 R. C. Easter, S. J. Ghan, Y. Zhang, R. D. Saylor, E. G. Chapman, N. S. Laulainen, H. Abdul-Razzak, H. R. Leung, X. D. Bian and R. A. Zaveri, "MIRAGE: Model Description and Evaluation of Aerosols and Trace Gases," Journal of Geophysical Research, Vol. 109, No. D20, 2004, Article ID: D20210.

25 N. Mahowald, J.-F. Lamarque, X. X. Tie and E. Wolff, "Sea Salt Aerosol Response to Climate Change: Last Glacial Maximum, Preindustrial, and Doubled Carbon Dioxide Climates," Journal of Geophysical Research, Vol. 111, No. D5, 2006, Article ID: D05303.

26 S. Tamm and M. Schulz, "Open-Ocean Aerosol Composition Obtained during 15 Months on a North Sea Ferry," Atmospheric Environment, Vol. 37, Suppl. 1, 2003, pp. 133-143.

27 E. M. Knipping and D. Dabdub, "Impact of Chlorine Emissions from Sea-Salt Aerosol on Coastal Urban Ozone," Environmental Science & Technology, Vol. 37, No. 2, 2003, pp. 275-284.

28 Cohan, W. Chang, M. Carreras-Sospedra and D. Dabdub, "Influence of Sea-Salt Activated Chlorine and Surface-Mediated Renoxification on the Weekend Effect in the South Coast Air Basin of California," Atmospheric Environment, Vol. 42, No. 13, 2008, pp. 3115-3129.

29 S. C. Pryor and R. J. Barthelmie, "Particle Dry Deposition to Water Surfaces: Processes and Consequences," Marine Pollution Bulletin, Vol. 41, No. 1-6, 2000, pp. 220-231.

30 M. Van Loon, R. Vautard, M. Schaap, R. Bergström, B. Bessagnet, J. Brandt, P. Builtjes, J. H. Christensen, J. H. Cuvelier, A. Graf, J. E. Jonson, M. Krol, J. Langner, P. Roberts, L. Rouil, R. Stern, L. Tarrasón, P. Thunis, E. Vignati, L. White and P. Wind, "Evaluation of LongTerm Ozone Simulations from Seven Regional Air Quality Models and Their Ensemble Average," Atmospheric Environment, Vol. 41, No. 10, 2007, pp. 2083-2097.

31 T. H. Muster and I. S. Cole, "Attachment Efficiencies of Salt Aerosols onto Infrastructure and Implications for Atmospheric Corrosion," Journal of the Electrochemical Society, Vol. 152, No. 3, 2005, pp. B125-B131.

32 Y. J. Yoon and P. Brimblecombe, "Modelling the Contribution of Sea Salt and Dimethyl Sulfide Derived Aerosol to Marine CCN," Atmospheric Chemistry and Physics, Vol. 2, No. 1, 2002, pp. 17-30.

33 V. Vinoj and S. K. Satheesh, "Direct and Indirect Radiative Effects of Sea-Salt Aerosols over Arabian Sea," Current Science (India), Vol. 86, No. 10, 2004, pp. 1381- 1390.

34 S. L. Gong, L. A. Bartie, J.-P. Blanchet, "Modeling SeaSalt Aerosols in the Atmosphere—1. Model Development," Journal of Geophysical Research, Vol. 102, No. D3, 1997, pp. 3805-3818.

35 E. Athanasopoulou, M. Tombrou, S. N. Pandis and A. G. Russell, "The Role of Sea-Salt Emissions and Heterogeneous Chemistry in the Air Quality of Polluted Coastal Areas," Atmospheric Chemistry and Physics, Vol. 8, No. 19, 2008, pp. 5755-5769.

36 B. Finlayson-Pitts and J. N. Pitts Jr., "Chemistry of the Upper and Lower Atmosphere, Theory, Experiments and Applications," Academic Press, Waltham, 1999.

37 H. Hass, M. van Loon, C. Kessler, R. Stern, J. Matthijsen, F. Sauter, Z. Zlatev, J. Langner, V. Foltescu and M. Schaap, "Aerosol Modelling: Results and Intercomparison from European Regional-Scale Modelling Systems," Special Rep. EUROTRAC-2 ISS, Munchen, 2003.

38 B. Bessagnet, A. Hodzic, H. Vautard, M. Beekmann, S. Cheinet, C. Honoré, C. Liousse and L. Rouil, "Aerosol Modeling with CHIMERE: Preliminary Evaluation at the Continental Scale," Atmospheric Environment, Vol. 38, No. 18, 2004, pp. 2803-2817.

39 S. L. Gong, L. A. Barrie, J.-P. Blanchet, K. von Salzen, U. Lohmann, G. Lesins, L. Spacek, L. M. Zhang, E. Girard, H. Lin, R. Leaitch, H. Leighton, P. Chylek and P. Huang, "Canadian Aerosol Module: A Size-Segregated Simulation of Atmospheric Aerosol Processes for Climate and Air Quality Models 1. Module Development," Journal of Geophysical Research, Vol. 108, No. D1, 2003, pp. AAC 3-1-AAC 3-16.

40 W. Guelle, M. Schulz, Y. Balkanski and F. Dentener, "Influence of the Source Formulation on Modeling the Atmospheric Global Distribution of Sea Salt Aerosol," Journal of Geophysical Research, Vol. 106, No. D21, 2001, pp. 27509-27524.

41 P. Stier, J. Feichter, S. Kinne, S. Kloster, E. Vignati, J. Wilson, L. Ganzeveld, I. Tegen, M. Werner, Y. Balkanski, M. Schulz, O. Boucher, A. Minikin and A. Petzold, "The Aerosol-Climate Model ECHAM5-HAM," Atmospheric Chemistry and Physics, Vol. 5, No. 4, 2005, pp. 1125- 1156.

42 P. Jiménez-Guerrero, O. Jorba, M. T. Pay, J. P. Montávez, S. Jerez, J. J. Gómez-Navarro and J. M. Baldasano, "Comparison of Two Different

Sea-Salt Aerosol Schemes as Implemented in Air Quality Models Applied to the Mediterranean Basin," Atmospheric Chemistry and Physics, Vol. 11, No. 10, 2011, pp. 4833-4850.

43 V. L. Foltescu, S. C. Pryor and C. Bennet, "Sea Salt Generation, Dispersion and Removal on the Regional Scale," Atmospheric Environment, Vol. 39, No. 11, 2005, pp. 2123-2133.

44 B. Langmann, S. Varghese, E. Marmer, E. Vignati, J. Wilson, P. Stier and C. O'Dowd, "Aerosol Distribution over Europe: A Model Evaluation Study with Detailed Aerosol Microphysics," Atmospheric Chemistry and Physics, Vol. 8, No. 6, 2008, pp. 1591-1607.

45 M. Schaap, R. M. A. Timmermans, F. J. Sauter, M. Roemer, G. J. M. Velders, G. A. C. Boersen, J. P. Beck and P. J. H. Builtjes, "The LOTOS-EUROS Model: Description, Validation and Latest Developments," International Journal of Environment and Pollution, Vol. 32, No. 2, 2008, pp. 270-290.

46 S. Zakey, F. Giorgi and X. Bi, "Modeling of Sea Salt in a Regional Climate Model: Fluxes and Radiative Forcing," Journal of Geophysical Research, Vol. 113, No. D14, 2008.

47 Bhugwant, H. Cachier, M. Bessafi and J. Leveau, "Impact of Traffic on Black Carbon Aerosol Concentration at La Reunion Island (Southern Indian Ocean)," Atmospheric Environment, Vol. 34, No. 20, 2000, pp. 3464- 3473.

48 Bhugwant and P. Brémaud, "Simultaneous Measurements of Black Carbon, PM10, Ozone and NO_x Variability at a Locally Polluted Island in the Southern Tropics," Journal of Atmospheric Chemistry, Vol. 39, No. 3, 2001, pp. 261-280.

49 Bhugwant, M. Bessafi and B. Siéja, "The Potential Impact of Marine Aerosols via the Swell and the Oceanic Waves on the PM10 Concentration Measurements at Urban Marine Locations, Air Pollution Emissions," Nova Publishers, New York, 2012.

50 INSEE, "TER (Tableau Economique de La Réunion)," 2010, p. 3014. in French.

51 W. E. Wilson, B. D. Grover, R. W. Long, N. L. Eatough and D. J. Eatough, "The Measurement of Fine-Particulate Semivolatile Material in Urban Aerosols," Journal of the Air & Waste Management Association, Vol. 56, No. 4, 2006, pp. 384-397.

52 B. D. Grover, N. L. Eatough, D. J. Eatough, J. C. Chow, J. G. Watson, J. L. Ambs, M. B. Meyer, P. K. Hopke, R. Al-Horr, D. W. Later and W. E. Wilson, "Measurement of Both Nonvolatile and Semi-Volatile Fractions

of Fine Particulate Matter in Fresno, CA," Aerosol Science and Technology, Vol. 40, No. 10, 2006, pp. 811-826.

53 J. Wanjura, B. Shaw, C. Parnell, R. Lacey and S. Capareda, "Comparisons of Continuous Monitor (TEOM) and Gravimetric Sampler Particulate Matter Concentrations," American Society of Agricultural and Biological Engineers, Vol. 51, No. 1, 2008, pp. 251-257.

54 C. Bhugwant and J.-L. Hoareau, "Variability of NO_2 in Different Environments at a Moderately Polluted Island over the Southwestern Indian Ocean," Atmospheric Research, Vol. 66, No. 4, 2003, pp. 241-259.

55 F. Cavalli, M. Viana, K. E. Yttri, J. Genberg and J.-P. Putaud, "Toward a Standardised Thermal-Optical Protocol for Measuring Atmospheric Organic and Elemental Carbon: The EUSAAR Protocol," Atmospheric Measurement Techniques, Vol. 3, No. 1, 2010, pp 79-89.

56 M. E. Birch and R. A. Cary, "Elemental Carbon-Based Method for Monitoring Occupational Exposures to Particulate Diesel Exhaust," Aerosol Science and Technology, Vol. 25, No. 3, 1996, pp. 221-241.

57 O. S. Météo-France, "Atlas Climatique de La Réunion," Bureau d'étude Climatologique, Direction Interrégionale de La Réunion, Annual Report No. 1657, 2000.

58 N. Pérez, J. Pey, M. Cusack, C. Reche, X. Querol, A. Alastuey and M. Viana, "Variability of Particle Number, Black Carbon, and PM10, PM2.5, and PM1 Levels and Speciation: Influence of Road Traffic Emissions on Urban Air Quality," Aerosol Science and Technology, Vol. 44, No. 7, 2010, pp. 487-499.

59 M. T. Limon-Sanchez, P. Carbajal-Romero, L. Hernandez-Mena, H. Saldarriaga-Norena, A. Lopez-Lopez, R. Cosio-Ramirez, J. L. Arriaga-Colina and W. Smith, "Black Carbon in PM2.5, Data from Two Urban Sites in Guadalajara, Mexico during 2008," Atmospheric Pollution Research, Vol. 2, No. 3, 2011, pp. 358-365.

60 V. Grondin and C. Bhugwant, "Internship Report, ORA, Etude de l'Impact des Embruns Marins Sur la Qualité de l'Air en Milieu Urbain à la Réunion," 2008, in French.

61 B. J. Turpin, P. Saxena and E. Andrews, "Measuring and Simulating Particle Organics in the Atmosphere: Problems and Prospects," Atmospheric Environment, Vol. 34, No. 18, 2000, pp. 2983-3013.

62 J. C. Cabada, S. N. Pandis, R. Subramanian, A. L. Robinson, A. Polidori and B. Turpin, "Estimating the Secondary Organic Aerosol Contribution

to PM2.5 Using the EC Tracer Method Special Issue of Aerosol Science and Technology on Findings from the Fine Particulate Matter Super Sites Program," Aerosol Science and Technology, Vol. 38, Suppl. 1, 2004, pp. 140-155.

63 E. Friese and A. Ebel, "Temperature Dependent Thermodynamic Model of the System H^+-$NH4^+$-Na^+-SO_4^{2-} - NO_3^- -Cl^--H_2O," Journal of Physical Chemistry A, Vol. 114, No. 43, 2010, pp. 11595-11631.

CHAPTER 6

The Chemistry of Dicarboxylic Acids in the Atmospheric Aerosols

Mohd Zul Helmi Rozaini

[1] School of Environmental Sciences, University of East Anglia, Norwich, Norfolk,, UK

[2] Department of Chemical Sciences, University Malaysia Terengganu, Kuala Terengganu, Terengganu,, Malaysia

1. INTRODUCTION

Atmospheric chemistry is a branch of atmospheric science in which the chemistry of the Earth's atmosphere and that of other planets is studied. It is a multidisciplinary field of research and draws on environmental chemistry, physics, meteorology, computer modeling, oceanography, geology and volcanology and other disciplines. It also deals with chemical compounds in the atmosphere, their distribution, origin, chemical transformation into other compounds and finally their removal from the atmospheric domain. These substances may occur as gasses, liquids or solid. The composition of the atmosphere is dominated by the gasses nitrogen and oxygen in proportions that have been found to be invariable in time and space at altitudes up to 100 km. All other compounds are minor ones, with many of them occurring only in traces.

The composition and chemistry of the atmosphere is of importance for several reasons, but primarily because of the interactions between the atmosphere and living organisms. The composition of the Earth's atmosphere (Figure 1) has been changed by human activity and some of these changes are harmful to human health, crops and ecosystems. Examples of problems which have been addressed by atmospheric chemistry include acid rain, photochemical smog and global warming. Atmospheric chemistry seeks to understand the causes of these problems, and by obtaining a theoretical understanding of them, allow possible solutions to be tested and the effects of changes in government policy evaluated.

Observations, lab measurements and modeling are the three important methodologies in atmospheric chemistry. Progress in atmospheric chemistry is often driven by the interactions between these components and they form an integrated whole. For example observations may tell us that more of a chemical compound exists than previously thought possible. This will stimulate new modelling and laboratory studies which will increase our scientific understanding to a point where the observations can be explained. Measurements made in the laboratory are essential to our understanding of the sources and sinks of pollutants and naturally occurring compounds. Lab studies tell us which gases react with each other and how fast they react. Measurements of interest include reactions in the gas phase, on surfaces and in water. Also of high importance is photochemistry which quantifies how quickly molecules are split apart by sunlight and what the products are plus thermodynamic data such as Henry's law coefficients.

Modelling for instance is important to synthesize and test theoretical understanding of atmospheric chemistry. Computer models (such as chemical transport models) are used. Numerical models solve the differential equations governing the concentrations of chemicals in the atmosphere. They can be very simple or very complicated. One common trade off in numerical models is between the number of chemical compounds and chemical reactions modelled versus the representation of transport and mixing in the atmosphere. For example, a box model might include hundreds or even thousands of chemical reactions but will only have a very crude representation of mixing in the atmosphere. In contrast, 3D models represent many of the physical processes of the atmosphere but due to constraints on computer resources will have far fewer chemical reactions and compounds. Models can be used to interpret observations, test understanding of chemical reactions and predict future concentrations of chemical compounds in the atmosphere. One important current trend is for atmospheric chemistry modules to become one part of earth system models in which the links between climate, atmospheric composition and the biosphere can be studied.

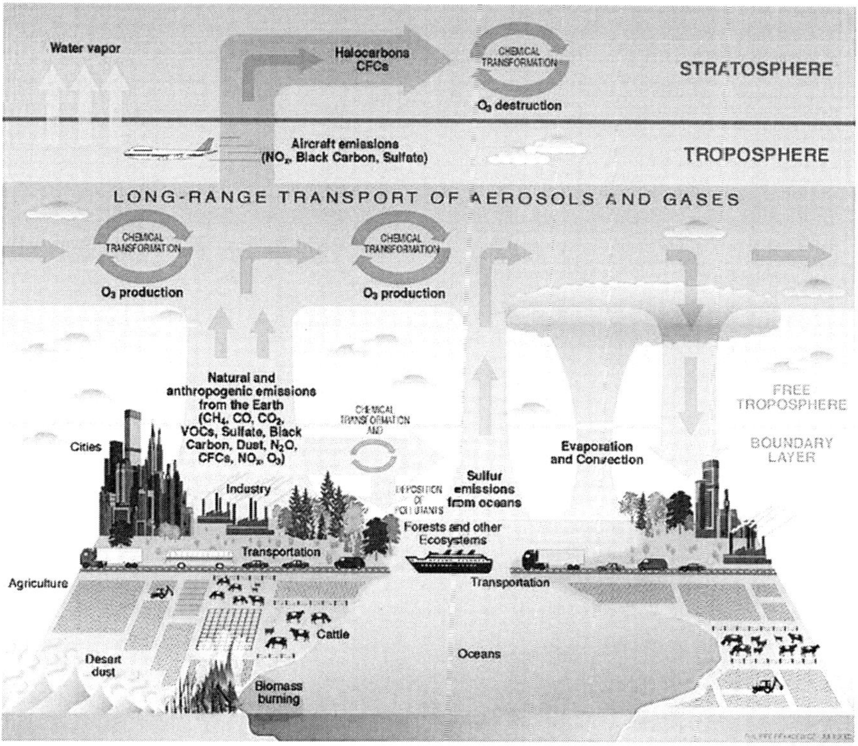

Figure 1. Schematic of chemical and transport processes related to atmospheric composition.

2. BACKGROUND KNOWLEDGE

2.1. Aerosol

An aerosol is a system (in the sense of a system as used in thermodynamics or chemistry) comprising liquid and/or solid particles in a carrier gas. It is generally defined as a suspension of liquid or solid particles in a gas, with particle diameters in the range of 10^{-9}-10^{-4} m (lower limit: molecules and molecular clusters: upper limit: rapid sedimentation). The most evident examples of aerosols in the atmosphere are clouds, which consist primarily of condensed water. The suspension of the particles in the gas must be significantly stable and homogenous. Hence the assumptions of stability and homogeneity, and consequently the possibilities to use statistical descriptors, are limited to understand and to predict the system, the particle properties, i.e. their size, shapes, chemical compositions, their surfaces, their optical properties, their volumes and masses must be known (Preining, 1993).

Aerosol particles scatter and absorb solar and terrestrial's radiation, they are involved in the formation of clouds and precipitation as cloud condensation and ice nuclei, and they affect the abundance and distribution of atmospheric traces gases by heterogeneous chemical reactions and other multiphase processes.

2.2. Aerosol Types

The atmospheric aerosol in the boundary layer and the lower troposphere is different for different regions, the main types are:

a. continental aerosol - a main component of which is mineral dust;

b. maritime aerosol - a main component of which is sea salt;

c. background aerosol - aged accumulation mode aerosol.

Chemically or photochemically produced from precursor gases, continental or oceanic biosphere or from anthropogenic releases including sulphates, nitrates, hydrocarbons, soot and so on. The continental aerosols are strongly influenced by man's activities and include urban and rural aerosols. Dust storms produce another type of continental aerosol. Aerosols with a lifetime of up to several years exist in the stratosphere, the sources of which are volcanic injections, and particles or gases entering the stratosphere via diffusion from the troposphere as well as interplanetary dust entering from space. The most important source is volcanic injection. Due to their long lifetime, these aerosols are distributed relatively homogeneously throughout the whole stratosphere and the size distribution is unimodal with only the accumulation mode present.

2.3. The Study Of Atmospheric Aerosols

Atmospheric aerosol particles are a ubiquitous part of earth's atmosphere, present in very lungful of air breathed. They are produced in vast numbers by both human activity (anthropogenic) and natural sources and subsequently modified by a multitude processes. They are known to be crucially important in many issues that directly affect everyday life which include respiratory health, visibility, clouds, rainfall, atmospheric chemistry and global regional climate but they are also one of the more poorly understood aspects of the atmosphere. These shortcomings in understanding are partly due to their small size, which is typically of the order of microns or less, making them difficult to study and also the fact that the processes involved are complex. The description of the organic chemistry in atmospheric aerosol is by no means straightforward, but the addition of the solubility variables, aerosol thermodynamic, hygroscopic properties, deliquescence behaviour makes understanding the atmosphere and its effect is even more challenging, requiring the application of wide spectrum of scientific disciplines including chemistry, physics, mechanics, biology and medicine.

2.4. Aerosols and Effect on Quality of Life

The effects of aerosols on the atmosphere, climate and public health are among the central topics in current environmental research. Urban areas have always been known to be a major source of particulate pollution (Finlayson-Pitss, 2000) which is expected to continue to increase due to world population growth and increasing industrialization and energy use, especially in developing countries (Fenger, 1999). The most obvious effects are the contributions to unsightly smogs and visible deterioration of the building materials (Grossi, 2002). In addition, the fact that urban particulate pollution impact directly on human health has been known for centuries (Brimblecombe, 1987) and has been the subject of much research (Adam et al., 1999).

In an attempt to reduce the health burden of atmospheric particulate pollution, regulatory authorities have attempted to place controls on the emission and the magnitude of pollution episodes within conurbations. The monitoring of particulate air pollution has traditionally focused on particles of less than 10 μm in aerodynamic diameter (the PM_{10} standard), as these are more likely to pass the throat when inhaled (DEFRA, 2005; Larrsen, 1999) but it has become apparent that the smaller particles are more significant, as these particles will penetrate deeper into the lungs and potentially cause more physiological distress or damage. This has lead to the use of the $PM_{2.5}$ standard in countries such Malaysia, where the total mass of particulate matter less than 2.5 μm in diameter is monitored (MOSTI, 2000).

2.5. Composition of Atmospheric Aerosol

The atmospheric aerosol consists of a complex mixture of organic and inorganic compounds (Cruz, 1998). The typical composition of fine continental aerosol will usually contain various sulphates (mostly ammonium and calcium), nitrates (mostly ammonium), chlorides (mostly sodium), elemental carbon (EC) and organic carbon (OC), especially traffic-related soot, biological materials and other organic compounds, iron compounds, trace metals, and mineral derived from rocks, soil and various human activities. Aerosol composition also can be influenced by local geology, geographic location and climate (Moreno et al., 2003).

2.5.1. Organic and Elemental Carbon of Aerosol

Several studies have shown that over 30% of aerosol is organic carbon, and carbon containg matter can account for as much as 50%. Typically, two classes of carbonaceous aerosol are commonly present in ambient air: organic carbon (OC) and elemental carbon (EC), which are the largest contributors to the fine particle burden in urban atmospheres and heavily industrialised areas (Cachier et al., 1989).

Field measurements also shown a significant mass fraction of atmospheric aerosol consist of organic compounds (Rogge et al., 1991). Around 5 to 10% of the known fraction is often limited to low molecular weight species, which are identified by standard analytical techniques, using gas chromatography coupled with mass spectrometry. A significant fraction of the organic mass in tropospheric aerosol, is comprised of high molecular weight, oxygenated species which remain unidentified (Decesari et al., 2002).

Organic compounds are emitted into the atmosphere from various anthropopgenic and biogenic sources. These include primary emission, mainly from combustion and biogenic sources and secondary organic aerosol resulting from the reaction of primary volatile organic compounds in the atmosphere (Fisseha et al., 2004). In urban areas, a number of emission sources are responsible for the presence of organic aerosol in the atmosphere among which are road traffic, industrial processes, waste incineration, wastewater treatment processes and domestic heating. Some of these are pure organic aerosols, which may be formed by primary particle emissions (primary organic carbon) or produced from atmospheric reactions involving gaseous organic precursors (secondary OC)(Cruz and Pandis, 1998).

Organic material is important in controlling the aerosol physico-chemical properties (Cornell et al., 2003). They also found that the uptake of liquid water in aerosol was enhanced by the presence of organic carbon compounds. Organic carbon is also an effective light scatter and may contribute significantly to both visibility degradation and direct aerosol climate forcing (Heintzenberg., 1989). Elemental carbon (often named black carbon or soot) may be the second most important elemental in global warming in terms of direct forcing, after CO_2 due to specific surface properties. Elemental carbon provides a good adsorbtion site for many semi-volatile compounds such as poly-aromatic hydrocarbon (PAH) and offers a large specific surface area for interactions with reactive trace gases such as ozone. Annually, about 13 Tg black carbons are emitted into the atmosphere, mainly through fossil fuel combustion and biomass burning (Jacob, 1999).

As for other aerosols, the removal of particulate carbon is likely to occur via two main scavenging processes: the in-cloud process, whereby particles are directly incorporated into cloud droplets; and the below-cloud process, where particles are washed out by precipitation itself. The physico-chemical atmospheric processes which transform young combustion particles, expected to be hydrophobic, into a water soluble aerosol phase remains a major unknown. The atmospheric behaviour of the carbonaceous particles is likely to be dictated by the chemical nature of their surfaces (Cachier et al., 1989). If the surface is hydrophobic, the particle remains inactive. However, if it is coated with hygroscopic substances, it may be activated enough to be incorporated into water droplets (Charlson and Heintzenberg, 1995).

2.5.2. Water Soluble Organic Compounds

A significant fraction of the particulate organic carbon is water soluble, ranging from 20% to 70% of the total soluble mass, thus making it important to various aerosol-cloud interactions (Decesari et al., 2000; Facchini et al., 2000). Water soluble organic compounds (WSOC) contribute to the ability of the particles to act as cloud condensation nuclei (CCN) (Novokov and Penner, 1993).

WSOC have been postulated to be partially responsible for the water uptake of airbone particulate matter, which can substantially affect the physical and chemical properties of atmospheric aerosols (Yu et al., 2005). Decesari et al. (2001) have suggested that WSOC are composed of higly oxidised species with residual aromatic nuclei and aliphatic chains. The current understanding of atmospheric particles describes their WSOC fraction as a complex mixture of very soluble organic compounds, slightly soluble organic compounds, and some undetermined macromolecular compounds (MMCs)(Saxena and Hildemann, 1996).

The composition of WSOC varies among sampling regions. It was found to constitute between 20 and 67% of the total organic carbon present in aerosol samples collected in Tokyo (Sempere andKawamura, 1994). The percentage is ranged from 65 to 75% in aerosol samples collected in Hungary, Italy and Sweeden (Zappoli et al., 1999). The study also found that the percentage of WSOC species with respect to the total soluble mass was much higher at the background site (Aspvreten, Central Sweeden) (c.a. 50%) compared to the polluted site (San Pietro Copofiume, Po Valley, Italy) (c.a 25%). A very high fraction (over 70%) of organic compounds in the aerosol consisted of polar species. A study by Wang et al. (2002) showed that most water soluble carbon is total organic carbon (TOC) and range between 20.53 to 35.58 μg m^{-3} in PM_{10} and $PM_{2.5}$. A further study by (Narukawa et al., 1999) concluded that individual haze particles over Kalimantan of Indonesia were mainly composed of water soluble organic materials and inorganic salt such as ammonium sulphate.

The ionic organic compounds (including carboxylic, dicarboxylic and ketoacids) were distributed between both sub-micron and super micron mode, indicating origins in both gas-to-particle conversion and heterogeneous reaction on pre-existing particles. WSOC in atmospheric aerosols and droplets can be divided by their functional groups into three classes which are neutral, mono- and dicarboxylic acid and also polycarboxylic acid, which were found to account on average for 87% of total fine aerosol WSOC (Decesari et al., 2000). The most frequently determined WSOC are the low molecular weight (LMW) carboxylic and dicarboxylic acids (Yu, 2000). Most of carboxylic acids compound are a secondary oxidation products of atmospheric organic compounds and also found in remote marine as well as continental rural and urban areas (Simoneit and Mazurek, 1982). Among these dicarboxylic acids (DCA's), oxalic acid is the most abundant, followed by succinic and malonic in atmospheric aerosol especially during summer season.

In the aqueous phase, organic oxidation also can be initiated by various radical anions in the atmosphere (e.g. OH^- ,NO^-_3 ,SO^2_4 ,Cl^-). Among these species, it is very likely that OH is the most efficient iniating organic oxidation (Dutot et al., 2003). The DCA's are the late products in the photochemistry of aliphatic and aromatic hydrocarbons, and due to the low vapour pressure, it is almost entirely partitioned to the particulate phase. They also constitute an important fraction of the water soluble part of particulate organic matter (POM) in atmospheric aerosol particles at remote and urban areas (Rohrl and Lammel, 2001).

3. DICARBOXYLIC ACIDS

During the past decade, much attention has been paid to the low molecular weight dicarboxylic acids and related polar compounds which are ubiquitous water-soluble organic compounds that have been detected in a variety of environmental samples including atmospheric aerosols, rainwaters, snow packs, ice cores, meteorites, marine sediments, hypersaline brines and freshwaters (Kawamura and Ikushima, 1993; Tedetti et al., 2006). In the atmosphere, dicarboxylic acids originate from incomplete combustion of fossil fuels (Kawamura and Ikushima, 1993; Kawamura and Kaplan, 1987), biomass burning (Narukawa et al., 1999), direct biogenic emission and ozonolysis and photo-oxidation of organic compound (Sempere and Kawamura, 2003).

Low molecular weight (LMW) dicarboxylic acids have also been identified in cloud water samples collected at a high mountain range in central europe (Puxbaum and Limbeck, 2000), in the condensed phase at a semi-urban site in the northeastern US (Khwaja, 1995) and in Arctic aerosol (Kawamura et al., 1996). As a result of their hygroscopic properties, dicarboxylic acids can act as cloud condensation nuclei and have an impact on the radiative forcing at earth's surface (Kerminen et al., 2000). Dicarboxylic acids also participate in many biological processes. They are important intermediates in the tricarboxylic acid and glyoxylate cycles and the catabolism and anabolism of amino acids (Tedetti et al., 2006).

Photochemical reactions are also an important source of atmospheric dicarboxylic acids. For example, glutaric acids photooxidation is likely the dominant pathway formation, as measured atmospheric concentrations of dicarboxylic acids in Los Angeles far surpasses contributions from direct emissions and seasonal trends suggest that dicarboxylic acids are largely produced in photochemical smog (Puxbaum and Limbeck, 2000; Rogge et al., 1993).

Aliphatic dicarboxylic acids (or diacids) can be described by the following general formula:

$HOOC\text{-}(CH_2)_n\text{-}COOH$

According to IUPAC nomenclature, dicarboxylic acids are named by adding the suffix dioic acid to the name of the hydrocarbon with the same number of carbon atoms, e.g., nonanedioic acid for $n = 7$. The older literature often uses another system based on the hydrocarbon for the $(CH_2)n$ carbon segment and the suffix dicarboxylic acid, e.g., heptanedicarboxylic acid for $n = 7$. However, trivial names are commonly used for the saturated linear aliphatic dicarboxylic acids from $n = 0$ (oxalic acid) to $n = 8$ (sebacic acid) and for the simple unsaturated aliphatic dicarboxylic acids; these names are generally derived from the natural substance in which the acid occurs or from which it was first isolated.

Aliphatic dicarboxylic acids are found in nature both as free acids and as salts. For example, malonic acid is present in small amounts in sugar beet and in the green parts of the wheat plant; oxalic acid occurs in many plants and in some minerals as the calcium salt. However, natural sources are no longer used to recover these acids.

The main industrial process employed for manufacturing dicarboxylic acids is the ring-opening oxidation of cyclic compounds.

Oxalic acid is the most important dicarboxylic acid. Adipic, malonic, suberic, azelaic, sebacic, and 1,12-dodecanedioic acids, as well as maleic and fumaric acids, are also manufactured on an industrial scale.

Physical properties: Dicarboxylic acids are colorless, odorless crystalline substances at room temperature. Table 1 lists the major physical properties of some saturated aliphatic dicarboxylic acids.

The lower dicarboxylic acids are stronger acids than the corresponding monocarboxylic ones. The first dissociation constant is considerably greater than the second. Density and dissociation constants decrease steadily with increasing chain length. By contrast, melting point and water solubility alternate: Dicarboxylic acids with an even number of carbon atoms have higher melting points than the next higher odd-numbered dicarboxylic acid. In the $n = 0 - 8$ range, dicarboxylic acids with an even number of carbon atoms are slightly soluble in water, while the next higher homologues with an odd number of carbon atoms are more readily soluble. As chain length increases, the influence of the hydrophilic carboxyl groups diminishes; from $n = 5$ (pimelic acid) onward, solubility in water decreases rapidly. The alternating solubility of dicarboxylic acids can be exploited to separate acid mixtures. Most dicarboxylic acids dissolve easily in lower alcohols; at room temperature, the lower dicarboxylic acids are practically insoluble in benzene and other aromatic solvents.

Chemical properties: The chemical behavior of dicarboxylic acids is determined principally by the two carboxyl groups. The neighboring methylene groups are activated generally to only a minor degree. Thermal decomposition of dicarboxylic acids gives different products depending on the chain length. Acids with an even number of carbon atoms require higher decarboxylation temperatures than the next higher odd-numbered homologues; lower dicarboxylic acids decompose more easily than higher ones. To avoid undesired decomposition reactions, aliphatic dicarboxylic acids should only be distilled in vacuum. When heated above 190 °C, oxalic acid decomposes to carbon monoxide, carbon dioxide, and water. Malonic acid is decarboxylated to acetic acid at temperatures above 150 C:

$$HOOC\text{-}(CH_2)_n\text{-}COOH\text{-}CH_3COOH + CO_2$$

When malonic acid is heated in the presence of P_2O_5 at ca. 150 °C, small amounts of carbon suboxide (C_3O_2) are also formed. Succinic and glutaric acids are converted into cyclic anhydrides on heating:

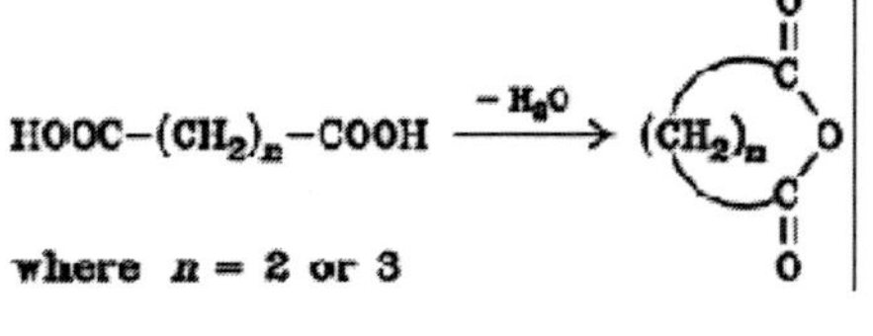

Table 1. Physical properties of saturated dicarboxylic acid (Clarke, 1986)

IUPAC name	Common name	Formula	Molecular weight, M_r	Melting point, M_p, °C	Boiling point, B_p at 13.3 kPa, °C	Density, ℓ at 25°C g/cm³	Solubility in mol/kg	Ionisation constant	
								K_1	K_2
Ethanedioc acid	Oxalic acid	HOOC-COOH	90.03	189.5	-	1.653	1.131	5.29×10^{-2}	5.33×10^{-5}
Propanedioc acid	Malonic acid	HOOC-CH2-COOH	104.06	135	-	1.619	16.03	1.42×10^{-3}	2.01×10^{-6}
Butanedioic acid	Succinic acid	HOOC-(CH2)2-COOH	118.08	188	235	1.572	0.748	6.16×10^{-5}	2.31×10^{-6}
Pentanedioic acid	Glutaric acid	HOOC-(CH2)3-COOH	132.11	99	200	1.424	8.468	4.57×10^{-5}	3.89×10^{-6}
Hexanedioic acid	Adipic acid	HOOC-(CH2)4-COOH	146.14	153	265	1.360	0.171	3.85×10^{-5}	3.89×10^{-6}
Heptanedioic acid	Pimelic acid	HOOC-(CH2)5-COOH	160.17	106	272	1.329	0.423	3.19×10^{-5}	3.74×10^{-6}
Octanedioic acid	Suberic acid	HOOC-(CH2)6-COOH	174.19	144	279	1.266	0.0139	3.05×10^{-5}	$3.85x^{-6}$
Nonanedioic acid	Azelaic acid	HOOC-(CH2)7-COOH	188.22	108	287	1.225	0.00946	2.88×10^{-5}	3.86×10^{-6}
Decanedioic acid	Sebacic acid	HOOC-(CH2)8-COOH	202.25	134.5	-	1.207	0.00012	3.1×10^{-5}	3.6×10^{-6}

Scheme 1. Succinic and glutaric acids are converted into cyclic anhydrides on heating

When the ammonium salt of succinic acid is distilled rapidly, succinimide is formed, with the release of water and ammonia.

Higher dicarboxylic acids from $n = 4$ (adipic acid) to $n = 6$ (suberic acid) split off carbon dioxide and water to form cyclic ketones:

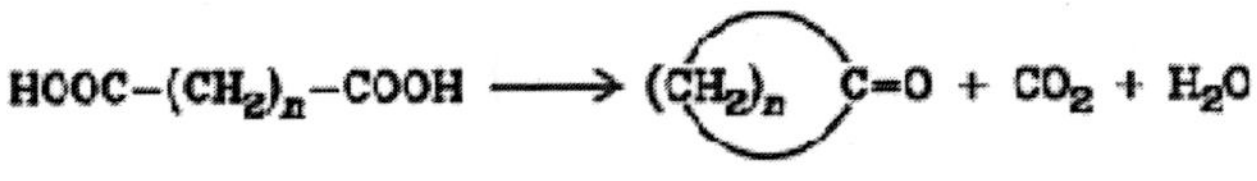

where $n = 4-6$

Scheme 2. Higher dicarboxylic acids from n = 4 (adipic acid) to n = 6 (suberic acid) split off carbon dioxide and water to form cyclic ketones

The decomposition of still higher dicarboxylic acids leads to complex mixtures. With the exception of oxalic acid, dicarboxylic acids are resistant to oxidation. Oxalic acid is used as a reducing agent for both commercial and analytical purposes. Dicarboxylic acids react with dialcohols to form polyesters and with diamines to form polyamides. They also serve as starting materials for the production of the corresponding diamines. Reaction with monoalcohols yields esters. All of these reactions are commercially important. Several reactions with malonic and glutaric acids are of interest in organic syntheses: the Knoevenagel condensation, Michael addition, and malonic ester synthesis (Clarke, 1986)

Succinic acid ester reacts with aldehydes or ketones in the presence of sodium ethoxide or potassium*tert*-butoxide to form alkylidenesuccinic acid monoesters (Stobbe condensation). These can subsequently be converted into monocarboxylic acids by hydrolysis, decarboxylation, and hydrogenation (Clarke, 1986).

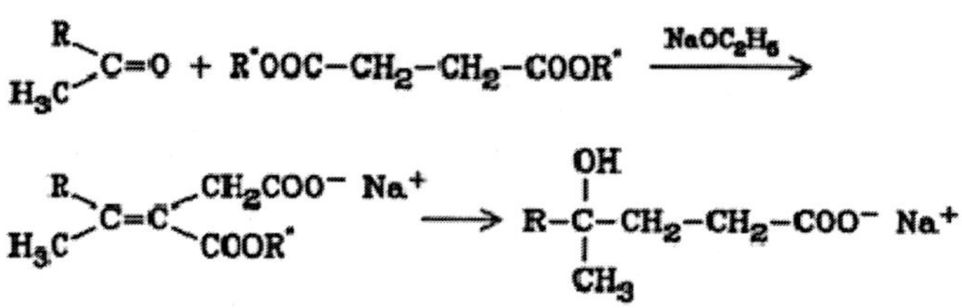

where R = H, alkyl

R' = C_2H_5

Scheme 3. Production number of straight-chain aliphatic dicarboxylic acids and their derivatives occur in nature

Production: A number of straight-chain aliphatic dicarboxylic acids and their derivatives occur in nature. However, isolation from natural substances has no commercial significance. Although many syntheses for the production of aliphatic dicarboxylic acids are known, only a few have found industrial application. This is due partly to the shortage of raw materials.

Individual saturated dicarboxylic acids: Dicarboxylic acids are used mainly as intermediates in the manufacture of esters and polyamides. Esters derived from monofunctional alcohols serve as plasticizers or lubricants. Polyesters are obtained by reaction with dialcohols. In addition, dicarboxylic acids are employed in the manufacture of hydraulic fluids, agricultural chemicals, pharmaceuticals, dyes, complexing agents for heavy-metal salts, and lubricant additives (as metal salts).

3.1. Oxalic Acid

Oxalic acid (ethanedioic acid, acidum oxalicum) is the simplest saturated dicarboxylic acid (Clarke, 1986). The compound exists in anhydrous form [144-62-7] or as a dihydrate [6153-56-6]. The anhydrous acid is not found in nature and must be prepared from the dihydrate even when produced industrially. Oxalic acid is widely distributed in the plant and animal kingdom (nearly always in the form of its salts) and has various industrial applications.

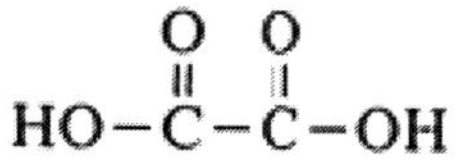

Scheme 4. Chemical structure of oxalic acid

The acidic potassium salt of oxalic acid is found in common sorrel (Latin: oxalis acetosella) and the name oxalic acid is derived from that plant. Table 2 shows examples of plants in which oxalic acid occurs (in the form of potassium, sodium, calcium, magnesium salts, or iron complex salts) are given below (oxalic acid content in milligrams per 100 g dry weight):(Tsu-Ning Tsao G., 1963)

Table 2. Oxalic acid content in milligrams per 100 g dry weight

Spinach	460 – 3200
Rhubarb	500 – 2400
Chard	690
Parsley	190
Beets	340
Cocoa	4500
Tea	3700
Beet leaves	up to 12 000

Oxalic acid is formed in plants through incomplete oxidation of carbohydrates, e.g., by fungi (*Aspergillus niger*) or bacteria (*acetobacter*) and in the animal kingdom through carbohydrate metabolism via the tricarboxylic acid cycle. The urine of humans and of most mammals also contains a small amount of calcium oxalate. In pathological cases, an increased calcium oxalate content in urine leads to the formation of kidney stones (Clarke, 1986). Calcium and iron(II) oxalates are also found as minerals. Both the anhydrous and dihydrated forms of oxalic acid form colorless and odorless crystals.

3.1.1 Anhydrous oxalic acid

Anhydrous oxalic acid [144-62-7] exists as rhombic crystals in the *a*-form and as monoclinic crystals in the *b*-form (West, 1980). These forms differ mainly in their melting points. The slightly stable *b*-form changes into the *a*-form at 97 °C and 0.2 barr. Anhydrous oxalic acid is prepared by dehydration of the dihydrate through careful heating to 100 °C. It is then sublimated in a dry air stream. The sublimation is fast at 125 °C and can be carried out at temperatures up to 157 °C without decomposition. The dehydration can also be accomplished by azeotropic distillation with benzene or toluene. Anhydrous oxalic acid is slightly hygroscopic; it absorbs water from moist air ("weathers") to form the dihydrate again. The hydration occurs very slowly because of surface caking.

3.1.2 Oxalic acid dihydrate

Oxalic acid dihydrate [6153-56-6], HOOC–COOH 2 H_2O is the industrially produced and usual commercial form of oxalic acid. The compound forms colorless and odorless prisms or granules that contain 71.42 wt % oxalic acid and 28.58 wt % water. Oxalic acid dihydrate is stable at room temperature and under normal storage conditions. The most important physical properties are as follows:

The solubility in water and the density of these solutions are presented in Table.1. Oxalic acid is readily soluble in polar solvents such as alcohols (although partial esterification occurs), acetone, dioxane, tetrahydrofuran, and furfural. Oxalic acid is sparingly soluble in diethyl ether (1.5 g oxalic acid dihydrate in 100 g ether at 25 °C), and insoluble in benzene, chloroform, and petroleum ether. The ionization constants show that oxalic acid is a strong acid. The value of K_1 is comparable to that of mineral acids and the value of K_2 corresponds to ionization constants of strong organic acids, for example, benzoic acid.

In the homologous series of dicarboxylic acids, oxalic acid, the first member, shows unique behavior because of the interaction of the neighboring carboxylate groups. This results in an increase in the value of the dissociation constant and in the ease of decarboxylation. Upon rapid heating to 100 °C oxalic acid decomposes into carbon monoxide, carbon dioxide, and water with formic acid as an isolable intermediate.

In aqueous solution decomposition is induced by light and to a much greater extent by g- or X-rays (to carbon monoxide, carbon dioxide, formic acid, and occasionally hydrogen). This decomposition is catalyzed by the salts of heavy metals, for example, by uranyl salts. Oxalic acid cannot form an intramolecular anhydride. Upon heating to over 190 °C or warming in concentrated sulfuric or phosphoric acid, oxalic acid decomposes to carbon monoxide, carbon dioxide, and water: this decomposition is not exothermic.

The reducing properties of oxalic acid (which itself is oxidized to the harmless end products carbon dioxide and water) form the basis for the variety of practical applications. Oxalic acid is also oxidized relatively easily to carbon dioxide by many other oxidizing agents in addition to air, especially in the presence of the salts of heavy metals. Oxalic acid is easily esterified, whereby two types, the acidic mono or neutral diesters can result. These esters are applied as intermediates in chemical syntheses. They react relatively easily with water, ammonia, or amines to afford the corresponding acyl derivatives.

Important chemical characteristics are also demonstrated by the metal salts of oxalic acid. These exist in two types-the acidic and neutral salts. The alkali metal and iron (III) salts are readily soluble in water. All other salts are sparingly soluble in water. The near complete insolubility of the alkaline-earth salts of oxalic acid, especially of calcium oxalate, finds some applications in quantitative analysis. When heated all these metal salts lose carbon monoxide. Other salts which are easier decomposable lose carbon dioxide in addition. The alkali and alkaline-earth salts form carbonates under these conditions. Manganese, zinc, and tin salts form oxides; iron, cadmium, mercury, and copper salts form mixtures of oxides and metals. Nickel, cobalt, and silver salts afford pure metals. Anhydrous fusion of oxalates with alkali yield carbonates and hydrogen. For a review see Dollimore (1987).

3.2. Malonic Acid

Three-carbon 1,3-dicarboxylic acid derivatives (malonic acid, malonates, cyanoacetic acid, cyanoacetates, and malononitrile) are widely used in industry for the manufacture of pharmaceuticals, agrochemicals, vitamins, dyes, adhesives, and fragrances. The common feature of malonic acid and its derivatives is the high reactivity of the central methylene group. Due to the increasingly electron-withdrawing character of the substituents, the acidity of the hydrogen atoms in the 2-position increases in the order malonates < cyanoacetates < malononitrile. Therefore, all these compounds undergo reactions typical of 1,3-dicarbonyl compounds. For example they are easily alkylated or arylated, undergo aldol and Knoevenagel condensations, and they can be used for the synthesis of pyrimidines and other nitrogen heterocycles.

Physical Properties: Important physical properties of malonic acid (propanedioic acid, methanedicarboxylic acid) are listed in Table 1. Its pK_a values are 2.83 and 5.70. Malonic acid forms a colorless hygroscopic solid which sublimes in vacuum with some decomposition. It's really soluble in the water; but slightly soluble in ethanol and diethyl ether, and is completely insoluble in benzene.

Chemical Properties: Malonic acid is found in small amounts in sugar beet and green wheat, being formed by oxidative degradation of malic acid. Reaction with sulfuryl chloride or bromine gives mono- and dihalogenated malonic acid, whereas treatment with thionyl chloride or phosphorus pentachloride leads to mono- or diacyl chloride. When heated with phosphorus pentoxide, malonic acid does not form an anhydride but rather carbon suboxide, a toxic gas that reacts violently with water to reform malonic acid. On heating the free acid above 130 °C, or an aqueous solution above 70 °C, decomposition to acetic acid and carbon dioxide takes place. The mono- and dianion of malonic acid are more stable. In aqueous solution the monosodium salt decomposes above 90 °C and the disodium salt above 130 °C (Bolton, 1995).

3.3. Succinic Acid

Succinic acid is found in amber, in numerous plants (e.g., algae, lichens, rhubarb, and tomatoes), and in many lignites.

Production: A large number of syntheses are used to manufacture succinic acid. Hydrogenation of maleic acid, maleic anhydride, or fumaric acid produces good yields of succinic acid; the standard catalysts are Raney nickel, Cu, NiO, or CuZnCr, Pd – Al_2O_3, Pd – $CaCO_3$, or Ni – diatomite. 1,4-Butanediol can be oxidized to succinic acid in several ways: (1) with O_2 in an aqueous solution of an alkaline-earth hydroxide at 90 – 110 °C in the presence of Pd – C; (2) by ozonolysis in aqueous acetic acid; or (3) by reaction with N_2O_4 at low temperature. Succinic acid or its esters are also obtained by Reppe carbonylation of ethylene glycol, catalyzed with $RhCl_3$ – pentachlorothiophenol; Pd-catalyzed methoxycarbonylation of ethylene; and carbonylation of acetylene, acrylic acid, dioxane, or β- propiolactone (Bolton, 1995).

Acid mixtures containing succinic acid are obtained in various oxidation processes. Examples include the manufacture of adipic acid; the oxidation of enanthic acid and the ozonolysis of palmitic acid. Succinic acid can also be obtained by phase-transfer-catalyzed reaction of 2-haloacetates, electrolytic dimerization of bromoacetic acid or ester, oxidation of 3-cyanopropanal, and fermentation of *n*-alkanes.

Uses: Succinic acid is used as a starting material in the manufacture of alkyd resins, dyes, pharmaceuticals, and pesticides. Reaction with glycols gives polyesters; esters formed by reaction with monoalcohols are important plasticizers and lubricants (Bolton, 1995).

3.4. Glutaric Acid

Glutaric acid occurs in washings from fleece and, together with malonic acid, in the juice of unripened sugar beet.

Production: Glutaric acid is obtained from cyclopentane by oxidation with oxygen and cobalt (III) catalysts or by ozonolysis; and from cyclopentanol – cyclopentanone by oxidation with oxygen and $Co(CH_3CO_2)_2$, with potassium peroxide in benzene, or with N_2O_4 or nitric acid. Like succinic acid, glutaric acid is formed as a byproduct during oxidation of cyclohexanol – cyclohexanone. Other production methods include reaction of malonic ester with acrylic acid ester, carbonylation of *Y*-butyrolactone, oxidation of 1,5-pentanediol with N_2O_4, and oxidative cleavage of *Y*-caprolactone.

Uses: The applications of glutaric acid, e.g., as an intermediate, are limited. Its use as a starting material in the manufacture of maleic acid has no commercial importance.

3.5. Adipic Acid

Adipic acid, hexanedioic acid, 1,4-butanedicarboxylic acid, $C_6H_{10}O_4$, M_r 146.14, $HOOCCH_2CH_2CH_2CH_2COOH$ [124-04-9], is the most commercially important aliphatic dicarboxylic acid. It appears only sparingly in nature but is manufactured worldwide on a large scale. The historical development of adipic acid was reviewed in 1997 (Luedeke, 1997)

Physical properties: Adipic acid is isolated as colorless, odorless crystals having an acidic taste. It is very soluble in methanol and ethanol, soluble in water and acetone, and very slightly soluble in cyclohexane and benzene. Adipic acid crystallizes as monoclinic prisms from water, ethyl acetate, or acetone/petroleum ether.

Chemical properties: Adipic acid is stable in air under most conditions, but heating of the molten acid above 230 – 250 °C results in some decarboxylation to give cyclopentanone [120-92-3], *bp* 131 °C. The reaction is markedly catalyzed by salts of metals, including iron, calcium, and barium. The tendency of adipic acid to form a cyclic anhydride by loss of water is much less pronounced compared to glutaric or succinic acids.

Adipic acid readily reacts at one or both carboxylic acid groups to form salts, esters, amides, nitriles, etc. The acid is quite stable to most oxidizing agents, as evidenced by its production in nitric acid. However, nitric acid will attack adipic acid autocatalytically above 180 °C, producing carbon dioxide, water, and nitrogen oxides.

Use: Adipic acid has been used in the manufacture of mono- and diesters as well as polyamides. Nylon 6,8 is obtained by reaction of suberic acid with hexamethylenediamine, and nylon 8,8 by reaction with octamethylenediamine. Polyamides of adipic acid with diamines such as 1,3-bis(aminomethyl)benzene, 1,4(bisaminomethyl)cyclohexane, and bis(4-aminocyclohexyl)methane are also of commercial interest. Esters of adipic acid with mono- and bifunctional alcohols are used as lubricants.

4. DICARBOXLIC ACIDS DISTRIBUTIONS IN THE ATMOSPHERE

Numerous organic compounds significantly contribute to the aerosol load of the atmosphere and thus to the radiative forcing of climate. Among others the influence of organic aerosol on cloud droplet formation is a key point in evaluating effects of anthropogenic emissions on climate. In contrast to sulfate more uncertainties exist about organics and in particular for secondary organic aerosol species which are more oxygenated and hygroscopic than primary organic species (Saxena and Hildemann, 1996). Among oxygenated organic species, dicarboxylic acids are probably the best quantified species, though they represent a small fraction of the total organic mass (Kawamura and Ikushima, 1993). Glutaric and malonic acid the atmosphere have potential to increase the cloud condensation nuclei (CCN) activation of major inorganic aerosol such as ammonium sulfate (Cruz and Pandis, 1998). These findings suggest a potentially important role played by dicarboxylic acids on radiative forcing and stimulate their studies since the sources of diacids in the atmosphere remain poorly understood and quantified.

Whatever the region; urban and continental, or remote marine (see Figure 1 which carried out fromTable 3), oxalic acid (C_2: HOOCCOOH) is always found to be the most abundant diacid followed by succinic (C_4: $HOOC(CH_2)_2COOH$) and/or malonic (C_3: $HOOCCH_2COOH$) acid with concentrations of several hundreds of nanograms per cubic meter in urban and continental regions (Kawamura and Ikushima, 1993; Kawamura and Kaplan, 1987) to a few tens of nanograms per cubic meter in remote marine boundary layer (Kawamura and Sakagushi, 1999; Sempere and Kawamura, 2003). In Europe, the most continuous study of diacids was conducted over one year by Limbeck et al., (2005) at Vienna, Austria. Although available data on diacids are more sparse at midlatitudes in Europe, they tend to show that oxalic acid levels at nonurban or rural sites are not considerably different from those at urban sites (Limbeck and Puxbaum, 1999; Rohrl and Lammel, 2001).

Motor exhausts have been proposed to be primary sources of oxalic, malonic, succinic, and glutaric (C_5: $HOOC(CH_2)_3COOH$) acids (Grosjean et al., 1978; Kawamura and Kaplan, 1987). Some of these diacids are also emitted by wood burning, particularly malonic acid (pine wood) and succinic acid (oak wood) (Rogge et al., 1991; Rogge et al., 1993). Note that until now no direct source of malic (hydroxysuccinic: hC_4: $HOOCCH_2CHOHCOOH$) and tartaric (dihydroxysuccinic: dhC_4: $HOOC(CHOH)_2COOH$) acids has been identified.

Glutaric, succinic, and adipic (C_6: $HOOC(CH_2)_4COOH$) acids have been identified in laboratory studies (Hatakeyama et al., 1985) as secondary organic aerosol products of the reaction of O_3 with cyclohexene, a symmetrical alkene molecule similar to monoterpenes emitted by the biosphere.Hatakeyama et al. (1985) also suggested that malonic and oxalic acids are also produced in the cyclohexene-ozone system.

Unsaturated fatty acids with a double bond at the C_9 position like cis-9-octadecenoic (oleic) acid are oxidized into C_9 diacid (azelaic acid) and other products hereafter mainly oxidized into shorter diacids hahah(Kawamura and Ikushima, 1994; Kawamura and Kaplan, 1987; Kawamura et al., 1985). These unsaturated acids which are abundant in marine phytoplankton and terrestrial higher plant leaves are also emitted by anthropogenic sources such as meat cooking (Rogge, 1991; Rogge et al., 1998) and wood burning processes (Rogge et al., 1998).

Warneck suggested that in the marine atmosphere clouds generate oxalic acid from glyoxal formed by oxidation of acetylene and glycolaldehyde formed by oxidation of ethane (Warneck, 2000). Note that along these processes glyoxylic acid (CHOCOOH) represents a key intermediate (see figure 3) whereas diacids other than oxalic acid are not produced. The formation of dicarboxylic acids in the continental atmosphere (Ervens et al., 2004a) involves production of glyoxal from toluene and of glycolaldehyde from isoprene as well as aqueous phase reactions of adipic and glutaric acids produced by oxidation of cyclohexene. Recently more literature has become available on the formation of oxalic acid that includes also the oxidation of methylglyoxal, an oxidation product of toluene and isoprene, via intermediate steps involving pyruvic and acetic acids (Lim et al., 2005). Since this diacid production pathway also forms oligomers, the knowledge of the sources of diacids is also of importance for the understanding of secondary organic aerosol formation.

The relative contribution of primary and secondary sources of diacids in the atmosphere remains poorly understood. Even though it is agreed that they are likely to be mainly secondary in origin it is not known in which proportion their precursors come from anthropogenic and biogenic sources.

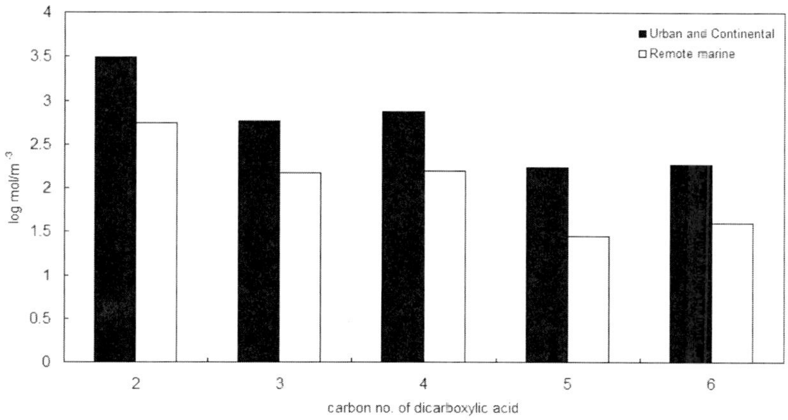

Figure 2. Comparison of dicarboxylic acids distribution in urban/continental and remote marine based on the data collection on table 3

Table 3. Summary of aerosol dicarboxylate concentration (ng m^{-3}) in urban/continental (a) remote marine (b) locations

References	Location	Oxalic	Malonic	Succinic	Glutaric	Adipic
(Grosjean et al., 1978)	New York	0	3.6	21.8	17.2	13.2
(Grosjean et al., 1978)	New York	0	3.9	24.9	23.2	11.6
(Kawamura and Kaplan, 1987)	West LA	6.38	1.58	1.96	0.6	2.22
(Kawamura and Kaplan, 1987)	West LA	2.12	0.4	0.66	0.22	0.94
(Kawamura and Kaplan, 1987)	West LA	8.13	0.72	2.34	0.66	3.31
(Kawamura and Kaplan, 1987)	West LA	8.65	1.45	2.37	0.74	0.49
(Kawamura and Kaplan, 1987)	Down Town LA	6.21	0.71	1.19	0.52	0.1
(Kawamura and Kaplan, 1987)	Down Town LA	6.6	0.76	1.84	0.52	0.2
(Kawamura and Kaplan, 1987)	Down Town LA	8.31	1.22	2.13	0.83	0.63
(Sempere and Kawamura, 1994)	Tokyo	29.65	6.69	13.18	3.72	6.66
(Sempere and Kawamura, 1994)	Tokyo	58.89	20.29	28.82	7.54	6.79
(Sempere and Kawamura, 1994)	Tokyo	330	141.3	161.1	4.15	2.91
(Limbeck and Puxbaum, 1999)	South Africa	193	142	58	8.8	7.9
(Limbeck and Puxbaum, 1999)	Sonblick Observatory	153	22	14	2.7	4.4
(Limbeck and Puxbaum, 1999)	Vienna	340	244	117	26	117
(Kawamura and Watanabe, 2004)	Tokyo	357	71.4	73.4	23.1	25.8
(Kawamura and Watanabe, 2004)	Tokyo	157	44	41	11	13
(Kawamura and Watanabe, 2004)	Tokyo	186	40.5	47.4	18.2	14.2
(Rohrl and Lammel, 2001)	Helsinki	0	0	30	0	0
(Ho et al., 2006)	Hong Kong (Road)	478	89.1	71.88	20	10.7

(Ho et al., 2006)	Hong Kong (Road)	268	47.6	33	6.95	12.7
(Hsieh et al., 2007)	Tainan,Taiwan	574	65.8	101	43	13.2
(Hsieh et al., 2007)	Tainan,Taiwan	432	34.2	87.9	10.3	8.8
(Limbeck et al., 2005)	Vienna, Austria	99.6	34	37	7.7	3.3
(Limbeck et al., 2005)	Vienna, Austria	66.2	38.6	30.8	6.6	3.2
(Limbeck et al., 2005)	Vienna, Austria	63.1	21.5	31.2	5.6	2.5
(Limbeck et al., 2005)	Mt Rax, Austria	34.5	9.1	16.4	2.3	0.8
(Limbeck et al., 2005)	Mt Rax, Austria	26.4	6.9	14.9	2.3	4.3
(Limbeck et al., 2005)	Mt Rax, Austria	32.6	16.4	22.4	3	1.7
(Decesari et al., 2006)	Rondonia, Brazil	194.7	73.1	123.5	23.5	14.5
(Decesari et al., 2006)	Rondonia, Brazil	793.3	56.8	210.2	32.1	12.6
(Decesari et al., 2006)	Rondonia, Brazil	937.9	128.5	423.9	34.7	21.2
(Decesari et al., 2006)	Rondonia, Brazil	1260	476.5	667.2	121.1	97.4
(Wang et al., 2006)	Hong Kong (Tunnel)	505	69.4	85.2	20.9	26.4
(Wang et al., 2006)	Hong Kong (Tunnel)	221	34.5	32.7	14.7	13.5
(Wang et al., 2006)	Hong Kong (Tunnel)	234	42	51.4	17.1	24.7
(Wang et al., 2006)	Hong Kong (Tunnel)	312	59.7	62.9	16.7	15.5
(Wang et al., 2006)	Hong Kong (Tunnel)	633	59.3	95.1	30.3	25.9

(a)

References	Location	Oxalic	Malonic	Succinic	Glutaric	Adipic
(Kawamura and Kaplan, 1987)	Green House LA	1.31	0.3	0.29	0.04	0.1
(Kawamura and Kaplan, 1987)	Green House LA	2.83	0.14	0.86	0	0.22
(Kawamura and Sakagushi, 1999)	North Pacific	44.7	23.2	19.5	2.57	3.08
(Kawamura and Sakagushi, 1999)	North Pacific	8.73	2.18	2.16	0.61	1.26
(Kawamura and Sakagushi, 1999)	North Pacific	10.6	1.98	2.22	0.23	2.12
(Kawamura and Sakagushi, 1999)	North Pacific	28.6	12.8	13	1.84	1.34
(Kawamura and Sakagushi, 1999)	North Pacific	667	189	93	20.1	4.9
(Kawamura and Sakagushi, 1999)	North Pacific	190	38.6	16.7	10.2	2.76
(Kawamura and Sakagushi, 1999)	North Pacific	88.5	34.5	21.6	4.72	6.04
(Kawamura and Sakagushi, 1999)	North Pacific	24.9	5.66	10.1	1.87	1.67
(Kawamura and Sakagushi, 1999)	North Pacific	10	2.12	1.52	0.32	0.43
(Kawamura and Sakagushi, 1999)	North Pacific	18.3	3.45	4.02	0.62	0.46
(Kawamura and Sakagushi, 1999)	North Pacific	25.5	5.93	2.99	0.65	0.4
Kawamura (1996)	Antarctic	1.59	0.13	0.63	0.31	0.49
Kawamura (1996)	Antarctic	3.12	0.38	5.77	0.58	0.85
Kawamura (1996)	Antarctic	3.26	0.52	1.18	0.34	0.33
Kawamura (1996)	Antarctic	10.29	2.69	61.53	2.26	1.81
Narukawa(1999)	Indonesia	2200	800.3	1090	310	350
Narukawa(1999)	Indonesia	225	18.4	123	30	40
Khwaja (1994)	semi urban site NY	308	84	55	12	89

Khwaja (1994)	semi urban site NY	245	92	106	16.3	101
Khwaja (1994)	semi urban site NY	118	165	107	15	40
Khwaja (1994)	semi urban site NY	58	81	129	20	21
Khwaja (1994)	semi urban site NY	298	96	90	23	31
Khwaja (1994)	semi urban site NY	1	43	0.5	39	20
Khwaja (1994)	semi urban site NY	360	88	167	46	50
Sempere (2003)	Western Pacific	428.5	78.6	33.4	7.6	7.2
Rohrl (2002)	rural(I)	0	0	14	0	0
Rohrl (2002)	rural(II)	0	0	8.8	0	0
Rohrl (2002)	rural(III)	0	0	18	0	0
(Kawamura et al., 2007)	Canadian arctic	9.89	2.74	2.16	0.54	0.51
(Kawamura et al., 2007)	Canadian arctic	8.3	2.87	1.44	0.37	0.26
(Kawamura et al., 2007)	Canadian arctic	5.26	1.67	1.08	0.22	0.27
(Narukawa et al., 2002)	Arctic,Alert	23.5	5.03	3.21	1.21	0.54
(Narukawa et al., 2002)	Arctic,Alert	40.09	11.6	15.67	2.16	0.55
(Mochida et al., 2007)	North Pacific,ACE	600	110	52	8.9	2
		(b)				

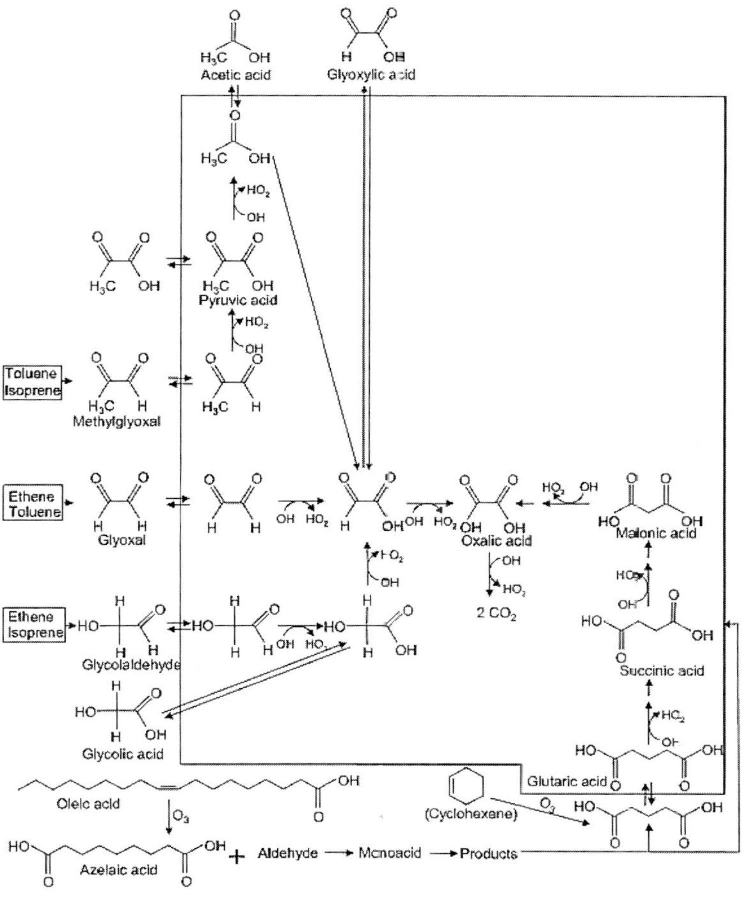

Figure 3. Multiphase organic chemistry producing C_2–C_5 diacids from key biogenic and anthropogenic precursors. The box refers to the aqueous phase. The figure is mainly adapted from (Ervens et al., 2004b) with modifications to account for the reaction pathway methylglyoxal/pyruvic acid/acetic acid/glyoxylic acid suggested by (Lim et al., 2005). In addition to cyclohexene used by (Ervens et al., 2004b)as a model compound for symmetrical alkenes, following (Legrand et al., 2007) we also report the oleic acid degradation into azelaic, C_4 and C_5 diacids.

REFERENCES

1. P. J. Adam, J. H. Seinfeld, D. M. Koch, 1999 Global concentration of tropospheric sulfate, nitrate and ammonium aerosol simulated in a general circulation model. J. Geophys. Res, 104: 13791.
2. G. L. Bolton, 1995 Encyclopedia of Reagents for Organic Synthesis, 5. John Wiley & Sons, New York, 32133222 . pp.
3. P. Brimblecombe, 1987 The big smoke: A history of air pollution in London since medieval times. Methuen, New York.
4. H. Cachier, M. P. Bremond, P. Buatmenard, 1989 Carboneceous Aerosols from Different Tropical Biomass Burning Sources. Nature, 340 371373 .
5. R. J. Charlson, Heintzenberg, 1995 Aerosols as acaouse of uncertainty in climate forecast: Report of Dahlem Workshop on Aerosl Forcing of Climate. Wiley & Sons, Chichester, 110 .
6. H. Clarke, D. T.a, A. W., 1986 Organic Synthesis. J. Wiley and sons, New York, 421425 .
7. S. E. Cornell, T. D. Jickells, J. N. Cape, 2003 Organic nitrogen deposition on land and coastal environments; a review of methods and data. Atmospheric Envionment, 37 21732191 .
8. C. N. Cruz, S. N. Pandis, 1998 The effect of organic coatings on the cloud condensation nuclei activation of inorganic atmosphere aerosol. Journal of Geophys. Res., 103(D11): 13111-13123.
9. C. N. Cruz, S. Pandis, N., 1998 The effect of organic coatings on the cloud condensation nuclei activation of inorganic atmosphere aerosol. Journal of Geophys. Res., 103(D11): 13111-13123.
10. S. Decesari, M. C. Facchini, S. Fuzzi, E. Tagliavini, 2000 Characterization of water-soluble compounds in atmospheric aerosol: A new approach. Journal of Geophysiscal research, 105 14811489 .

11. S. Decesari, et al. 2002 Water soluble organic compounds formed by oxidation of soot. Atmospheric Environment, 36 11 18271832 .

12. DEFRA, 2005 UK Department for Environment Food & Rural Afffairs, The Quality Strategy for England, Scotland, Wales and Northern Island (Rep. 01EP0538), London,UK.

13. D. Dollimore, 1987 Thermochimica Acta, 117 331363 .

14. L. Dutot, J. Rude, B. Aumont, 2003 Neutral network method to estimate teh aqueous rate constant for the OH reactions with organic compounds. Atmospheric Envionment, 37 269276 .

15. Ervens, G. Feingold, S. L. Clegg, S. M. Kreidenweis, 2004a A modeling study of aqueous production of dicarboxylic acids: 2. Implications for cloud microphysics. Journal of Geophysical Research D: Atmospheres, 109(15).

16. Ervens, G. Feingold, G. J. Frost, S. M. Kreidenweis, 2004b A modeling of study of aqueous production of dicarboxylic acids: 1. Chemical pathways and speciated organic mass production. Journal of Geophysical Research D: Atmospheres, 109(15).

17. J. Fenger, 1999 Urban air quality. Atmospheric Environment, 33 29 48774900 .

18. R. Fisseha, J. Dommen, M. Sax, 2004 idenfification of organic aerosol and the acids in secondary corresponding gas phase from chamber experiments. Analytical Chemistry, 76 63356540 .

19. D. Grosjean, K. Van Cauwenberghe, J. P. Schmid, 1978 Identification of C3-C10 aliphatic dicarboxylic acids in airborne particulate matter. Environmental Science and Technology, 12 3 313317 .

20. C. M. Grossi, P. Brimblecombe, 2002 The effect of atmospheric pollution on building materials. Journal De Physique Iv, 12(PR10): 197-210.

21. S. Hatakeyama, T., et al. 1985 Ozone-cyclohexene reaction in air: Quantitative analyses of particulate products and the reaction mechanism. Environ. Sci. Technol, 19,: 935-942.

22. D. J. Jacob, 1999 Introduction to a atmospheric chemistry. Princeton University, New Jersy.

23. K. Kawamura, F. Sakagushi, 1999 Molecular distribution of water soluble dicarboxylic acids in marine aerosols over the pacific ocean including tropics. J. Geophys. Res, 104 350135C9 .

24. K. Kawamura, K. Ikushima, 1993 Seasonal changes in the distribution of dicarboxylic acids in the urban atmosphere. Environmental Science & Technology, 27 10 22272235 .

25. K. Kawamura, K. Ikushima, 1994 Seasonal changes in the distribution of dicarboxylic acids in the urban atmosphere. Environ. Sci. Technol, 27 22272235 .

26. K. Kawamura, I. R. Kaplan, 1987 Dicarboxylic acids generated by thermal alteration of kerogen and humic acids. Geochimica et Cosmochimica Acta, 51 32013207 .

27. K. Kawamura, H. Kasukabe, L. A. Barrie, 1996 Source and reaction pathways of dicarboxylic acids, ketoacids and dicarbonyls in arctic aerosols: One year of observations. Atmospheric Environment

28. K. Kawamura, L. L. Ng, I. R. Kaplan, 1985 Determination of organic acids (C1-C10) in the atmosphere, motor exhausts, and engine oils. Environmental Science and Technology, 19 11 10821086 .

29. V. Kerminen, M. et, al, 2000 Low molecular weight dicarboxylic acids in an urban and rural atmosphere. Journal of Aerosol Science, 31 3 349362 .

30. H. A. Khwaja, 1995 Atmospheric concentrations of carboxylic acids and related compounds at a semiurban site. Atmospheric Environment, 29 1 127139 .

31. M. Legrand, et al. 2007 Origin of C2 -C5 dicarboxylic acids in the European atmosphere inferred from year-round aerosol study conducted at a west-east transect. J. Geophys. Res.,, 112: D23S07, doi:10.1029/2006JD008019.

32. H. Lim, J. , A. G. Carlton, a. , B. J. Turpin, 2005 Isoprene forms secondary organic aerosol through cloud processing: Model simulations. Environ. Sci. Technol., 39 44414446 .

33. Limbeck, Y. Kraxner, H. Puxbaum, 2005 Gas to particle distribution of low molecular weight dicarboxylic acids at two different sites in central Europe (Austria). Journal of Aerosol Science, 36 8 9911005 .

34. Limbeck, H. Puxbaum, 1999 Organic acids in continental background aerosols. Atmospheric Environment, 33 12 18471852 .

35. V. Luedeke, 1997 Encyclopedia of Chemical Processing and Design. In: W.C. J. McKetta (Editor). Marcel Dekker Inc, New York, 128146 .

36. T. Moreno, W. Gibbons, T. Jones, R. Richards, 2003 The geology of ambient aerosols: Characterising urban and rural/coastal silicate PM10-

2.5 and PM2.5 using high-volume cascade collection and scanning electron microscopy. Atmospheric Environment, 37 30 42654276 .

37. M. O. S. T. I. M. O. S.a, I. M. , 2000 Environmental Quality (Standard for Particulate Matter) Regulations, Kuala Lumpur.

38. M. Narukawa, K. Kawamura, N. Takeuchi, T. Nakajima, 1999 Distribution of dicarboxylic acids and carbon isotopic compositions in aerosols from 1997 Indonesian forest fires. Geophysical Research Letters, 26 20 31013104 .

39. T. Novokov, J. E. Penner, 1993 Large Contribution of Organic Aerosols to Cloud-Condensation-Nuclei Concentration. Nature, 365 823826 .

40. O. Preining, 1993 Global Climate Change Due to Aerosols. In: H.C. N. (Editor), Global Atmospheric Chemical Change. Elsevier Applied science, London and New York, 93122 .

41. H. Puxbaum, A. Limbeck, 2000 Scavenging of Organic aerosol Constituents in Supercooled clouds. 13th International Conference on Cloud Precipitation, Reno,Nevada.

42. W. F. Rogge, M. Lynn, Monica. A. Hildemann, Glen. R. Mazurek, a. Cass, B. R. T. Simoneit, 1991 Sources of fine organic aerosol. 1. Charbroilers and meat cooking operations. Environ. Sci. Technol, 25 11121125 .

43. W. F. Rogge, M. Lynn, Monica. A. Hildemann, Glen. R. Mazurek, Cass, R. T. Bernd, Simoneit, 1991 Sources of fine organic aerosol. 1. Charbroilers and meat cooking operations. Environ. Sci. Technol, 25 11121125 .

44. W. F. Rogge, M. A. Mazurek, L. M. Hildemann, G. R. Cass, B. R. T. Simoneit, 1993 Quantification of urban organic aerosols at a molecular level: Identification, abundance and seasonal variation. Atmospheric Environment- Part A General Topics, 27 A(8): 1309-1330.

45. W. F. Rogge, M. A. Mazurek, L. M. Hildemann, G. R. Cass, B. R. T. Simoneit, 1998 Quantification of urban organic aerosols at a molecular level: Identification, abundance and seasonal variation. Atmospheric Environment- Part A General Topics, 27 A(8): 1309-1330.

46. Rohrl, G. Lammel, 2001 Low-Molecular Weight Dicarboxylic Acids and Glyoxylic Acid: Seasonal and Air Mass Characteristics. Environ. Sci. Technol., 35 1 95101 .

47. J. Rudloff, H. Cölfen, 2004 Langmuir, 20: 991.

48. P. Saxena, L. M. Hildemann, 1996 Water-Soluble Organics in Atmospheric Particles: A critical Review of the Literature and

Application of Thermodynamics to Identify Candidate Compounds. J. Atmos. Chem, 24 57109 .

49. P. Saxena, L. M. Hildemann, 1997 Water Absorption by Organics: Survey of Laboratory Evidence and Evaluation of UNIFAC for Estimating Water Activity. Environ. Sci. Technol., 31 11 33183324 .

50. R. Sempere, K. Kawamura, 2003 Trans-hemispheric contribution of C2-C10 a,w-dicarboxylic acids, and related polar compounds to water-soluble organic carbon in the western Pacific aerosols in relation to photochemical oxidation reactions. Global Biogeochemical Cycles, 17 2 381 .

51. R. T. Simoneit, D. A. Mazurek, 1982 Organic matter of the troposphere II: Natural background of biogenic lipid matter in aerosols over the rural western United States. Atmos Environment, 16 21392159 .

52. M. Tedetti, K. Kawamura, B. Charriere, N. Chevalier, R. Sempere, 2006 Determination of Low Molecular Weight Dicarboxylic and Ketocarboxylic Acids in Seawater Samples. Anal. Chem., 78 17 60126018 .

53. P. Warneck, 2000 Chemistry of the Natural Atmosphere. Second Edition, Oxford, AP publication.

54. L. Yu, M. Shulman, R. Kopperud, L. Hildemann, 2005 Characterization of Organic Compounds Collected during Southeastern Aerosol and Visibility Study: Water-Soluble Organic Species. Environ. Sci. Technol., 39 3 707715 .

55. S. Yu, 2000 Role of organic acids (formic, acetic, pyruvic and oxalic) in the formation of cloud condensation nuclei (CCN): a review. Atmospheric Research, 53 4 185217 .

56. Yusunov, S. Tukhtaev, L. N. Semenova, 1972 Deposited Doc.,, VINITI: 461272 .

57. S. Zappoli, et al. 1999 Inorganic, organic and macromolecular components of fine aerosol in different areas of Europe in relation to their water solubility. Atmospheric Environment, 33 27332743 .

CHAPTER 7

Aerosol Optical Properties over South Asia from Ground-Based Observations and Remote Sensing: A Review

S. Ramachandran[1,*] and Sumita Kedia [2]

[1] Space and Atmospheric Sciences Division, Physical Research Laboratory, Navrangpura, Ahmedabad 380009, India

[2] Computational Earth Sciences Group, Centre for Development of Advanced Computing, Pune 411007, India

ABSTRACT

Seasonal and inter-annual variabilities in aerosol optical depth (AOD) and aerosol size distribution are investigated using ground-based measurements (sun photometers and sun/sky radiometers), and MODIS (MODerate Imaging Spectroradiometer) and MISR (Multiangle Imaging SpectroRadiometer) satellites over Ahmedabad, Gurushikhar, Karachi, Kanpur and Gandhi College in South Asia during 2006–2008. An analysis and a review on the comparison between aerosol optical depths measured from ground-based observations and remote sensing over South Asia is performed. Karachi and Ahmedabad AODs are two times higher than Gurushikhar, a high altitude remote site. AODs over Kanpur and Gandhi College in the Indo-Gangetic basin are higher than those measured over Ahmedabad, Gurushikhar and Karachi. Summer monsoon AODs are higher over Ahmedabad and Karachi, while winter AODs are higher over Kanpur and Gandhi College. AOD ratio, ratio of AODs obtained at 0.38 μm and 0.87 μm, is higher during postmonsoon and winter than premonsoon and monsoon ratio suggesting the abundance of fine mode aerosols during postmonsoon and winter. Monsoon AOD ratios are lowest owing to the dominance of coarse mode (mainly sea salt) particles. Ångstr wavelength exponent (α) during postmonsoon and winter are higher than that of premonsoon and monsoon values. Lower α values over Gurushikhar and Karachi indicate the dominance of coarse mode aerosols (dust in Gurushikhar, and dust and sea salt in Karachi). Dominance of fine mode aerosols due to anthropogenic activities give rise to higher α values over urban locations (e.g., Ahmedabad and Kanpur). Comparison between ground-based and MODIS (Terra and Aqua) retrieved AODs show that aerosol optical depths do not change significantly in an hour and much of the diurnal AOD variation is captured well by the two MODIS instruments. The temporal difference (about an hour) between the ground-based and remote sensing measurements contributes negligibly to the observed differences in AODs. The differences between ground-based and remote sensing (MODIS and MISR) AODs vary on spatial scales. During the study period (2006–2008) MODIS underestimates AODs over western India by about 0.04, while over the Indo-Gangetic Plain MODIS overestimates AOD by 0.06 in (Kanpur) and underestimates by 0.07 (Gandhi College) with respect to AODs measured by hand held sun photometer and Microtops sun photometer (Ahmedabad and Gurushikhar), and AERONET sun/sky radiometers (Karachi, Kanpur and Gandhi College) respectively. During the same period MISR underestimates AODs in the range of 0.02–0.17 over Ahmedabad, Gurushikhar, Karachi and Kanpur, while in Gandhi College MISR overestimates AOD by 0.2 when compared to ground-based AODs. Results on spatial, seasonal and inter-annual variations in aerosol characteristics will be useful in improving the aerosol retrieval algorithms in remote sensing, and in regional and global estimates of aerosol radiative forcing.

KEYWORDS

aerosols; optical properties; observations; remote sensing; South Asia; review

1. INTRODUCTION

Atmospheric aerosols influence the global energy balance by scattering and absorbing both solar and terrestrial radiation (direct effect), as well as by acting as cloud condensation nuclei and altering the cloud microphysical properties (indirect effect) [1]. Aerosols are produced in the atmosphere through various natural and anthropogenic processes, and they get dispersed horizontally and vertically through prevailing atmospheric circulation. Aerosols are short lived with a residence time of about a week in the lower troposphere. Due to their short lifetimes and widely distributed sources, aerosols exhibit large spatial and temporal variabilities. In recent years, there has been a substantial increase in the interest in studying natural and anthropogenic aerosols because of their influence on climate through direct and indirect radiative effects [1]. Despite this, aerosols are still a major source of uncertainty in the prediction of climate change due to inadequate information on the variabilities of aerosol characteristics on regional and temporal scales [1].

In this regard, aerosol characteristics retrieved from MODerate Imaging Spectroradiometer (MODIS) and Multiangle Imaging SpectroRadiometer (MISR) on board Terra and Aqua satellites can provide a detailed knowledge over a larger spatial range. MODIS (Terra, Aqua) provides near-global daily observations of the Earth in a wide spectral range which are used to derive aerosol optical properties over land and ocean [2]. MISR observes the Earth's environment globally and makes near simultaneous measurements at nine view angles to provide greater sensitivity while deriving aerosol optical depths, however, the frequency of global coverage is lower (9 days) when compared to MODIS [3,4]. Validation of satellite derived data through collocated ground-based direct measurement is very essential [2,5], as the satellite retrieval involves*apriori* assumptions including the atmospheric composition over different regions across the globe. Furthermore, satellite retrievals are based on the radiant energy reflected and emitted by the Earth [2]. In contrast, to determine aerosol optical depths ground-based instruments [6] rely on the measurements of attenuation of solar radiation reaching the surface. Aerosol optical depth (AOD) retrieved from ground-based measurements, is considered the most reliable because of lower uncertainties as compared to satellite retrievals. In addition, ground-based measurements can reliably be used to obtain continuous data on a variety of aerosol characteristics over a particular location [7]. Several recent studies are based on the comparison between independent aerosol retrieval both from satellite and ground-based measurements over different regions [2,4,8,9,10,11,12,13,14].

A seven year (2000–2006) analysis of global aerosol climatology obtained using MODIS AODs revealed that on annual mean scale AODs over

South Asia were the highest ([2], Table 2) when compared to the rest of the world. South Asia comprises India, Pakistan, Nepal, Bhutan, Bangladesh and Sri Lanka. It is densely populated as more than one-fifth of the world population lives in this region, and faces serious environmental threats in terms of air pollution, monsoon floods, droughts and associated climate change. Increases in aerosol loading due to growing population and industrialization in recent decades have resulted in an increase in health-related problems, and impacted air quality, agriculture and water resources in Asia [1,15]. Anthropogenic aerosols over South and East Asia can significantly change the energy balance of the Earth-atmosphere system on regional scales [1,15,16]. A climatology of aerosol optical and microphysical properties over the Indian subcontinent using 9 years of MISR data indicated that the winter to premonsoon changes in aerosol properties are not dominated by dust, but also by an increase in anthropogenic components, particularly biomass burning [17]. Aerosol properties over the The Indian subcontinent were found to display strong seasonal and spatial heterogeneity. It was reported that both natural and anthropogenic aerosols decrease over large areas of the landmass in winter compared to postmonsoon, and in postmonsoon when compared to monsoon. It was shown that AODs are high over the densely populated rural areas, thus, emphasizing that aerosols over rural areas in India need to investigated. Many regional hot spots (mainly urban centers and densely populated rural areas) over India exhibit statistically significant seasonal linear trends in AOD, with an increase of about 0.1–0.4 in the last decade [18].

In this context, it is important to examine aerosol characteristics over a regional aerosol hot spot which can be helpful not only in regional scale, but also in global air quality and climate. Earlier studies on the comparison of ground-based and satellite (mainly MODIS) retrieved AODs over the South Asian region were mostly over India and in particular over Kanpur, an AERONET (Aerosol Robotic Network [6]) location; to name a few [19,20,21,22,23]. All the above mentioned studies used the monthly mean MODIS and AERONET AODs. The absolute difference between MODIS and AERONET AODs during January-May and August-December for a 5-year period (2001–2005) over Kanpur was 0.08 ± 0.07, while during June-July the absolute difference in the 5-year period was estimated to be 0.24 ± 0.10 [21]. These errors were within the predicted retrieval expected error bounds of $\Delta AOD = \pm 0.05 \pm 0.2AOD$ over land [2]. Results on validation of MODIS and MISR AODs using the AERONET AODs over Kanpur during winter and summer seasons showed that MISR performs better than MODIS during both the seasons [20]. This better performance by MISR was attributed to its design in terms of viewing and spectral capability. MODIS was found to overestimate AODs during summer where as both the sensors were found to underestimate AOD during winter [20].

Over the South Asian region, analysis and validation of aerosol optical properties have been conducted over a particular location as was seen above. In the present study, for the first time we analyze aerosol characteristics from five locations spread over a large region in South Asia during different seasons with an aim to examine the space-time variation in aerosol characteristics, and the effect of long range transport. AODs obtained from ground-based measurements over these locations are compared with AODs retrieved from two satellites (MODIS and MISR) during 2006–2008. To get further insight into the aerosol size distribution Ångström wavelength exponent (α) is retrieved from the spectral AODs. Daily measurements of aerosol optical depths in the 0.38–0.87 μm wavelength region made during 2006–2008 over Ahmedabad, Gurushikhar, Karachi, Kanpur and Gandhi College are used in the study. Daily mean AODs measured in situ are compared with daily mean MODIS and MISR AODs over each location and inferences are drawn. An analysis and documentation on the spatial and seasonal variabilities in aerosol optical characteristics over South Asia, attain significance in the context of air quality, air pollution and regional and global climate. The results obtained on the inter-comparison between ground-based and remote sensing AODs will be useful to improve the retrieval algorithms and the trend analysis of AODs.

2. TOPOGRAPHY OF STUDY LOCATIONS

The South Asian region is a land of many contrasts. Its landscape varies from the mighty peaks of Himalayas in the north, through vast plain and arid desert, to tropical forest and coastal regions in the south. The South Asian region experiences tropical and subtropical climatic conditions resulting in extreme temperatures, rainfall and relative humidity which introduces large spatial and temporal variability in aerosol characteristics over this region. The region is a potential source for many types of both natural and anthropogenic aerosols such as mineral dust, sea salt, black carbon, sulfate, nitrates, and other organic water soluble substances. This region, in general, and India, in particular, due to rapidly growing industrialization and expanding urbanization in recent years has become a major regional aerosol hot spot.

The present study is conducted over five different locations in South Asia, namely, Ahmedabad (23.03°N, 72.55°E, 55 m above mean sea level (AMSL)), Gurushikhar (24.65°N, 72.78°E, 1,680 m AMSL), Karachi (24.87°N, 67.03°E, 49 m AMSL), Kanpur (26.51°N, 80.23°E, 123 m AMSL), and Gandhi College (25.87°N, 84.13°E, 60 m AMSL) (Figure 1). Ahmedabad is an industrialized urban location in western India with several small and large scale industries including two power plants [24].

The city has a population of ~5.8 million and is located in the southeast direction of Thar desert. Gurushikhar, Mount Abu is a relatively pristine site situated at a height of about 1.7 km above the mean sea level and is ~300 km northeast of the That desert. The site is situated at the highest peak of Aravalli range of mountains in western India with a cleaner and stable atmosphere where the effect of local pollution is minimum. During winter Gurushikhar exhibits free tropospheric characteristics as the boundary layer height is lower than the altitude of Gurushikhar; while during summer Gurushikhar lies within the boundary layer. Karachi, an urban, coastal city located in the southeastern part of Pakistan on the Arabian Sea experiences scanty rainfall and subtropical climate [25,26]. Karachi has a large industrial base which includes oil-fired power plants, cement factories, steel mill, textile plants, heavy petrochemical industries, and several small industries. Karachi is densely populated (population of ~14 million) [26]. Kanpur and Gandhi College are situated in the Indo-Gangetic basin which is one of the largest river basins in the world, and is one of the densely populated and highly polluted regions in India. This region is bordered by the Himalayas to the north and Vindhyan Satpura ranges to the south. Kanpur is an urban, industrial city with a population of more than four million [27], while Gandhi College is a rural village location in Ganga basin.

3. WIND PATTERNS AND METEOROLOGICAL CONDITIONS

Surface level daily mean temperature, relative humidity (RH), wind speed and wind direction data during 2006–2008 are obtained from National Center for Environmental Prediction (NCEP) reanalysis. NCEP reanalysis data is available at 2.5° × 2.5° latitude–longitude resolution. To examine the spatial variation in synoptic winds (Figure 1) the NCEP data over the study domain are utilized. However, temperature, winds and relative humidity for each study location (data for Figure 2,Figure 3, Figure 4, Figure 5 and Figure 6) are obtained at the respective latitude–longitude of each location. Since the objective of the study is to examine the seasonal and spatial variations in aerosol characteristics, the coarse resolution meteorological data used in the study is not expected to alter the inferences and overall conclusions. The daily mean temperature, RH and wind speed are further used to calculate the monthly mean value and standard deviation. Daily rainfall corresponding to each location are obtained from Tropical Rainfall Measuring Mission (TRMM) satellite 1° × 1°latitude-longitude rainfall data, from which monthly accumulated rainfall are calculated. Surface level synoptic winds over the study region during winter (December-January-February, DJF) are calm, north/northeasterly and are from the polluted northern hemisphere (Figure 1(a)); temperatures are colder and the atmosphere is dry (low RH). In premonsoon (March–April–May, MAM) winds originate and travel from/through a less polluted west (arid/marine) (Figure 1(b)); temperatures increase slightly and RH is lower than winter. In this season mineral dust gets transported from adjacent Thar desert and the Middle East to western India [24]. During monsoon (June-July-August-September, JJAS) the winds (NCEP reanalysis) are stronger, moist and are from the marine and western regions (Figure 1(c)); both temperature and RH are high during this season. During postmonsoon (October-November, ON) wind patterns start shifting in direction from southwest to northeast (Figure 1(d)); temperature and RH reduce. On the basis of meteorology, the aerosol data are grouped into four major seasons of winter, premonsoon, monsoon, and postmonsoon, and discussed. Seasonal variation in meteorological parameters during 2006–2008 for each location are drawn subsequently in Figure 2, Figure 3, Figure 4, Figure 5 and Figure 6.

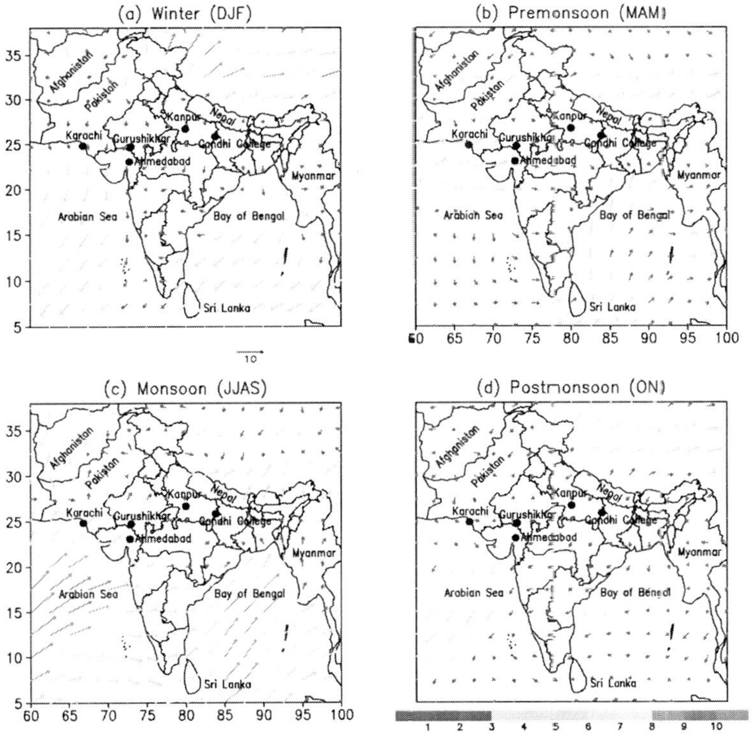

Figure 1. Synoptic surface level wind speed (ms−1) over South Asia during (**a**) winter (DJF); (**b**) premonsoon (MAM); (**c**) monsoon (JJAS); and (**d**) postmonsoon (ON). Study locations—Ahmedabad, Gurushikhar, Karachi, Kanpur, and Gandhi College are marked in the figure.

The sources of aerosols which include natural and anthropogenic vary on regional basis. Ahmedabad, Karachi and Kanpur are urban centers, industrialized and have large number of automobiles. The gas to particle reaction products of the exhausts from the industry and automobiles can contribute to the fine mode aerosols in these locations, while during monsoon and favorable wind conditions sea salt and dust particles in the coarse mode (Figure 2, Figure 3, Figure 4, Figure 5and Figure 6) can be present in these locations. Gurushikhar, a remote, high altitude site is observed to be influenced by long range transport of aerosols. Gandhi College, a rural site located downwind of major urban centers, namely, Delhi, Lucknow, Kanpur, *etc.* in the Indo-Gangetic basin will have a mix of rural and urban aerosol emissions. Over the west (Karachi, Ahmedabad and Gurushikhar) the meteorological features are similar between Ahmedabad and Gurushikhar, while they are different in Karachi. The seasonal rainfall, for example, is much lower over Karachi when compared to Ahmedabad and Gurushikhar (Figure 2, Figure 3 and Figure 4). Kanpur and Gandhi College in the north exhibit similar meteorological features; however, there exist differences in winds and rainfall (Figure 5 and Figure 6). The variation in the local sources of aerosols, long range transport, in addition to meteorology and atmospheric dynamics can influence and modulate the spatiotemporal variabilities in aerosol characteristics over a region which are investigated in detail.

4. MEASUREMENTS, DATA, AND ANALYSIS

4.1. Aerosol Optical Depths

4.1.1. Sun Photometers: Hand Held and Microtops II

Measurements of direct solar radiation intensities reaching the surface are conducted using two kinds of sun photometers—(i) a hand held sun photometer (October 2006–December 2007); and (ii) a set of Microtops-II sun photometers (January 2008–December 2008) over Ahmedabad and Gurushikhar. These measurements were conducted everyday at 1-hour time interval from 09:00 Local Standard Time (LST) to 17:00 LST under only clear sky conditions. The indigenously built hand held sun photometer consists of an interference filter, photo diode and necessary electronics. The sun photometer measures the solar intensity at five different wavelength bands centered at 0.38, 0.50, 0.65, 0.75 and 0.87 μm. The bandwidths of the filters used for the measurement in the present study are about 0.01 μm and the field of view of the sun photometer is about 4°. The indigenously built hand held sun photometer has been used extensively and successfully in several studies over oceans and continental locations [28].

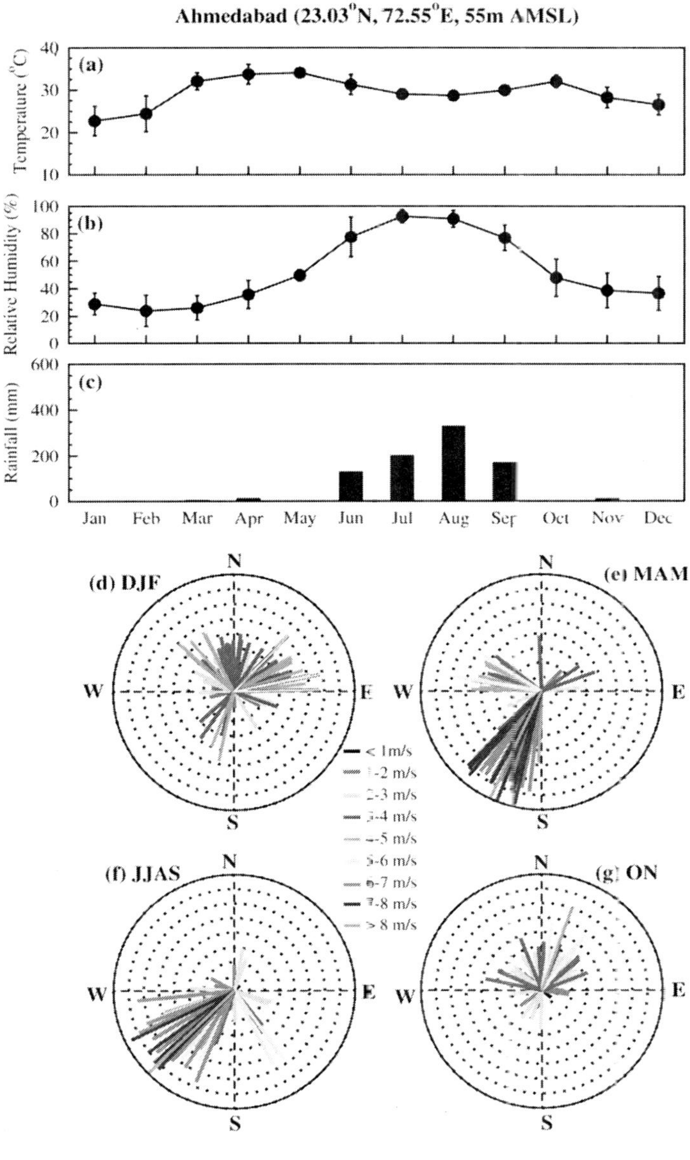

Figure 2. Monthly mean (**a**) temperature (∘C); (**b**) relative humidity (%); (**c**) rainfall (mm); and daily average wind speed and wind direction in (**d**) winter; (**e**) premonsoon; (**f**) monsoon; and (**g**) postmonsoon over Ahmedabad during 2008. Vertical bars indicate $\pm1\sigma$ variation from the mean. The wind speeds are shown as function of its magnitude varying from <1 to >8 ms−1 at an interval of 1 ms−1.

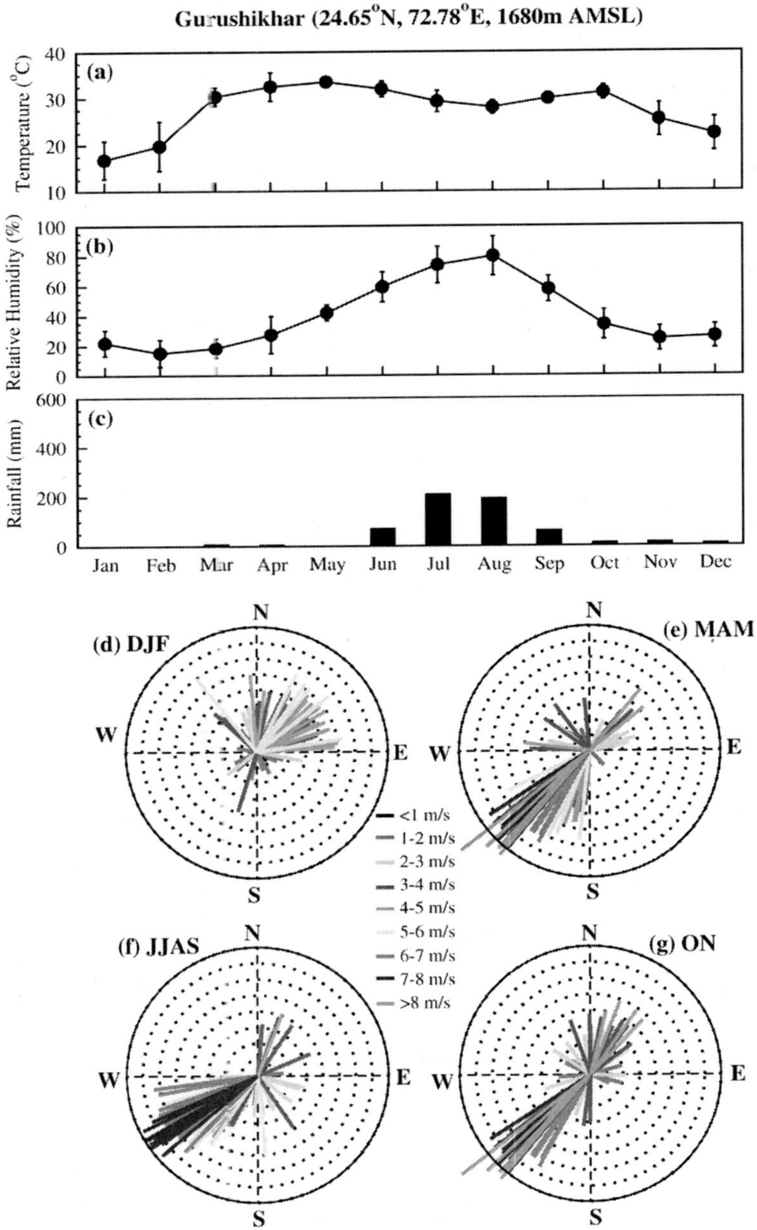

Figure 3. Monthly mean (**a**) temperature (∘C); (**b**) relative humidity (%); (**c**) rainfall (mm); and daily mean wind speed and wind direction during (**d**) winter; (**e**) premonsoon; (**f**) monsoon; and (**g**) postmonsoon over Gurushikhar in 2008. Vertical bars denote $\pm 1\sigma$ variation from the mean.

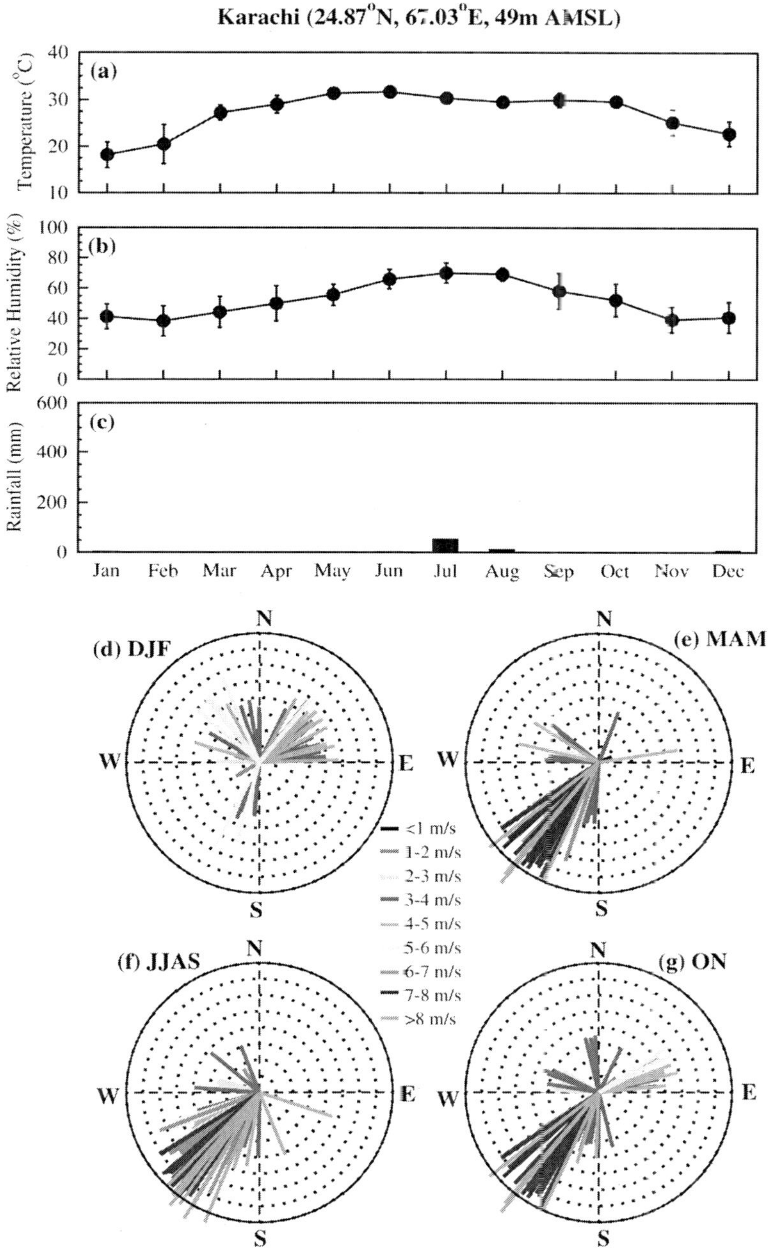

Figure 4. Monthly mean (**a**) temperature (∘C); (**b**) relative humidity (%); (**c**) rainfall (mm); and daily average wind speed and wind direction corresponding to (**d**) winter; (**e**) premonsoon; (**f**) monsoon; and (**g**) postmonsoon over Karachi in 2008. Vertical bars represent $\pm 1\sigma$ variation from the mean.

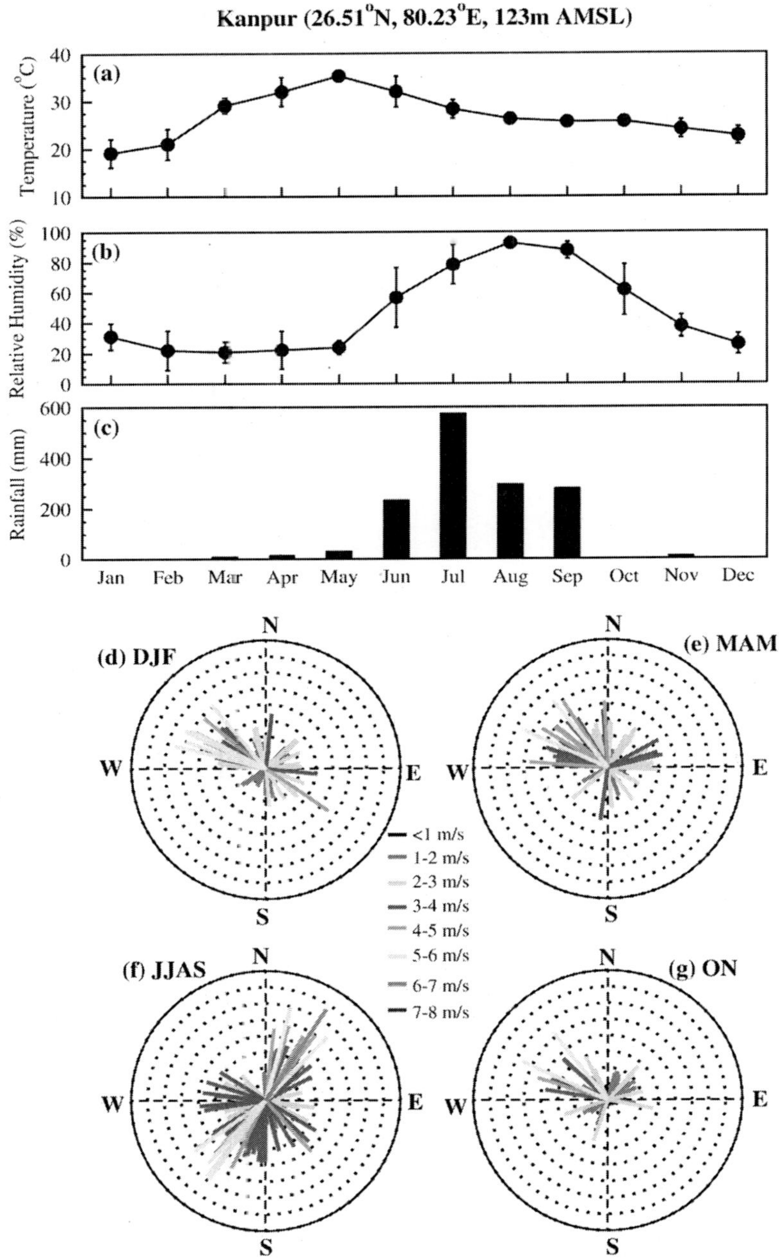

Figure 5. Monthly mean (**a**) temperature (∘C); (**b**) relative humidity (%); (**c**) rainfall (mm); and daily mean wind speed and wind direction in (**d**) winter; (**e**) premonsoon; (**f**) monsoon; and (**g**) postmonsoon over Kanpur during 2008. Vertical bars indicate $\pm 1\sigma$ variation from the mean.

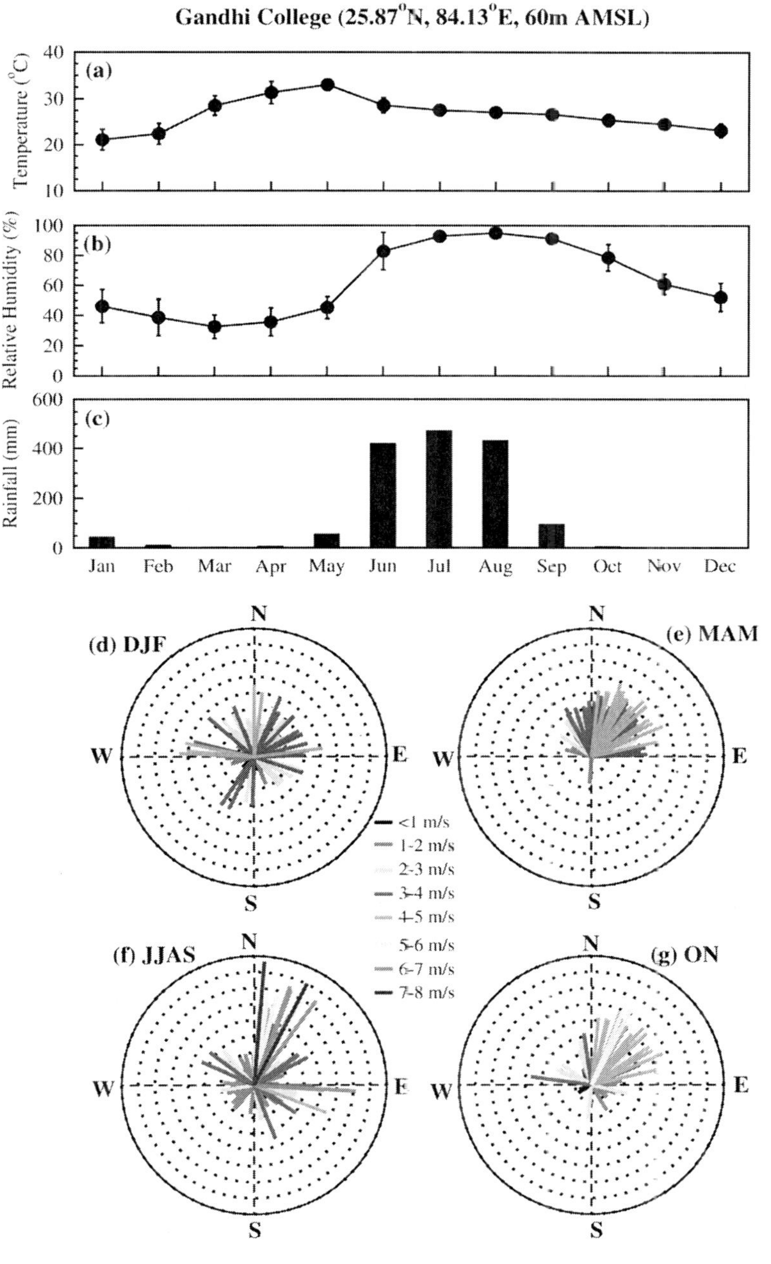

Figure 6. Monthly mean (**a**) temperature (∘C); (**b**) relative humidity (%); (**c**) rainfall (mm); and daily average wind speed and wind direction corresponding to (**d**) winter; (**e**) premonsoon; (**f**) monsoon; and (**g**) postmonsoon over Gandhi College in 2008. Vertical bars correspond to $\pm 1\sigma$ variation from the mean.

The governing equation of sun photometry measurements follows Beer-Lambert's law, which can be written as

$$\tau = \frac{-1}{m}\left[\ln\left(\frac{I}{I_o}\right) - 2\ln\left(\frac{r_o}{r}\right)\right] \quad (1)$$

where τ is the total integrated columnar optical depth of the atmosphere, I is the instantaneous solar radiation intensity measured by the photometer and Io is the solar radiation intensity obtained from Langley plot for zero air mass, m is the atmospheric air mass, r is the instantaneous value of the Sun-Earth distance and ro is the Sun-Earth distance when Io values are obtained. The total integrated columnar optical depth of the atmosphere comprises

$$\tau = \tau_{rs} + \tau_{aerosol} + \tau_{ma} \quad (2)$$

where τrs is the Rayleigh scattering optical depth (scattering due to air molecules), τaerosol is the aerosol optical depth (AOD) and τma is the optical depth due to molecular absorptions such as ozone, water vapor or nitrogen dioxide.

Io(λ) values are obtained using Langley plot technique. In this technique, the natural logarithm of the photometer outputs measured for various solar zenith angles are plotted with respect to air mass. Through the experimentally observed points a least square fitted straight line is drawn and is extrapolated to meet zero air mass in the abscissa. The intercept gives Io(λ) of a particular wavelength. Similarly, Langley plots are made to obtain Io(λ) for all wavelengths. The voltage Io(λ) obtained for zero air mass is a constant in time for a particular wavelength, if the response of the instrument is constant and if the value is corrected for the mean solar distance.

To draw Langley plots observations are to be taken in a cleaner and stable atmosphere where the effects of local pollutants are minimum. These conditions are best met at high altitude mountain sites. In the present case observations are taken from sunrise to sunset, at Gurushikhar in Mt. Abu, at a height of about 1,680 meters above the mean sea level. From the total columnar optical depths (Equation (1)), contribution due to Rayleigh scattering and absorption due to ozone and water vapor at the respective wavelengths are subtracted to calculate aerosol optical depths (AODs) (Equation (2)). An air column density of 2.16 × 1025 molecules per cm2 appropriate for the tropical region is used in the estimation of Rayleigh scattering contribution. Monthly mean columnar concentrations of ozone and water vapor over the tropics are used to estimate ozone and water vapor optical depths [28]. Water vapor contributes much less than 0.1% to the total optical depth in the wavelength range of 0.38–0.87 μm. Therefore, even during the wet season water vapor does not give rise to additional uncertainty in AODs. Contribution due to nitrogen dioxide is less than 0.5% in this wavelength region and is not considered in the present study. The uncertainties in the aerosol optical depth measurements arise from (i) instrumental error due to bias and precision; and (ii) ignoring the forward scattering contribution to the measured irradiance. The solar radiation intensities are measured with an accuracy better than 1%. The uncertainty in aerosol optical depths retrieved using hand held sun photometer is <15% [28].

The hand held Microtops II sun photometer (Solar Light Company, Glenside, PA, USA) can measure AODs at five wavelength bands centered around 0.38, 0.44, 0.50, 0.675 and 0.87 μm [29]. Field of view of Microtops instrument is 2.5° and the band width of each filter is about 0.01 μm. The indigenous hand held sun photometer and the Microtops sun photometer were periodically calibrated (3–6 month intervals) during 2006–2008 at Gurushikhar and the corresponding calibration constants obtained from Langley plot analysis have been used for the subsequent period. The calibration constants (Io) in the 0.38–0.87 μm wavelength range varied <2% during the study period. Aerosol optical depths from Microtops are derived following the same procedure outlined earlier for the sun photometer. The absolute uncertainty in AODs retrieved from Microtops measurements is 2%–5% in the 0.38–0.87 μm range. AODs measured using the hand held sun photometer and Microtops were found to show a good agreement (R2 ≥ 0.94) in the 0.38–0.87 μm wavelength range.

4.1.2. Aeronet

Aerosol Robotic Network (AERONET) is a ground-based Sun/sky radiometer which measures the total ambient aerosol characteristics at different wavelengths. AERONET derives the total column aerosol properties from the measurements of solar radiance. The field of view of the instrument is about 1.2° and the instrument makes direct Sun measurements at every 15 min [6]. AERONET Level 2, cloud screened and quality assured AOD data [30], at the wavelength bands of 0.38, 0.44, 0.5, 0.675 and 0.87 μm at Kanpur, Karachi and Gandhi College during 2006–2008 are used. The uncertainty in the calculation of AERONET AODs is estimated to be <±0.01 for wavelengths >0.44 μm and <±0.02 for ultraviolet wavelengths [31].

4.1.3. MODIS Terra/Aqua Satellites

The MODerate resolution Imaging Spectroradiometer (MODIS) is a remote sensor on board the two Earth Observing System (EOS) Terra and Aqua satellites. Terra and Aqua satellites operate at an altitude of 705 km with Terra spacecraft crossing the equator at about 10:30 LST (local standard time) (ascending northward) while Aqua spacecraft crosses the equator at around 13:30 LST (descending southward) [2]. In the present study, Level 3 MODIS Collection 5.1 atmosphere daily global aerosol optical depth at $1^{\circ} \times 1^{\circ}$ grid resolution are utilized. MODIS Terra and Aqua derived aerosol products over land and oceans are tested, validated, compared, and are being extensively used to investigate spatiotemporal variations in aerosol optical characteristics [2,32]. The algorithms developed to retrieve AODs from MODIS Terra and Aqua were tested and validated with data obtained from airborne imagers [2]. The results of these field tests combined with sensitivity studies suggested that 1σ (standard deviation) of AOD retrievals from MODIS would fall within ±(0.05 + 0.15AOD) over land [5]. These error bounds are referred to as expected error. A global evaluation of collection 5 MODIS AODs over land revealed that >66% (one standard deviation) of MODIS-retrieved aerosol optical depth (AOD) values compare to AERONET observed values within an expected error envelope of (0.05 + 15%), with a high correlation of 0.9 [5].

Daily aerosol optical depth at 0.55 μm from both Terra and Aqua satellites from January 2006–December 2008 are obtained over each location, and daily mean (Terra + Aqua) AODs are calculated. The daily mean AODs are further used to calculate the seasonal mean AODs for all the five locations during 2006–2008. The LST 10:30 and 13:30 of MODIS Terra and Aqua are the local times at the equatorial crossing point of the sub-satellite track. The local time at any other observation location can differ by as much as about ±42 min depending on the longitudinal distance from the sub-satellite track. An additional correction which arises due to the latitude of the observation point, is small in case of MODIS, as its orbit is nearly polar [33].

For the day to day intercomparison of MODIS and sun photometer AODs, in this study, the sun photometer measured AODs between 09:48 and

11:12 LST (corresponding to Terra), and 12:48 and 14:12 LST (corresponding to Aqua) taking into account the longitudinal dependence, are averaged and plotted. Since, MODIS Terra and Aqua AODs are available only twice a day it may not be able to adequately represent the daily mean over many places. In this study, we use the ground based sun photometer AOD data to characterize the diurnal cycle seasonally and examine how the diurnal variation is captured by the two MODIS overpass times during the day. The results (discussed later) suggest clearly that the AODs obtained at two times are able to capture the diurnal cycle seasonally in all the study locations adequately well. Details on the absolute differences between sun photometer and Terra, and sun photometer and Aqua, and their mean are discussed inSection 5.3.

4.1.4. MISR

Multiangle Imaging SpectroRadiometer (MISR) was launched into polar Earth orbit aboard the Terra spacecraft in December 1999 [34]. MISR images the Earth at nine discrete view angles ranging from 70∘ aftward to 70∘ forward, and at four different wavelengths (0.443, 0.555, 0.670 and 0.865 μm). Daily MISR Level-3 0.555 μm AOD data at 17.6 km resolution are obtained for all the five study locations during 2006–2008 and the seasonal means are estimated by averaging all the available AODs. Comparison between daily MISR AODs with ground-based AODs is discussed in Section 5.3. A comparison between AODs from MISR and MODIS retrievals over collocated AERONET sites in land showed that MODIS AODs at 0.47 and 0.66 μm are larger than those obtained from MISR by about 35% and 10% on an average, when all land surface types are considered [10]. MISR was found to show a good agreement with AERONET within the maximum of ±0.05 or ±0.2AOD over land [10]. A global assessment of comparison between MISR and AERONET AODs showed that about 70% to 75% of MISR AOD retrievals are within 0.05% or 20% AOD of the paired validation data [14]. When AERONET AOD was ≥0.4, the MISR retrieved AOD was frequently underestimated over land [14].

4.2. Ångström Exponent (α)

Analysis of the spectral variation of aerosol optical depths can provide information on aerosol size distribution. The spectral variation of aerosol optical depth can be represented by Ångström power law, which is expressed as

$$\tau = \beta\lambda^{-\alpha} \quad (3)$$

where λ is the wavelength in μm, α and β are Ångström parameters. The value of α depends on the ratio of concentration of larger to smaller aerosols in the aerosol size distribution and β represents the columnar aerosol loading in the atmosphere. The Ångström power law is a useful tool for extrapolating AOD throughout the shortwave spectral region. The value of the Ångström exponent qualitatively indicates the aerosol particle size. Values of $\alpha \leq 1$ indicate size distributions that are dominated by coarse mode aerosols such as dust and sea salt, and values of $\alpha \geq 2$ indicate size distributions dominated by fine mode aerosols typically produced by urban pollution and biomass burning [31,35]. Such characterization is useful in the retrieval of aerosol parameters and applying correction factors in remote sensing, and in modeling the radiative effects of aerosols [35]. The α values are estimated for an individual data set of AODs by least square fitting AODs against the respective wavelength on a log-log scale for all the five locations during 2006–2008, and seasonal means are calculated.

5. RESULTS AND DISCUSSION

5.1. Aerosol Optical Depths: Ground-Based Sun Photometer, Microtops and AERONET

Seasonal mean AODs measured at 0.38, 0.50 and 0.87 μm using hand held sun photometer, Microtops sun photometer, AERONET sun/sky radiometers during 2006–2008 are shown over the five study locations (Figure 7). The 0.38, 0.50 μm AODs represent the short wavelength region and are expected to be dominated by fine mode particles, while 0.87 μm AODs represent the long wavelength region and will be influenced by coarse mode particles. At the outset, AODs exhibit strong seasonal and wavelength dependence. AODs during winter and premonsoon are lower than that of monsoon season. AODs decrease gradually as wavelength increases (Figure 7); the gradient in AODs is steeper in Kanpur and Gandhi College. 0.38 and 0.50 μm AODs are higher than that of 0.87 μm AODs during the entire study period (Figure 7) at all the locations except Gurushikhar. AODs in Ahmedabad are about a factor of two higher than Gurushikhar. AODs in the Indo-Gangetic Plain are higher than that of Ahmedabad, while Karachi AODs are comparable. Higher AODs at lower wavelengths indicate the abundance of smaller size particles. During postmonsoon and winter the boundary layer is shallow and holds the pollutants in a smaller volume near the Earth's surface when compared to summer. Over the Indian subcontinent the dust activity is found to peak in the spring (March-May) which decreases in summer with the onset of southwest monsoon [36]. The increase in dust activity accompanied with southwesterly winds contributes to the increase in AODs during premonsoon over Ahmedabad (Figure 7(a)). Wet deposition by precipitation or wet removal is one of the main mechanisms for removal of aerosols from the atmosphere. Though wet removal is most effective in reducing the near surface aerosol concentration, it has been shown that AOD increases significantly when RH increases mainly due to the hygroscopic growth of water soluble aerosols [37]. This increase in AOD due to the hygroscopic growth was found to overwhelm the removal of aerosols and addition of sea salt aerosols in the atmosphere during summer monsoon.

In contrast over Gurushikhar, the difference in AODs at shorter and longer wavelengths is not significant, suggesting lower concentration of fine mode aerosols. AODs could not be measured over Gurushikhar due to cloudy, overcast conditions during monsoon. AODs in the visible wavelength region are more influenced by fine mode aerosols, while when coarse mode aerosols are dominant AODs in the longer wavelength are affected. AODs over Gurushikhar are lowest during winter season as the measurement site remains above the boundary layer. Highest value of AODs are observed during premonsoon, when the measurement site is affected by long range transport of dust particles.

Over Karachi winter, premonsoon and postmonsoon AODs at 0.38 and 0.50 μm are more or less the same in 2007, while in 2008 winter AODs are lower than premonsoon and postmonsoon AODs. AODs during monsoon are higher by about 1.5 times over Karachi (Figure 7(c)). Seasonal and inter-annual variation is similar in short and long wavelength regions in Karachi which indicates that the abundance of fine and coarse mode particles showed similar variabilities. Higher AODs during premonsoon and monsoon occur due to winds coming from the southwest with higher speed (Figure 4(e),(f)) which increases the concentration of dust and sea salt in the atmosphere respectively. The seasonal cycle in AODs is consistent with aerosol sources and transport effects over Karachi. Over Afghanistan, Iran, Pakistan and China, which are mountainous regions dust sources are found in closed intermountain basins [36]. The dust activity starts in April–May, becomes strong in June–July, and weakens considerably by September over Pakistan basin [36]. Karachi, located on the coast of Arabian Sea, witnesses considerable dust activity during premonsoon and monsoon seasons. In addition, during the monsoon season (Figure 1(a)) winds transport sea salt aerosols over Karachi.

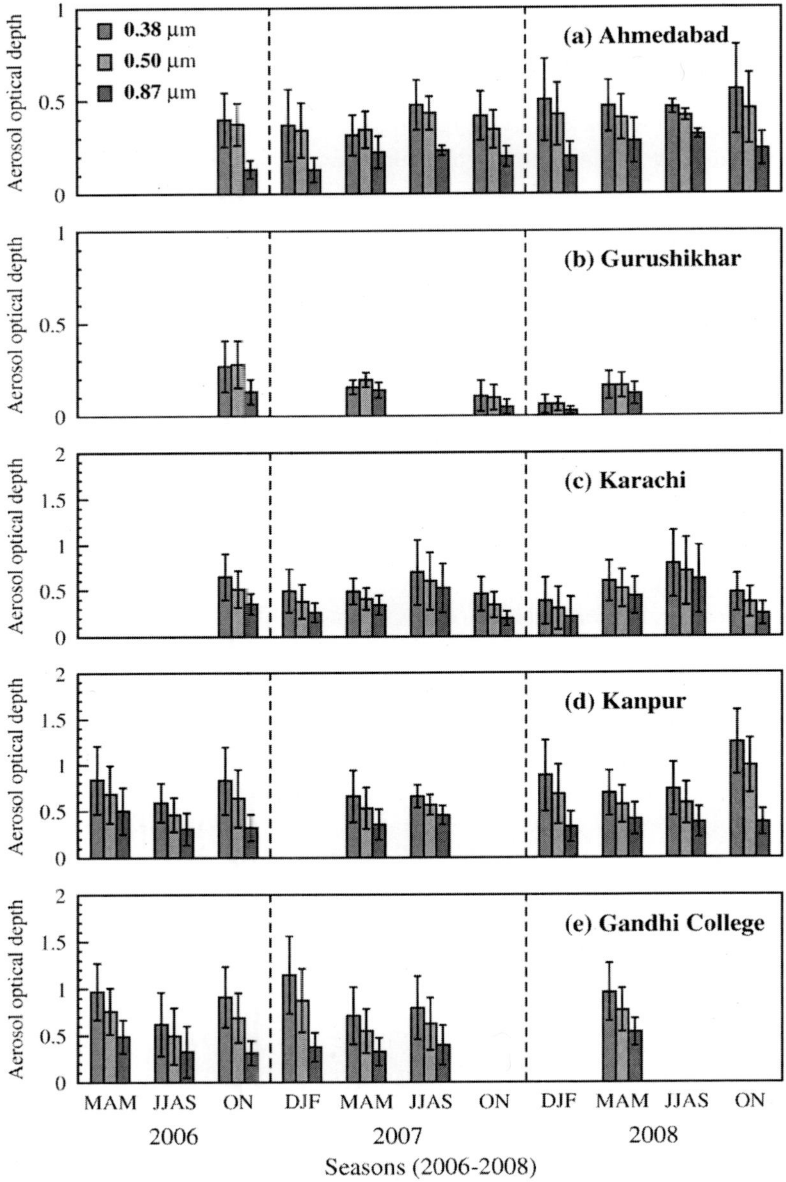

Figure 7. Seasonal mean aerosol optical depths at 0.38, 0.50 and 0.87 μm during 2006–2008 over (**a**) Ahmedabad; (**b**) Gurushikhar; (**c**) Karachi; (**d**) Kanpur and (**e**) Gandhi College. Gaps in the figure denote the absence or non availability of sun photometer, Microtops and AERONET sun/sky radiometer aerosol optical depth data. Vertical bars indicate $\pm 1\sigma$ deviation from the seasonal mean.

Unlike the other three locations seasonal variation in Kanpur AODs are not so strong while there exists a large inter-annual variability. AERONET AODs during 2001–2003 also showed distinct inter-annual variations over Kanpur [27] as seen here. Gandhi College being a rural, downwind location south of Kanpur exhibits similar features as that of Kanpur. Higher AODs during postmonsoon and winter is a characteristic feature of the locations in Indo-Gangetic basin. Over much of northern India the temperatures are colder in the winter. During postmonsoon and winter the boundary layer is shallow and holds the pollutants in a smaller volume, when compared to summer, which results in confinement of aerosols. The colder temperatures along with the trapping of pollutants give rise to hazy and foggy conditions over Indo-Gangetic basin [4,21]. In addition, the dense population in this region contributes to large aerosol sources (due to fossil fuel consumption and bio fuels used for domestic purposes) which produce higher AODs.

Ahmedabad, Gurushikhar and Karachi situated in the western region of South Asia are governed by more or less similar meteorological conditions except for rainfall over Karachi (a factor of ten less when compared to the other two locations), but are distinct in terms of the environment (urban, remote and high altitude, urban/coastal respectively) and therefore are influenced by different types of aerosols and sources. Over this region highest AODs are observed over Karachi; while the lowest AODs are obtained over Gurushikhar. AODs over Ahmedabad and Karachi are at least a factor of two higher than the AODs measured over Gurushikhar. This is due to the fact that Gurushikhar, being a high altitude remote location is dominated by coarse mode aerosols from natural sources (mainly mineral dust) in the atmosphere (Table 1); while both Ahmedabad and Karachi are densely populated and highly industrialized areas. Among the urban locations, Karachi AODs are higher in all the seasons during 2006–2008 when compared to Ahmedabad. Highest AODs over Karachi could be because of much higher population (~14 million, when compared to about 5.8 million over Ahmedabad, and about 50,000 in Gurushikhar), industrialization, and in addition, being a coastal station the contribution of sea salt cannot be excluded. Kanpur and Gandhi College in northern India situated in one of the the most polluted and densely populated regions, namely the Indo-Gangetic basin, are influenced by similar meteorological conditions. Kanpur is an urban, industrial city while Gandhi College is a rural location. AODs over the Indo-Gangetic basin are higher than those measured in the western region of South Asia. 0.38 and 0.50 μm AODs over Gandhi College are about 20% higher than Kanpur; while 0.87 μm AODs nearly the same for all the seasons over both the locations. This suggests that though both Kanpur and Gandhi College are located in the Indo-Gangetic basin, Gandhi College AODs can be higher as it is located downwind of major urban centers; also being a rural location the local production due to biomass emissions over Gandhi College could be contributing to the higher AODs.

Table 1. Seasonal mean 0.5 μm aerosol optical depths (along with $\pm 1\sigma$ standard deviation) measured using hand held sun photometer and Microtops over Ahmedabad and Gurushikhar, and from AERONET over Karachi, Kanpur and Gandhi College during 2006–2008. Seasonal mean AODs correspond to year 2007 unless otherwise mentioned.

Seasons	Ahmedabad	Gurushikhar	Karachi	Kanpur	Gandhi College
Winter	0.37 ± 0.15	0.06 ± 0.03 [a]	0.38 ± 0.18	0.88 ± 0.32 [b]	0.87 ± 0.34
Premonsoon	0.31 ± 0.10	0.19 ± 0.04	0.40 ± 0.12	0.66 ± 0.22	0.55 ± 0.24
Monsoon	0.46 ± 0.09		0.60 ± 0.31	0.65 ± 0.11	0.63 ± 0.29
Postmonsoon	0.42 ± 0.07	0.10 ± 0.06	0.34 ± 0.14	0.83 ± 0.31 [c]	0.69 ± 0.26 [d]

[a,b] 2008, and [c,d] 2006.

5.2. Ångström Coefficients and AOD Ratio

The α (estimated from the AODs in the 0.38–0.87 μm wavelength range) is found to show strong seasonal, inter-annual as well as spatial variability similar to AODs. Over Ahmedabad, α is found to be high during postmonsoon and winter season; while during premonsoon and monsoon α value is lower (Figure 8(a)) for all the three years. This feature of lower α value is consistent with the presence of mineral dust and hygroscopically grown water soluble aerosols, which are in the coarse mode, during premonsoon and monsoon respectively. Over Gurushikhar α is found to be lower than that of Ahmedabad throughout the study period indicating a higher concentration of coarse mode aerosols (Figure 8(b)). The intra seasonal variation in α is quite significant over Gurushikhar; premonsoon α is at least a factor of three lower when compared to postmonsoon and winter. The lower α during premonsoon highlights the dominance of long range transported dust particles in the atmosphere, while during postmonsoon and winter AODs are lower and fine mode aerosols are present giving rise to higher α values. α values over Karachi, an urban, coastal station are lower than those obtained over Ahmedabad (Figure 8(c)) suggesting the presence of coarse mode (sea salt) aerosols. The α values over Gurushikhar significantly lower during premonsoon (Figure 8(b)), due to the abundance of coarse mode dust particles. In Karachi, the αvalues during monsoon are lower during premonsoon and monsoon than during winter and postmonsoon. The monsoon αvalues are lower than even premonsoon because in addition to the dust particles that are present during monsoon, sea salt particles get transported (Figure 1). Thereby, coarse mode particles become more abundant during monsoon in Karachi and give rise to further lower values of α. Over the Indo-Gangetic basin (Kanpur and Gandhi College) α value is found to increase from premonsoon to winter (Figure 8(d,e)). Seasonal differences in the magnitudes of α over the Indo-Gangetic basin are not quite strong as compared to the other three locations in the west. Higher α during postmonsoon and winter over the Indo-Gangetic basin arise due to the increase in the abundance of fine mode aerosols contributed mainly by anthropogenic activities, and atmospheric dynamics (shallow boundary layer and colder temperatures help trap the aerosols).

The α values during postmonsoon and winter are higher than those obtained during premonsoon and monsoon in all the locations during 2006–2008 (Table 2). However, the magnitude of seasonal differences exhibit spatial variation. The higher α values ($\geq$1) suggest the dominance of fine mode aerosols, mainly from anthropogenic activities (fossil and bio fuel emissions); α values lower than 1 indicate the dominance and/or contribution of coarse mode particles from natural sources (dust and sea salt) [35]. While higher α values in the study locations can be attributed to the dominance of fine mode aerosols over urban areas (Ahmedabad and Kanpur), dust and sea salt aerosols also contribute and dominate (Gurushikhar, Kanpur and Karachi). The β is anti-correlated with α over Karachi; β decreases during post-monsoon and winter while α increases. This feature is seen in all the study locations, but this anti-correlation feature is more evident and significant over Karachi.

AOD ratio, indicative of the dominance of fine or coarse mode aerosols, is found to show seasonal and spatial variations corroborating the results obtained on AODs (Table 3). A higher AOD ratio over Ahmedabad when compared to Gurushikhar indicates that fine mode aerosols are less abundant over Gurushikhar. Postmonsoon and winter AOD ratios in the study locations are higher than those obtained during premonsoon and monsoon suggesting the abundance of fine mode aerosols during postmonsoon and winter. Monsoon AOD ratios are the lowest in all the locations corroborating the increase and the dominance of coarse mode aerosols (Table 3). The magnitudes of AOD ratios exhibit spatial differences indicating that the source strengths of fine and/or coarse mode aerosols over these locations vary (higher or lower) during the year (Table 3). For example, AOD ratios over Gandhi College are higher than Kanpur throughout the year indicating a higher amount of fine mode aerosols in Gandhi College, which is corroborated by higher α (Table 2) values. These results from the detailed analysis of spatial, seasonal and inter annual variations in aerosol characteristics (Table 1, Table 2 andTable 3) over five locations in South Asia highlight the spatial and seasonal variabilities in the contribution of fine and coarse mode particles from different aerosol sources, and the influence of meteorology including the effect of long range transport.

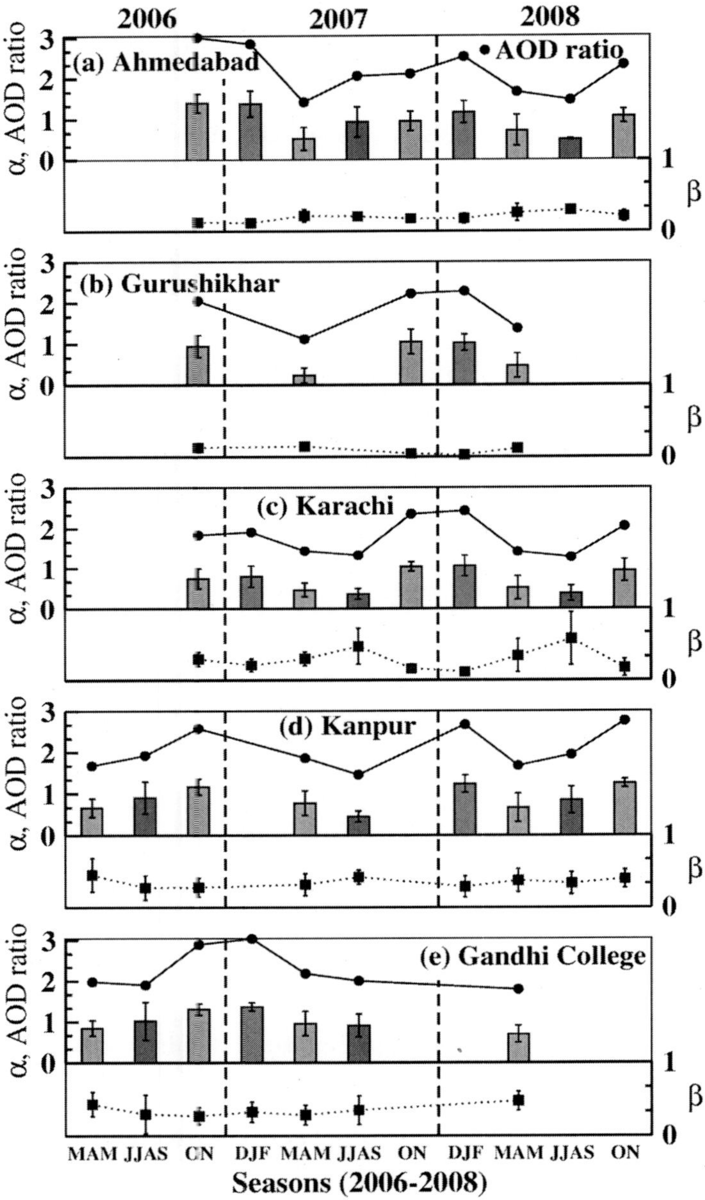

Figure 8. Ångström coefficients α and β (Equation (3)) derived from the aerosol optical depths measured in the 0.38–0.87 μm wavelength region over (**a**) Ahmedabad; and (**b**) Gurushikhar; (**c**) Karachi; (**d**) Kanpur; and (**e**) Gandhi College during 2006–2008 for different seasons. Vertical bars indicate $\pm 1\sigma$ deviation from the seasonal mean. Seasonal mean AOD ratio (=0.38 μm AOD/0.87 μm AOD) is also plotted.

Table 2. Ångström exponent (α) and $\pm 1\sigma$ standard deviation derived from spectral aerosol optical depths measured in the 0.38–0.87 µm wavelength region over Ahmedabad, Gurushikhar, Karachi, Kanpur and Gandhi College. α values correspond to 2007 unless otherwise stated.

Seasons	Ahmedabad	Gurushikhar	Karachi	Kanpur	Gandhi College
Winter	1.38 ± 0.32	1.03 ± 0.20 [a]	0.80 ± 0.26	1.23 ± 0.21 [b]	.37 ± 0.10
Premonsoon	0.52 ± 0.28	0.23 ± 0.18	0.46 ± 0.17	0.77 ± 0.29	0.96 ± 0.30
Monsoon	0.93 ± 0.37		0.36 ± 0.13	0.45 ± 0.13	0.91 ± 0.28
Postmonsoon	0.95 ± 0.24	1.05 ± 0.30	1.04 ± 0.12	1.16 ± 0.19 [c]	.31 ± 0.14 [d]

[a,b] 2008, and [c,d] 2006.

Table 3. Seasonal mean aerosol optical depth (AOD) ratio (=0.38 µm AOD/0.87 µm AOD) determined from aerosol optical depths measured using hand held sun photometer and Microtops over Ahmedabad and Gurushikhar, and from AERONET over Karachi, Kanpur and Gandhi College during 2006–2008. Seasonal mean AODs correspond to year 2007 unless otherwise mentioned.

Seasons	Ahmedabad	Gurushikhar	Karachi	Kanpur	Gandhi College
Winter	2.9	2.3 [a]	1.9	2.7 [b]	3.0
Premonsoon	1.7	1.1	1.4	1.9	2.2
Monsoon	1.5		1.3	1.4	2.0
Postmonsoon	2.1	2.2	2.4	2.6 [c]	2.9 [d]

[a,b] 2008, and [c,d] 2006.

5.3. Comparison of Ground-Based and Satellite AODs

Daily values of ground-based sun photometer AODs are compared with MODIS Terra and Aqua AODs in Figure 9 andFigure 10. Since the availability of ground-based AODs at the exact time of Terra and Aqua overpasses which are 10:30 and 13:30 LST respectively is difficult ground-based sun photometer AODs available within ±5 min of the satellite overpass are used. In Figure 11 and Figure 12 ground-based sun photometer AODs averaged during the ±42 min time frame of Terra and Aqua overpasses are compared with Terra and Aqua AODs as a function of season. At the outset, there are no visible differences in seasonal variations in the comparison figures between ground-based AODs measured within ±5 min interval and those that are averaged within the ±42 min with respect to Terra and Aqua AODs (Figure 9, Figure 10, Figure 11 andFigure 12, Table 4). This analysis confirms that (i) aerosol optical characteristics do not change significantly within an hour over the observation site (variation within an hour is quite less); (ii) much of the diurnal AOD variation is captured well by the two MODIS instruments during their overpass times and (iii) the temporal difference between the ground-based and remote sensing measurements contributes negligibly to the observed variations/differences in aerosol optical characteristics. The results show that mean of Terra + Aqua AOD may adequately represent the daily mean (Figure 13). Differences in MODIS Terra and Aqua, and MISR AOD values with respect to those obtained from the ground-based measurements are given in Table 4 and Table 5. Over Ahmedabad Terra, Aqua and MISR underestimate the AODs (~0.05) (Table 5); and the underestimation (~0.10) is more on seasonal time scales (Table 4). On an annual mean (2006–2008) scale, MODIS and MISR underestimate over Gurushikhar, a high altitude site (Table 5), while on a seasonal mean basis MODIS Terra underestimates, while Aqua overestimates during winter (Table 4). In Karachi, MODIS overestimates in different seasons (Table 4) and on annual mean basis by about 0.14, while MISR underestimates by about 0.06 (Table 5). In Kanpur and Gandhi College, located in the Indo-Gangetic Plain, the features differ; MODIS overestimates by about 0.06, while MISR underestimates by 0.17 (Table 5). In contrast, over Gandhi College, MODIS underestimates AOD by about 0.07, while MISR overestimates by about 0.2. The mean AODs over Kanpur and Gandhi College are >0.6, and the $\pm 1\sigma$ variation is about 20%. These results quantitatively document the differences between ground-based AOD measurements and remote sensing (MODIS and MISR) over distinctly different environments on a variety of temporal scales.

In Figure 14 seasonal means of sun photometer, Microtops and AERONET measured 0.5 μm AODs are compared with MODIS and MISR retrieved AODs at 0.55 μm and 0.555 μm respectively, over Ahmedabad, Gurushikhar, Karachi, Kanpur and Gandhi College. During the entire study period and for all the study locations MODIS and MISR AODs are broadly able to capture the seasonal variations (Figure 14), for example, the winter and postmonsoon low in AODs and the premonsoon and monsoon high in AODs. However, the magnitude of AODs are found to exhibit distinct behavior. MODIS underestimates AODs in Ahmedabad, Gurushikhar and Gandhi College by at least 0.04 in absolute magnitude, while it overestimates by about 0.06 in Kanpur and by 0.13 in Karachi. On the other hand, MISR underestimates AODs at all the locations except Gandhi College (Table 5). Over Kanpur and Gandhi College, separated by about 200-300 km, MODIS overestimates while MISR underestimates in Kanpur; MODIS underestimates while MISR overestimates over Gandhi College. Interestingly the over or underestimation in AODs by the two satellites over this region is almost the same in terms of the amount (Table 5).

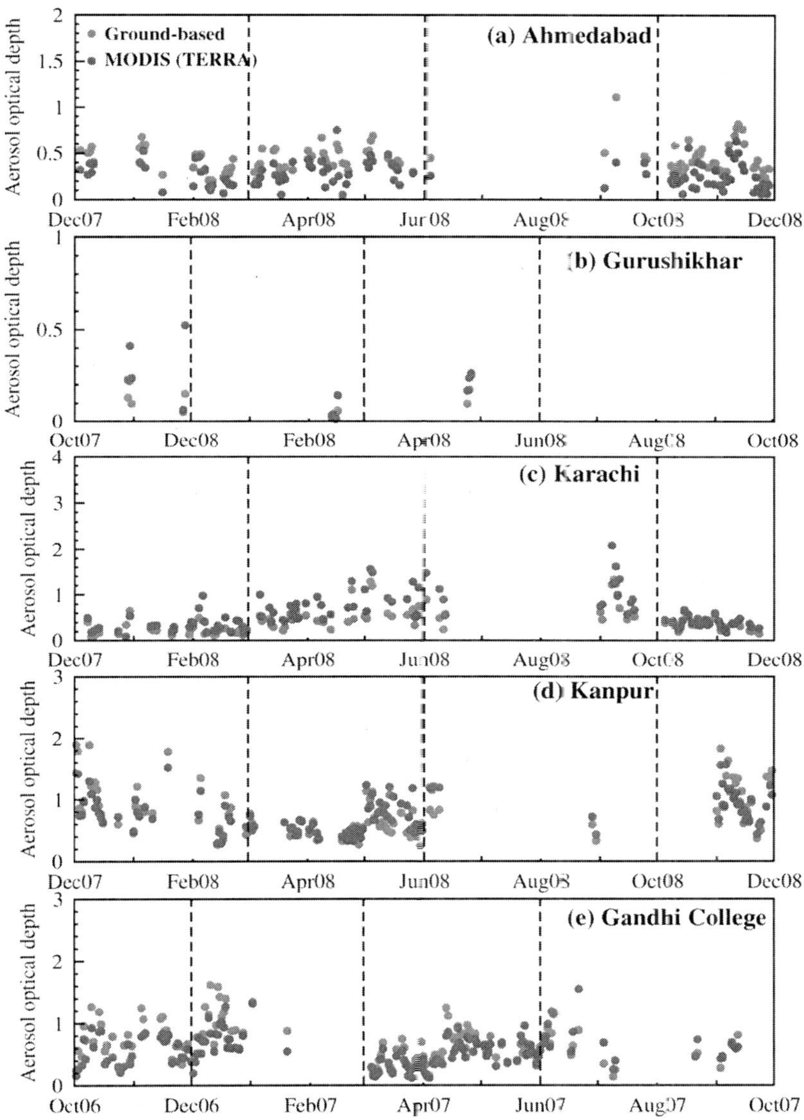

Figure 9. Aerosol optical depths measured using ground-based sun photometers compared with MODIS Terra retrieved AODs corresponding to 10:30 LST. AODs are measured using indigenously developed hand held sun photometer and Microtops sun photometer over (**a**) Ahmedabad and (**b**) Gurushikhar, while (**c**) Karachi, (**d**) Kanpur and (**e**) Gandhi College AODs are measured using AERONET sun/sky radiometers. For comparison AOD data measured within ±5 min of 10:30 LST are used.

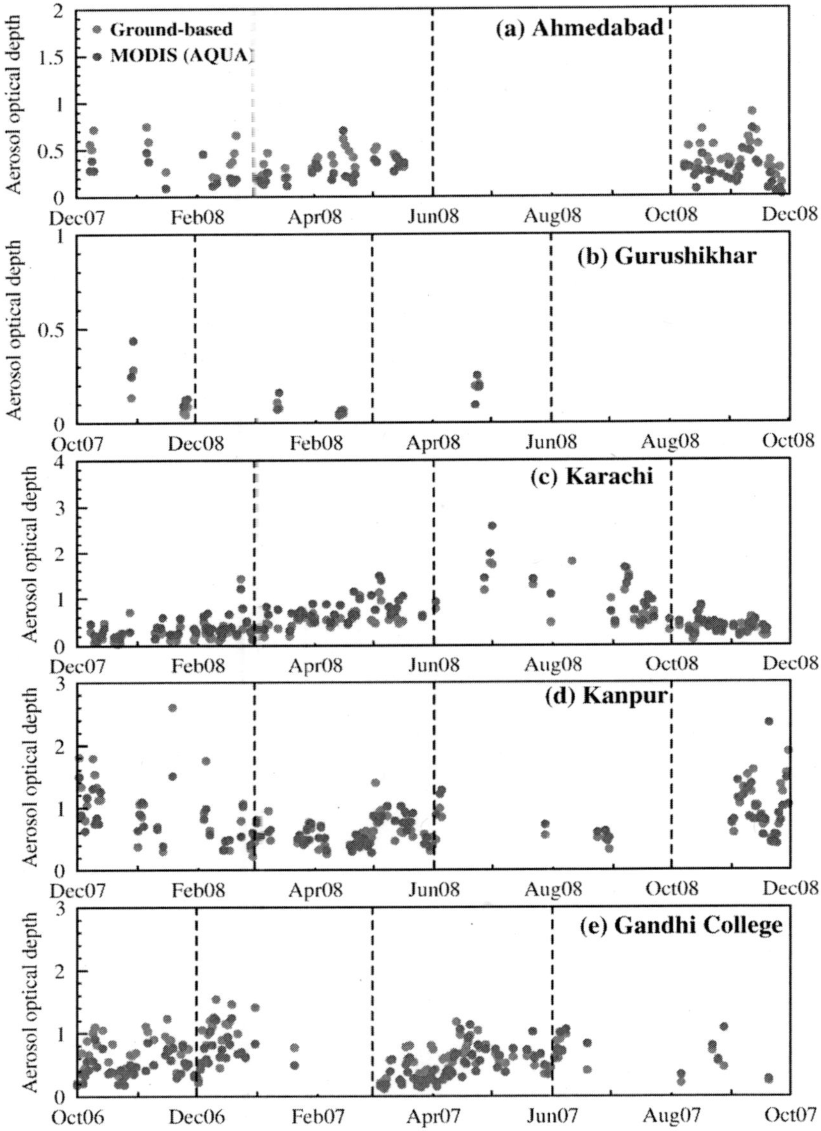

Figure 10. Aerosol optical depths measured using ground-based sun photometers compared with AODs retrieved from MODIS Aqua corresponding to 13:30 LST. AODs are measured using indigenously developed hand held sun photometer and Microtops sun photometer over (**a**) Ahmedabad and (**b**) Gurushikhar; while (**c**) Karachi, (**d**) Kanpur and (**e**) Gandhi College AODs are measured using AERONET sun/sky radiometers. For comparison AOD data measured within ±5 min of 13:30 LST are used.

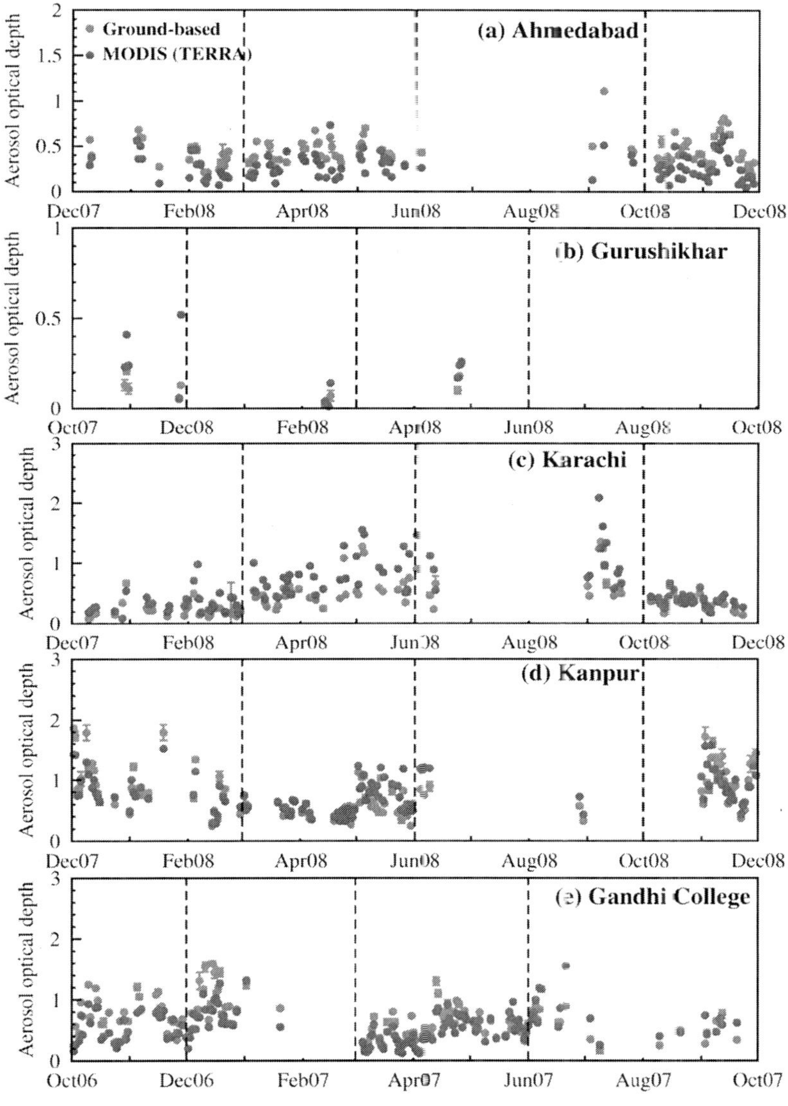

Figure 11. Daily aerosol optical depths measured using ground-based sun photometers compared with MODIS Terra AODs at each study location. AODs are measured using indigenously developed hand held sun photometer and Microtops sun photometer over (**a**) Ahmedabad and (**b**) Gurushikhar, while (**c**) Karachi, (**d**) Kanpur and (**e**) Gandhi College AODs are measured using AERONET sun/sky radiometers. For the intercomparison of MODIS and sun photometer AODs the ground based AODs measured between 09:48 and 11:12 LST (corresponding to Terra) taking into account the longitudinal dependence are averaged and drawn.

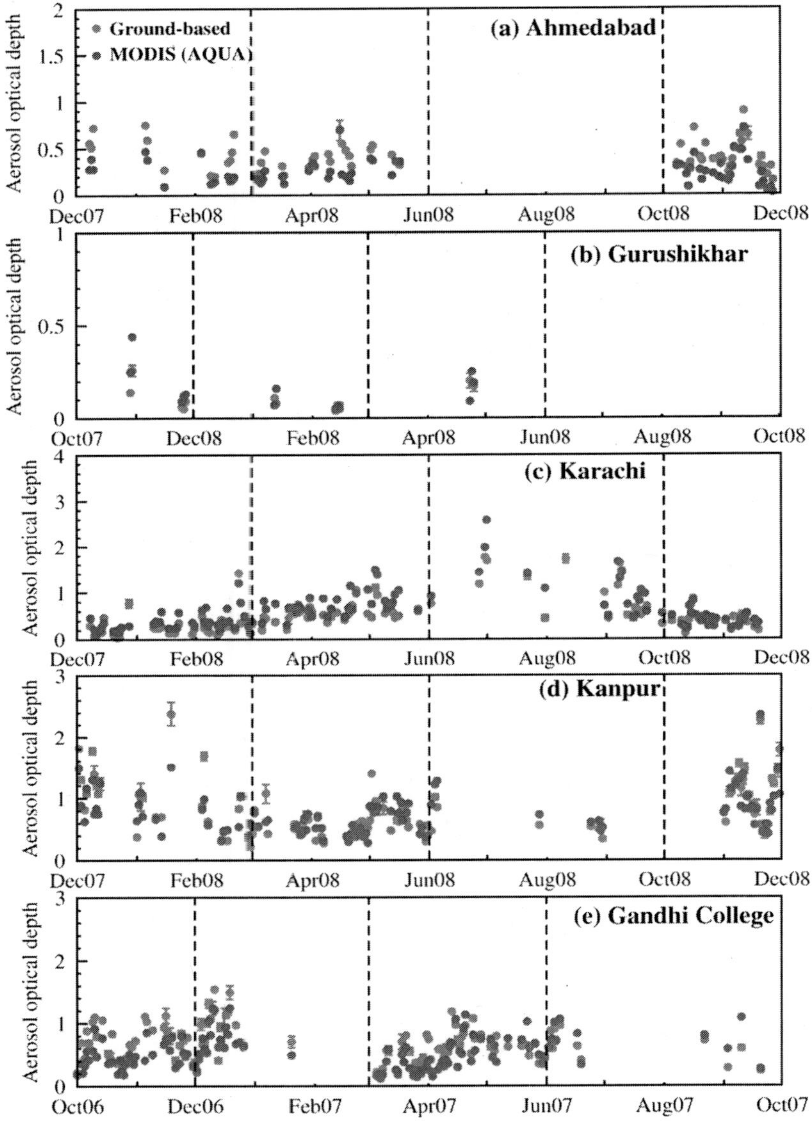

Figure 12. Daily aerosol optical depths measured using ground-based sun photometers compared with MODIS Aqua AODs at each study location. AODs are measured using indigenously developed hand held sun photometer and Microtops sun photometer over (**a**) Ahmedabad and (**b**) Gurushikhar, while (**c**) Karachi, (**d**) Kanpur and (**e**) Gandhi College AODs are measured using AERONET sun/sky radiometers. For the intercomparison of MODIS and sun photometer AODs the ground based AODs measured between 12:48 and 14:12 LST (corresponding to Aqua) taking into account the longitudinal dependence are averaged and drawn.

Table 4. Comparison of seasonal mean AODs in 2006–2008 obtained from ground-based measurements with AODs retrieved from MODIS Terra and Aqua. Ground-based AODs corresponding to within ±5 min of Terra and Aqua overpass times of 10:30 and 13:30 LST, and after averaging the measurements made during ±42 min corresponding to Terra and Aqua satellite pass tracks.

Source	Ahmedabad			
	DJF 2007–2008	MAM 2008	JJAS 2008	ON 2008
Ground-based (10:30 ± 5 min)	0.41	0.43	0.47	0.43
Ground-based (10:30 ± 42 min avg)	0.41	0.43	0.46	0.43
MODIS (Terra)	0.27	0.30	0.33	0.30
Ground-based (13:30 ± 5 min)	0.47	0.39		0.42
Ground-based (13:30 ± 42 min avg)	0.47	0.40		0.41
MODIS (Aqua)	0.27	0.27		0.28

	Gurushikhar			
	DJF 2008	MAM 2008	JJAS 2008	ON 2007
Ground-based (10:30 ± 5 min)	0.23	0.17		0.13
Ground-based (10:30 ± 42 min avg)	0.23	0.17		0.13
MODIS (Terra)	0.10	0.22		0.29
Ground-based (13:30 ± 5 min)	0.08	0.20		0.12
Ground-based (13:30 ± 42 min avg)	0.08	0.18		0.12
MODIS (Aqua)	0.08	0.18		0.20

	Karachi			
	DJF 2007–2008	MAM 2008	JJAS 2008	ON 2008
Ground-based (10:30 ± 5 min)	0.22	0.54	0.73	0.35
Ground-based (10:30 ± 42 min avg)	0.23	0.54	0.73	0.35
MODIS (Terra)	0.34	0.80	1.03	0.39
Ground-based (13:30 ± 5 min)	0.30	0.56	0.93	0.39
Ground-based (13:30 ± 42 min avg)	0.27	0.56	0.93	0.39
MODIS (Aqua)	0.37	0.75	1.21	0.40

	Kanpur			
	DJF 2007–2008	MAM 2008	JJAS 2008	ON 2008
Ground-based (10:30 ± 5 min)	0.79	0.57	0.75	1.03
Ground-based (10:30 ± 42 min avg)	0.81	0.57	0.76	1.02
MODIS (Terra)	0.82	0.65	0.99	0.95
Ground-based (13:30 ± 5 min)	0.81	0.59	0.60	1.09
Ground-based (13:30 ± 42 min avg)	0.83	0.59	0.62	1.06
MODIS (Aqua)	0.82	0.60	0.78	0.96

	Gandhi College			
	DJF 2006–2007	MAM 2007	JJAS 2007	ON 2006
Ground-based (10:30 ± 5 min)	0.94	0.54	0.61	0.74
Ground-based (10:30 ± 42 min avg)	0.94	0.55	0.61	0.74
MODIS (Terra)	0.73	0.46	0.71	0.56
Ground-based (13:30 ± 5 min)	0.92	0.56	0.60	0.68
Ground-based (13:30 ± 42 min avg)	0.89	0.56	0.62	0.68
MODIS (Aqua)	0.72	0.46	0.73	0.47

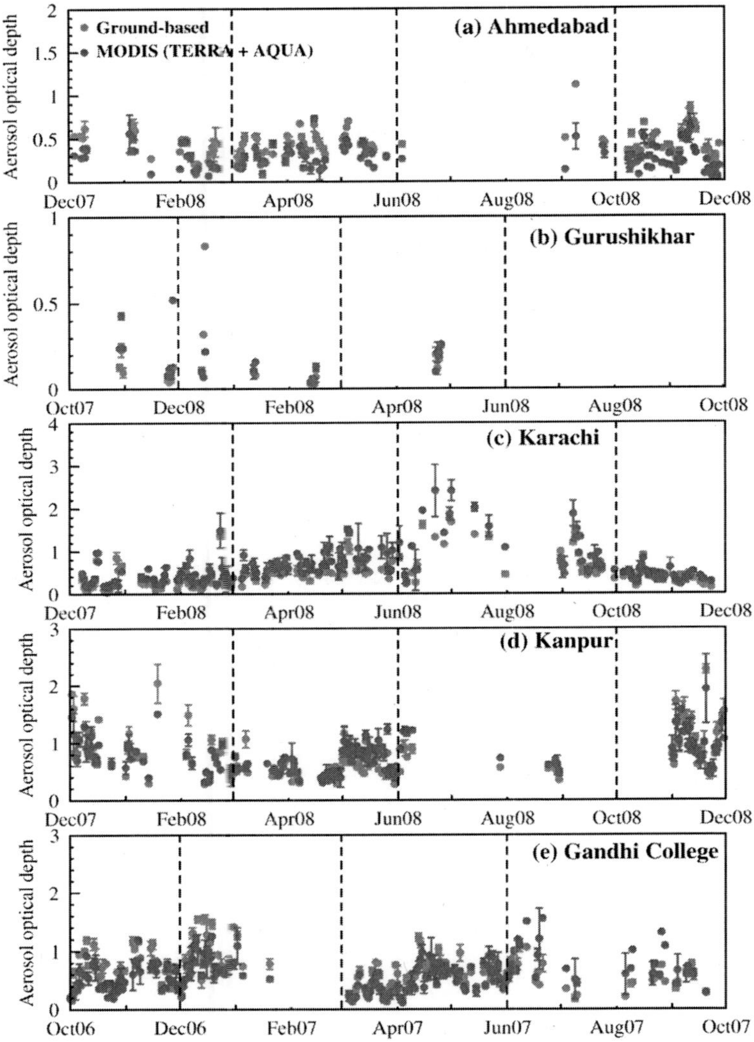

Figure 13. Diurnal mean aerosol optical depths measured using ground-based sun photometers compared with MODIS Terra and Aqua mean AODs at each study location. AODs are measured using indigenously developed hand held sun photometer and Microtops sun photometer over (**a**) Ahmedabad and (**b**) Gurushikhar, while (**c**) Karachi, (**d**) Kanpur and (**e**) Gandhi College AODs are measured using AERONET sun/sky radiometers. For the day to day intercomparison of MODIS and sun photometer AODs the ground based AODs measured between 09:48 and 11:12 LST (corresponding to Terra), and 12:48 and 14:12 LST (corresponding to Aqua) taking into account the longitudinal dependence are averaged and drawn.

Table 5. Comparison of average (2006–2008) AODs obtained from ground-based measurements with AODs retrieved from MODIS Terra, Aqua, their mean (Terra + Aqua) and MISR. Mean AODs from ground-based measurements corresponding to Terra and Aqua satellite passing times, and the overall mean for 2006–2008 are given. AODs obtained from ground-based measurements are subtracted from MODIS and MISR (2006–2008) mean AODs. A positive value indicates overestimate of AOD by that amount by MODIS/MISR with respect to the ground-based measurement, while a negative value indicates an underestimate.

Source	Ahmedabad	Gurushikhar	Karachi	Kanpur	Gandhi College
Ground-based (09:42-11:12 h mean)	0.44	0.18	0.41	0.78	0.67
MODIS (Terra)	−0.08	+0.01	+0.16	0.00	−0.10
Ground-based (12:48-14:12 h mean)	0.44	0.12	0.52	0.80	0.66
MODIS (Aqua)	−0.08	+0.03	+0.14	−0.05	−0.13
Ground-based (Mean)	0.39 ± 0.04	0.16 ± 0.08	0.46 ± 0.13	0.63 ± 0.15	0.68 ± 0.13
MODIS (Terra + Aqua)	−0.07	−0.04	+0.13	+0.06	−0.07
MISR	−0.05	−0.02	−0.06	−0.17	+0.20

A one-on-one correlation between the collocated daily mean AODs obtained over each study location from ground-based (sun photometer, Microtops, and sun/sky radiometers) and satellite (MODIS and MISR) measurements (Figure 15) is shown. Correlation coefficients lie in the range of 0.50–0.87 in the study locations (Figure 15). The correlations are not so strong due to the differences between the AODs retrieved from MODIS, MISR and ground-based instruments. The plausible causes that could contribute to the differences between ground-based and satellite retrieved AODs include: (1) spatial difference between the two measurements (point, *vs.* grid); (2) uncertainties associated in deriving AODs from both ground-based and satellite retrievals; and (3) differences in the wavelengths (0.50 μm in case of sun photometer/Microtops/sun/sky radiometers, 0.55 μm from MODIS, and 0.555 μm from MISR) [2,28]. It should be noted that while MODIS provides a near global coverage daily [38], MISR can provide the global coverage once in 9 days, which reduces the number of days on which AODs can be compared (Figure 15). A comparison of coincident MODIS, MISR and AERONET AODs revealed that instrument sampling, sampling differences, assumptions made in the MODIS and MISR standard algorithms about boundary conditions, missing particle property or mixing options, reflectances used in the retrievals, in addition to cloud screening could be contributing to differences in AODs between the ground-based and satellite measurements [39]. For example, it has been found that the dust models used in MISR retrievals are too absorbing, therefore, revised dust models including a wider range of bimodal, non-absorbing mixtures are being considered [40]. The analysis on the correlation between ground-based and remote sensing suggests that improvements to be made in retrievals including the surface reflectance and aerosol properties will enhance the correlation between ground-based and satellite retrieved AODs. The results on the spatio-temporal variations in aerosol optical characteristics over five locations governed by different sources, and the comparison between ground-based and remote sensing AODs will be useful in remote sensing to improve the retrieval algorithms, and in assessing the aerosol impact on climate.

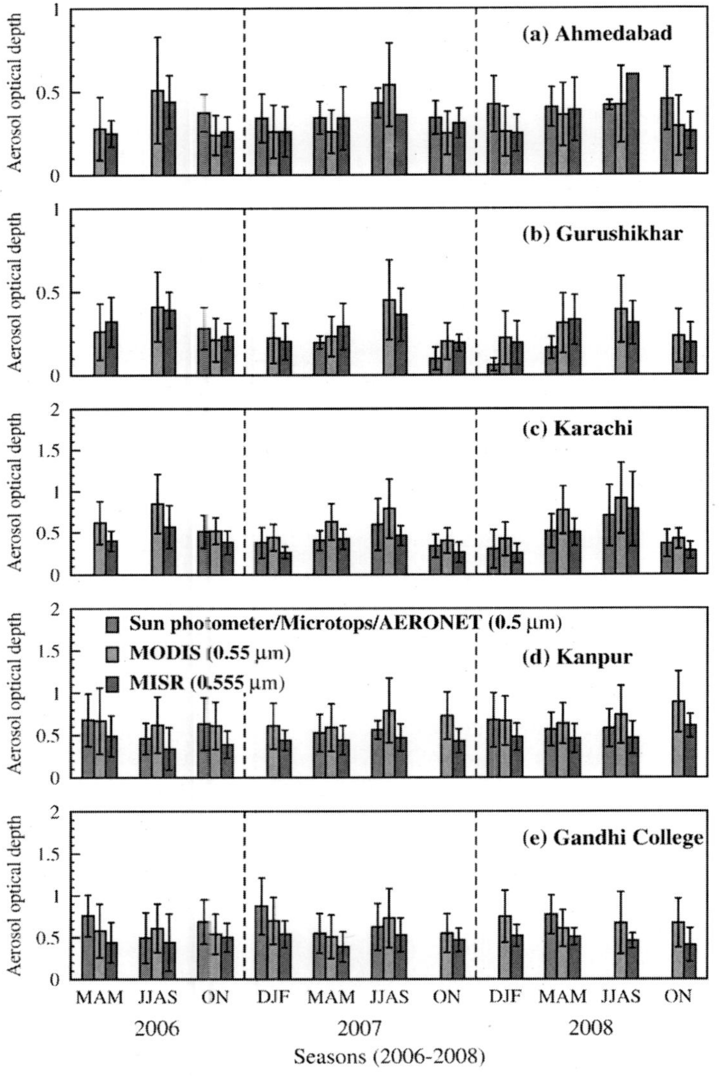

Figure 14. Comparison of seasonal mean aerosol optical depths obtained from ground-based (sun photometer, Microtops and AERONET sun/sky radiometers), and satellites (MODIS and MISR) during 2006–2008 over

(**a**) Ahmedabad; (**b**) Gurushikhar; (**c**) Karachi; (**d**) Kanpur; and (**e**) Gandhi College. Ground based aerosol optical depths correspond to 0.5 µm, while MODIS and MISR retrieved aerosol optical depths are at 0.55 and 0.555 µm respectively. Vertical bars denote $\pm 1\sigma$ deviation from the seasonal mean.

6. SUMMARY AND CONCLUSIONS

Seasonal and inter annual variabilities in aerosol optical depth (AOD) and size distribution are examined over five locations in South Asia during 2006–2008. Ahmedabad, Gurushikhar and Karachi in the West of South Asia are in the semi arid/arid region. Ahmedabad and Karachi are urban and industrialized locations. Karachi in addition is a coastal station. Gurushikhar, a remote, high altitude site is more influenced by long range transport of aerosols. Kanpur and Gandhi College in the Indo-Gangetic basin are urban, industrial city and a rural location respectively. AODs retrieved from hand held sun photometer and Microtops measurements in Ahmedabad and Gurushikhar, and from AERONET sun/sky radiometer measurements over Karachi, Kanpur and Gandhi College are utilized in the study. Ångström wavelength exponent (α) is determined from spectral AODs obtained in the wavelength region of 0.38–0.87 µm over Ahmedabad, Gurushikhar in western India, Karachi, Kanpur and Gandhi College in the Indo-Gangetic Plain. AODs obtained from the ground-based measurements (sun photometers and sun/sky radiometers) have been compared with satellite (MODIS and MISR) retrieved AODs during 2006–2008.

The major findings of the study are:

(1) Highest AODs are observed over Karachi; while the lowest AODs are obtained over Gurushikhar. AODs over Ahmedabad and Karachi are at least a factor of two higher than the AODs measured over Gurushikhar. AODs over the locations (Kanpur and Gandhi College) in the Indo-Gangetic basin are higher than those measured in the western region of South Asia. AODs in short wavelength (0.38 µm) over Gandhi College are about 20% higher than Kanpur; while AODs at 0.87 µm are nearly the same for all the seasons over both the locations. Higher AODs over Gandhi College occur as the location is downwind of major urban centers in the Indo-Gangetic basin, and being a rural location the local production from biomass emissions could contribute additionally to the increase in AODs.

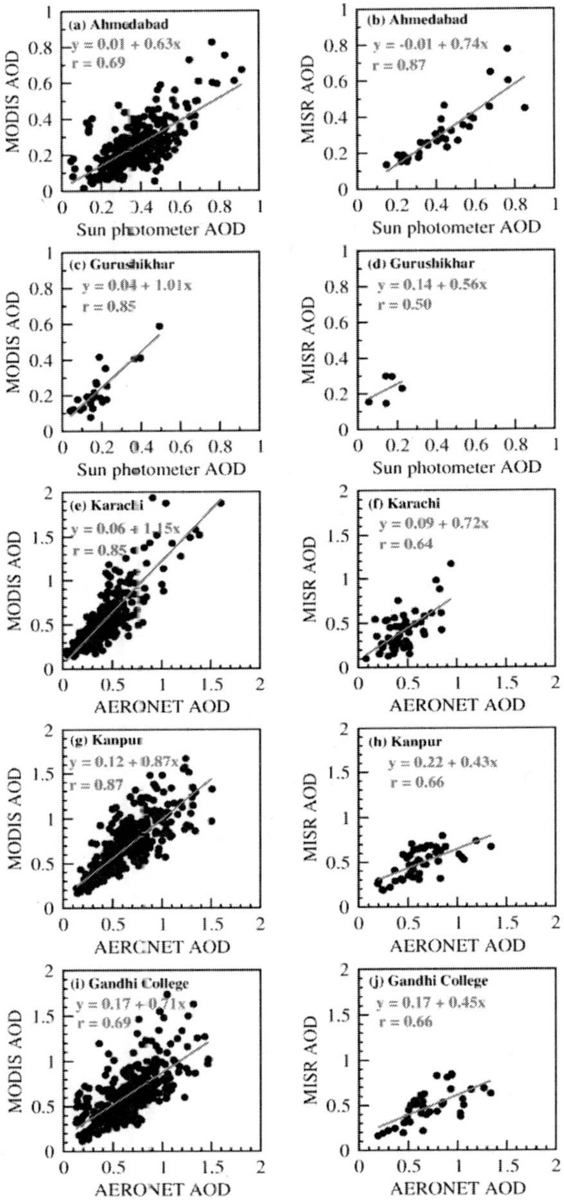

Figure 15. Correlation between ground-based sun photometer, Microtops and AERONET sun/sky radiometers measured AODs and satellite (MODIS and MISR) retrieved AODs during 2006–2008 over Ahmedabad (**a,b**); Gurushikhar (**c,d**); Karachi (**e,f**); Kanpur (**g,h**); and Gandhi College (**i,j**). Ground-based AODs correspond to 0.5 μm, while MODIS and MISR retrieved AODs are at 0.55 and 0.555 μm respectively.

(2) Over the study locations in west of South Asia, an increase in AODs is observed during monsoon because of higher wind speeds, relative humidity, hygroscopic growth of water soluble aerosols, and deeper boundary layer. Winter AODs are higher in the Indo-Gangetic basin, due to shallow boundary layer, increase in fossil fuel and bio fuel emissions which give rise to hazy and foggy conditions, and therefore higher AODs.

(3) AOD ratio, ratio of AODs at 0.38 μm and 0.87 μm, an indicator of whether sub-micron or super-micron aerosols dominate the aerosol size distribution, is found to show seasonal and spatial variations consistent with AOD variations. Postmonsoon and winter AOD ratios in the study locations are higher (≥2) than those obtained during premonsoon and monsoon suggesting the abundance of fine mode aerosols during postmonsoon and winter. Monsoon AOD ratios are the lowest because of the dominance of coarse mode (mainly sea salt) particles.

(4) Ångstr wavelength exponent (α) is found to exhibit large spatial, seasonal, and inter annual variations during 2006–2008. α values during postmonsoon and winter are higher than that of premonsoon and monsoon values in all the locations. α values are lower over Gurushikhar and Karachi indicating the dominance of coarse mode aerosols (dust in Gurushikhar, and sea salt in Karachi). Higher α values over the study locations can be attributed to the dominance of fine mode aerosols over urban locations (e.g., Ahmedabad and Kanpur).

(5) Comparison between ground-based and MODIS (Terra and Aqua) retrieved AODs show that aerosol optical characteristics do not change significantly in an hour and much of the diurnal AOD variation is captured well by the two MODIS instruments. The temporal difference (about an hour) between the ground-based and remote sensing measurements contributes negligibly to the observed differences in AODs. The differences between ground-based and remote sensing AODs vary on spatial scales. During the study period (2006–2008) MODIS underestimates AODs over western India by about 0.04, while over the Indo-Gangetic Plain MODIS overestimates AOD by 0.06 in (Kanpur) and underestimates by 0.07 (Gandhi College) when compared to AODs measured by hand held sun photometer and Microtops sun photometer (Ahmedabad and Gurushikhar), and AERONET sun/sky radiometers (Karachi, Kanpur and Gandhi College) respectively. During the same period MISR underestimates AODs in the range of 0.02–0.17 over Ahmedabad, Gurushikhar, Karachi and Kanpur, while in Gandhi College MISR overestimates AOD by 0.2 when compared to ground-based AODs.

(6) The aerosol characterization and the knowledge obtained on the correlation and the differences on different temporal scales between ground-based and remote sensing AODs over five different environments can serve as inputs to improve the remote sensing retrieval algorithms.

The spatial, seasonal and inter-annual variations in aerosol characteristics over South Asia will be useful in aerosol radiative forcing estimates on regional scale, and regional/global climate impact studies.

ACKNOWLEDGMENTS

MODIS and MISR aerosol optical depth data are downloaded from GES-DISC. Temperature, relative humidity, and winds are obtained from NCEP Reanalysis website. Rainfall data from TRMM are downloaded from GES-DISC. We thank Brent N. Holben, Ramesh P. Singh and Sachchida N. Tripathi for their efforts in establishing and maintaining the AERONET sun/sky radiometer at Karachi, Kanpur and Gandhi College the data of which are used in the present study.

CONFLICTS OF INTEREST

The authors declare no conflict of interest.

REFERENCES

1 Solomon, S.; Qin, D.; Manning, M.; Chen, Z.; Marquis, M.; Averyt, K.B.; Tignor, M.; Miller, H.L. Summary for Policymakers. In Climate Change 2007: The Physical Science Basis. Contribution of Working Group to the Fourth Assessment Report of the Intergovernmental Panel on Climate Change; Cambridge University Press: Cambridge, UK, New York, NY, USA, 2007; pp. 1–18.

2 Remer, L.A.; Kleidman, R.G.; Levy, R.C.; Kaufman, Y.J.; Tanré, D.; Mattoo, S.; Martins, J.V.; Ichoku, C.; Koren, I.; Yu, H.; et al. Global aerosol climatology from the MODIS satellite sensors. J. Geophys. Res. 2008, 113.

3 Kahn, R.A.; Li, W.-H.; Martonchik, J.V.; Bruegge, C.J.; Diner, D.J.; Gaitley, B.J.; Abdou, W. MISR calibration and implications for low-light-level aerosol retrieval over dark water. J. Atmos. Sci. 2005, 62, 1032–1052.

4 Di Girolamo, D.; Bond, T.C.; Bramer, D.; Diner, D.J.; Fettinger, F.; Kahn, R.A.; Martonchik, J.V.; Ramana, M.V.; Ramanathan, V.; Rasch, P.J. Analysis of Multi-angle Imaging SpectroRadiometer (MISR)

aerosol optical depths over greater India during winter 2001–2004. Geophys. Res. Lett. 2004, 31.

5 Levy, R.C.; Remer, L.A.; Kleidman, R.G.; Mattoo, S.; Ichoku, C.; Kahn, R.; Eck, T.F. Global evaluation of the collection 5 MODIS dark-target aerosol products over land. Atmos. Chem. Phys. 2010, 10, 10399–10410.

6 Holben, B.N.; Eck, T.F.; Slutsker, I.; Tanré, D.; Buis, J.P.; Setzer, A.; Vermote, E.; Reagen, J.A.; Kaufman, Y.J.; Nakajima, T.; et al. AERONET—A federated instrument network and data archive for aerosol characterization. Remote Sens. Environ. 1998, 66, 1–16.

7 Dubovik, O.; Holben, B.; Eck, T.F.; Smirnov, A.; Kaufman, Y.J.; King, M.D.; Tanré, D.; Slutsker, I. Variability of absorption and optical properties of key aerosol types observed in worldwide locations. J. Atmos. Sci. 2002, 59, 590–608.

8 Liu, Y.; Sarnat, J.A.; Coull, B.A.; Koutrakis, P.; Jacob, D.J. Validation of Multiangle Imaging Spectroradiometer (MISR) aerosol optical thickness measurements using Aerosol Robotic Network (AERONET) observations over the contiguous United States. J. Geophys. Res. 2004, 109.

9 Martonchik, J.V.; Diner, D.J.; Kahn, R.; Gaitley, B.; Holben, B.N. Comparison of MISR and AERONET aerosol optical depths over desert sites. Geophys. Res. Lett. 2004, 31.

10 Abdou, W.A.; Diner, D.J.; Martonchik, J.V.; Bruegge, C.J.; Kahn, R.A.; Gaitley, B.J.; Crean, K.A.; Remer, L.A.; Holben, B. Comparison of coincident Multiangle Imaging Spectroradiometer and Moderate Resolution Imaging Spectroradiometer aerosol optical depths over land and ocean scenes containing Aerosol Robotic Network sites. J. Geophys. Res. 2005, 110.

11 Kahn, R.A.; Garay, M.J.; Nelson, D.L.; Yau, K.K.; Bull, M.A.; Gaitley, B.J.; Martonchik, J.V.; Levy, R.C. Satellite-derived aerosol optical depth over dark water from MISR and MODIS: Comparisons with AERONET and implications for climatological studies. J. Geophys. Res. 2007, 112.

12 Li, Z.; Niu, F.; Lee, K.-H.; Xin, J.; Hao, W.-M.; Nordgren, B.; Wang, Y.; Wang, P. Validation and understanding of Moderate Resolution Imaging Spectroradiometer aerosol products (C5) using ground-based measurements from the handheld Sun photometer network in China. J. Geophys. Res. 2007, 112.

13 Xiao, N.; Shi, T.; Calder, C.A.; Munroe, D.K.; Berrett, C.; Wolfinbarger, S.; Li, D. Spatial characteristics of the difference between MISR and MODIS aerosol optical depth retrievals over mainland Southeast Asia. Remote Sens. Environ. 2009, 113, 1–9.

14 Kahn, R.A.; Gaitley, B.J.; Garay, M.J.; Diner, D.J.; Eck, T.F.; Smirnov, A.; Holben, B.N. Multiangle Imaging SpectroRadiometer global aerosol product assessment by comparison with Aerosol Robotic Network. J. Geophys. Res.2010, 115.

15 Lau, K.-M.; Ramanathan, V.; Wu, G.-X.; Li, Z.; Tsay, S.C.; Hsu, C.; Sikka, D.R.; Holben, B.; Lu, D.; Tartari, G.; et al. The joint aerosol-monsoon experiment. Bull. Am. Meteorol. Soc. 2008, 89, 369–383.

16 Ramachandran, S.; Srivastava, R.; Kedia, S.; Rajesh, T.A. Contribution of natural and anthropogenic aerosols to optical properties and radiative effects over an urban location. Environ. Res. Lett. 2012, 7, 034028.

17 Dey, S.; di Girolamo, L. A climatology of aerosol optical and microphysical properties over the Indian subcontinent from 9 years (2000–2008) of Multiangle Imaging Spectroradiometer (MISR) data. J. Geophys. Res. 2010, 115.

18 Dey, S.; di Girolamo, L. A decade of change in aerosol properties over the Indian subcontinent. Geophys. Res. Lett.2011, 38.

19 Tripathi, S.N.; Dey, S.; Chandel, A.; Srivastava, S.; Singh, R.P.; Holbeni, B.N. Comparison of MODIS and AERONET derived aerosol optical depth over the Ganga Basin, India. Ann. Geophys. 2005, 23, 1093–1101.

20 Prasad, A.K.; Singh, R.P. Comparison of MISR-MODIS aerosol optical depth over the Indo-Gangetic basin during the winter and summer seasons (2000–2005). Remote Sens. Environ. 2007, 107, 109–119.

21 Ramachandran, S. Aerosol optical depth and fine mode fraction variations deduced from Moderate Resolution Imaging Spectroradiometer (MODIS) over four urban areas in India. J. Geophys. Res. 2007, 112.

22 Misra, A.; Jayaraman, A.; Ganguly, D. Validation of MODIS derived aerosol optical depth over Western India. J. Geophys. Res. 2008, 113.

23 Choudhry, P.; Misra, A.; Tripathi, S.N. Study of MODIS derived AOD at three different locations in the Indo Gangetic Plain: Kanpur, Gandhi College and Nainital. Ann. Geophys. 2012, 30, 1479–1493.

24 Ramachandran, S.; Rajesh, T.A. Black carbon aerosol mass concentrations over Ahmedabad, an urban location in western India:

Comparison with urban sites in Asia, Europe, Canada, and the United States. J. Geophys. Res. 2007, 112.

25 Parekh, P. P.; Khwaja, H.A.; Khan, A.R.; Naqvi, R.R.; Malik, A.; Shah, S.A.; Khan, K.; Hussain, G. Ambient air quality of two metropolitan cities of Pakistan and its health implications. Atmos. Environ. 2001, 35, 5971–5978.

26 Dutkiewicz, V.A.; Alvi, S.; Ghauri, B.M.; Choudhary, M.I.; Husain, L. Black carbon aerosols in urban air in South Asia.Atmos. Environ. 2009, 43, 1737–1744.

27 Singh, R.P.; Dey, S.; Tripathi, S.N.; Tare, V.; Holben, B. Variability of aerosol parameters over Kanpur, northern India.J. Geophys. Res. 2004, 109.

28 Kedia, S.; Ramachandran, S. Latitudinal and longitudinal variation in aerosol characteristics from Sun photometer and MODIS over the Bay of Bengal and Arabian Sea during ICARB. J. Earth Syst. Sci. 2008, 117, 375–387.

29 Morys, M.; Mims, F.M., III; Hagerup, S.; Anderson, S.E.; Baker, A.; Kia, J.; Walkup, T. Design calibration, and performance of MICROTOPS II handheld ozone monitor and sun photometer. J. Geophys. Res. 2001, 106, 14573–14582.

30 Dubovik, O.; King, M.D. A flexible inversion algorithm for retrieval of aerosol optical properties from Sun and sky radiance measurements. J. Geophys. Res. 2000, 105, 20673–20696.

31 Eck, T.F.; Holben, B.N.; Reid, J.S.; Dubovik, O.; Smirnov, A.; O'Neill, N.T.; Slutsker, I.; Kinne, S. Wavelength dependence of the optical depth of biomass burning, urban, and desert dust aerosols. J. Geophys. Res. 1999, 104, 31333–31349.

32 Smirnov, A.; Holben, B.N.; Sakerin, S.M.; Kabanov, D.M.; Slutsker, I.; Chin, M.; Diehl, T.L.; Remer, L.A.; Kahn, R.; Ignatov, A.; et al. Ship-based aerosol optical depth measurements in the Atlantic Ocean: Comparison with satellite retrievals and GOCART model. Geophys. Res. Lett. 2006, 33.

33 Yu, H.; Dickinson, R.E.; Chin, M.; Kaufman, Y.J.; Zhou, M.; Zhou, L.; Tian, Y.; Dubovik, O.; Holben, B.N. Direct radiative effect of aerosols as determined from a combination of MODIS retrievals and GOCART simulations. J. Geophys. Res. 2004, 109.

34 Diner, D.J.; Abdou, W.A.; Bruegge, C.J.; Conel, J.E.; Crean, K.A.; Gaitley, B.J.; Helmlinger, M.C.; Kahn, R.A.; Martonchik, J.V.; Pilorz, S.H.; et al. MISR aerosol optical depth retrievals over southern Africa

during the SAFARI-2000 winter season campaign. Geophys. Res. Lett. 2001, 28.

35 Eck, T.F.; Holben, B.N.; Dubovik, O.; Smirnov, A.; Slutsker, I.; Lobert, J.M.; Ramanathan, V. Column-integrated aerosol optical properties over the Maldives during the northeast monsoon for 1998–2000. J. Geophys. Res. 2001, 106, 28555–28566.

36 Prospero, J.M.; Ginoux, P.; Torres, O.; Nicholson, S.E.; Gill, T.E. Environmental characterization of global sources of atmospheric soil dust identified with the Nimbus 7 Total Ozone Mapping Spectrometer (TOMS) absorbing aerosol index. Rev. Geophys. 2002, 40, 1002.

37 Srivastava, R.; Ramachandran, S.; Rajesh, T.A.; Kedia, S. Aerosol radiative forcing deduced from observations and models over an urban location and sensitivity to single scattering albedo. Atmos. Environ. 2011, 45, 6163–6171.

38 Levy, R.C.; Remer, L.A.; Dubovik, O. Global aerosol optical properties and application to Moderate Resolution Imaging Spectroradiometer aerosol retrieval over land. J. Geophys. Res. 2007, 112.

39 Kahn, R.A.; Nelson, D.L.; Garay, M.; Levy, R.C.; Bull, M.A.; Martonchik, J.V.; Diner, D.J.; Paradise, S.R.; Hansen, E.G.; Remer, L.A. MISR aerosol product attributes, and statistical comparisons with MODIS. IEEE Trans. Geosci. Remote Sens. 2009, 47, 4095–4114.

40 Kalashnikova, O.V.; Kahn, R. Ability of multiangle remote sensing observations to identify and distinguish mineral dust types: Part 2. Sensitivity over dark water. J. Geophys. Res. 2006, 111.

CHAPTER 8

Impact of Coupled NO_x/Aerosol Aircraft Emissions on Ozone Photochemistry and Radiative Forcing

Giovanni Pitari [1,*], Daniela Iachetti [1], Glauco Di Genova [1], Natalia De Luca [1], Ole Amund Søvde [2], Øivind Hodnebrog [2], David S. Lee [3] and Ling L. Lim [3]

[1] Department of Physical and Chemical Sciences, Università dell'Aquila, Dipartimento di Fisica, Via Vetoio, 67100 Coppito, L'Aquila, Italy

[2] Center for International Climate and Environmental Research—Oslo (CICERO), 0318 Oslo, Norway

[3] Dalton Research Institute, Manchester Metropolitan University (MMU), Manchester M1 5GD UK

ABSTRACT

Three global chemistry-transport models (CTM) are used to quantify the radiative forcing (RF) from aviation NO_x emissions, and the resultant reductions in RF from coupling NO_x to aerosols via heterogeneous chemistry. One of the models calculates the changes due to aviation black carbon (BC) and sulphate aerosols and their direct RF, as well as the BC indirect effect on cirrus cloudiness. The surface area density of sulphate aerosols is then passed to the other models to compare the resulting photochemical perturbations on NO_x through heterogeneous chemical reactions. The perturbation on O_3 and CH_4 (via OH) is finally evaluated, considering both short- and long-term O_3 responses. Ozone RF is calculated using the monthly averaged output of the three CTMs in two independent radiative transfer codes. According to the models, column ozone and CH_4 lifetime changes due to coupled NO_x/aerosol emissions are, on average, +0.56 Dobson Units (DU) and −1.1 months, respectively, for atmospheric conditions and aviation emissions representative of the year 2006, with an RF of +16.4 and −10.2 mW/m^2 for O_3 and CH_4, respectively. Sulphate aerosol induced changes on ozone column and CH_4 lifetime account for −0.028 DU and +0.04 months, respectively, with corresponding RFs of −0.63 and +0.36 mW/m^2. Soot-cirrus forcing is calculated to be 4.9 mW/m^2.

KEYWORDS

chemistry-transport models; aviation emissions; ozone photochemistry; sulphate and black carbon aerosols; soot-cirrus particles; radiative forcing from aviation NO_x and aerosols

1. INTRODUCTION

Aviation alters the composition of the atmosphere globally and can thus drive climate and ozone changes [1,2]. Aircraft emissions may impact the atmospheric composition both directly (mainly via emissions of CO_2, H_2O, NO_x, SO_2, soot) and indirectly (by increasing tropospheric O_3 and OH and decreasing the CH_4 lifetime; with formation of contrails and with indirect effects on upper tropospheric cirrus cloudiness, *i.e.*, contrail-cirrus and soot-cirrus). The majority of the emitted species produced by combustion of kerosene are gases and soot particles. The former may be directly relevant as greenhouse gases (CO_2 and stratospheric H_2O), while the latter may trigger cloud formation. In addition, the combustion process forms numerous trace species, such as NO, NO_2, SO_2, CO and hydrocarbons (HCs). NO_x emissions, in particular, play a key role in tropospheric and lower stratospheric chemistry, by enhancing the ozone production and the OH concentration [3,4]. The chemical reaction with OH, in turn, acts as the main sink for atmospheric CH_4, so that NO_x emission by the aircraft will decrease its lifetime. The lowering of

CH_4 atmospheric abundance induces a cooling that may partially counteract the warming due to the other greenhouse gases (CO_2, stratospheric H_2O, tropospheric O_3), as well as the warming due to upper tropospheric particle formation (contrails and aviation cirrus cloudiness) [5]. In addition, the aircraft impact on atmosphere and climate may act on very different time scales, inducing long-term responses with CO_2accumulation and CH_4 lifetime changes or producing short-term climate responses with additional O_3 production and by increasing aerosols and cloud particles. The aircraft impact on atmospheric ozone is particularly complex, because a superposition of short- and long-term effects takes place, via direct NO_x emissions (short-term) and via OH driven CH_4changes that can feedback, in turn, on HO_x chemistry and finally on ozone (long-term) [5]. The dilution of aircraft plumes increases the complexity of this problem. Kraabøl *et al.* [6] found ~20% reduction in O_3-changes caused by aircraft NO_x; Cariolle *et al.* [7] found the reduction to be in the range of 10%–25%. However, a possible impact on CH_4 is uncertain.

Most of the aircraft emitted $SO_2$is localized in the upper troposphere where is oxidized via OH radicals producing SO_3, which via reaction with water vapour, is rapidly converted to gaseous H_2SO_4; where gaseous H_2SO_4 represents an important aerosol precursor, and through heterogeneous reactions has an impact on ozone chemistry [8,9,10]. Among the emitted gases and gaseous products formed in the very young plume are condensable gases. These may undergo condensation in the young plume leading to volatile aerosol particles and volatile coating of soot. Small soot particles may combine with larger particles, and when large enough, they may act as condensation nuclei, which potentially play an important role in changing the occurrence of natural cirrus and its properties [11]. The aviation impact on upper tropospheric cirrus cloudiness may take place as line-shaped contrails that persist and spread into cirrus-like clouds. (*i.e.*, contrail-cirrus) or with heterogeneous freezing of adiabatically ascending water vapour on soot aerosol particles (*i.e.*, soot-cirrus, formed in the absence of spreading contrails). Aircraft soot emissions can possibly affect natural cloudiness by slightly reducing water vapour available for the formation of natural clouds. Burkhardt and Kärcher [12] have found that the change in the water budget due to contrail-cirrus formation could reduce natural cirrus by about 1%–2% in some regions at the main cruise altitudes.

The most important particles produced by aircraft emissions that affect atmospheric chemistry are those of sulphuric acid (H_2SO_4). They act through heterogeneous chemical reactions where NO_xconversion into nitric acid may take place on the surface of these aerosols: This reduces O_3 production via NO_x reactions with HO_2 and organic radicals RO_2. Because of the very small size of aircraft-generated particles and since emissions occur mostly at mid-latitudes (in the Northern Hemisphere), the relevant heterogeneous reactions are essentially hydrolysis of N_2O_5 and $BrONO_2$ [13]; their rates are treated following JPL recommendations [14]. The main effect of these two reactions is twofold: (a) NO_x enhancements from NO_x aircraft emissions are partly counter-balanced by additional heterogeneous conversion of NO_x into HNO_3 on these particles in the flight corridors, with decreased O_3 production caused by peroxy radicals and sunlight; (b) this reduced NO_xenhancement tends to increase the amount of reactive Cl and Br in the high latitude lowermost stratosphere, since less Cl/Br can be stored into Cl and Br nitrates, which enhances O_3 depletion via catalytic Cl and Br cycles.

Aircraft emissions may perturb the global amount and the size distribution of H_2SO_4 aerosols in two ways: (a) the direct emission of ultrafine particles (radius~5 nm) [15,16,17] and (b) the release of gas phase SO_2 [8]. Direct particle emissions are estimated to be 4%–15% of the overall aircraft emitted sulphur [16,18]. These direct particle emissions do not significantly change aerosol mass and extinction but they may substantially increase the sulphate aerosol surface area density (SAD) in the Northern Hemisphere upper troposphere and lower stratosphere (UTLS). Release of gas phase SO_2, on the other hand, may increase the net production of H_2SO_4, thus enhancing the sulphate mass in the accumulation mode and consequently, the direct radiative forcing (RF). It also may increase the gas phase contribution to SAD, in the range of 25% of the change produced by direct plume particle emission [8,9].

In this work, we focus on the impact of aircraft emitted NO_x on tropospheric and lower stratospheric O_3 formation and loss, also taking into account the effects that emitted sulphur has on UTLS NO_x, via heterogeneous reactions on sulphate aerosol SAD. The tropopause radiative forcing associated with ozone changes is also evaluated using monthly averaged three-dimensional output from three independent global models (ULAQ-CTM, Oslo CTM2, Oslo CTM3), feeding two well-tested and documented off-line radiative transfer models (ULAQ, Oslo). The RF due to sulphate and black carbon (BC) soot particles is evaluated in the ULAQ-CTM, taking into account both the direct forcing due to radiation scattering/absorption and also the indirect effects due to formation of soot-cirrus ice particles. First, we use the University of L'Aquila model ULAQ-CTM, with fully coupled aerosols and chemistry, to assess this impact. Then, the Oslo models CTM3 and CTM2 were also applied to robustly test the chemical impact of aerosols. As these models do not include the required coupled aerosol-chemistry, their heterogeneous reactions were described by monthly averages of aerosol SAD from the ULAQ-CTM. These models are briefly described in this paper, as well as the ULAQ-CTM results in terms of background atmospheric aerosols and their perturbations due to aviation emissions. Finally, a detailed discussion of the radiative calculations on the NO_x/aerosol coupled impact on ozone is presented.

The multi-model evaluation of the short-term ozone RF along with the estimated BC indirect effect on cirrus cloudiness (soot-cirrus) are two important advances in the present study with respect to Pitari *et al.* [9]. Soot-cirrus is integrated in the representation of ice clouds in the ULAQ-CTM, discriminating between natural cirrus cloudiness formed through homogeneous freezing of adiabatically ascending water vapour [19] and ice particles formed via heterogeneous freezing on non-hydrophobic BC particles [11]. Gettelman and Chen [20] have documented modeling results on direct and indirect effects of aviation aerosols on climate (BC, SO_4 and induced cloudiness). Another new insight from the present study is the evaluation of the long-term O_3 negative RF associated with CH_4 lifetime changes produced by the aviation induced tropospheric OH perturbation. In this case, we have followed an innovative approach, using not only IPCC-type parametric formulas to calculate long-term negative RFs associated with the CH_4 RF (*i.e.*, tropospheric O_3 and stratospheric H_2O), but designed an experiment (ULAQ_FLX) where surface CH_4 is calculated using a flux boundary conditions. In this way, the calculated OH change due to aircraft NO_x emissions produces a CH_4 lifetime change that may directly migrate onto the CH_4mixing ratio distribution. Therefore, the calculated O_3 RF results to be the sum of both the short- and long-term responses and the effects on stratospheric H_2O are explicitly calculated. Another recently published modeling work that has focused on NO_x/aerosol interaction is Unger [21]. The general effects of aviation NO_x have been widely discussed in the modeling community, e.g., by Köhler *et al.* [3], Hoor *et al.* [4], Grewe *et al.* [22], Holmes *et al.* [5], Gottschaldt *et al.* [23], with insights in many relevant aspects of the problem (photochemistry, sensitivity approach, climate, uncertainties).

2. EXPERIMENTAL SECTION

A short description of the chemistry-transport models used in this study and their basic setups are presented in the following subsections, along with a description of the adopted aviation emission inventories and an overall presentation of the numerical experiment setup. Details on the model adopted surface emissions of NO_x, CO and VOC are reported in Søvde *et al.* [24].

2.1. ULAQ-CTM

The University of L'Aquila model is a global scale climate-chemistry coupled model (ULAQ-CCM) extending from the surface to the mesosphere (0.04 hPa). In this study, the ULAQ-CCM was operated in

CTM mode. Dynamical data, *i.e.*, velocity stream-function and velocity potential, are provided by the background GCM run in a reference case, with no feedbacks of aviation-induced changes. The ULAQ-CCM has been fully described in Pitari *et al.* [25] and also in Eyring *et al.*[26] and Morgenstern *et al.* [27] for the SPARC (*i.e.*, Stratospheric Processes And their Role in Climate) climate-chemistry model validation (CCMVal) campaign. The most recent version of the ULAQ model has been fully documented in Pitari *et al.* [28], with some important updates with respect to the SPARC-CCMVal exercise. These are: (a) increase of horizontal and vertical resolution, now T21 with 126 log-pressure levels (with approximate pressure altitude increment of 568 m); (b) inclusion of a parameterization for the formation of upper tropospheric cirrus cloud ice particles [19]; (c) update to Sander *et al.* [14] recommendations for reaction rates of homogeneous and heterogeneous chemistry, cross sections of species, and the parameterization of Minschwaner *et al.* [29] for the Schumann-Runge bands, based on fixed-temperature opacity distribution function formulation; (d) new radiative transfer code for calculations of photolysis, solar heating rates and top-of-atmosphere radiative forcing. The oceanic surface temperature is assimilated from the Hadley Centre for Climatic Prediction and Research [30]. A parameterization of periodic natural forcings (solar cycle, quasi-biennial oscillation) is included on-line.

The chemistry module is organized by long-lived and surface-flux species (CH_4, N_2O, CFCs, HCFCs, CO, NO_x, NMVOC (*i.e.*, non-methane volatile organic compounds)) and by all medium and short-lived species grouped in the families O_x, NO_y, HO_x, CHO_x, Cl_y, Br_y, SO_x, and aerosols. In total, there are 40 transported species, 26 species at photochemical equilibrium, and 57 size categories for the aerosols. The photochemistry module is organized with 140 gas-phase reactions, 45 photolysis reactions and 30 heterogeneous reactions on the surface of sulphate aerosols, nitric acid trihydrate and ice particles [25,31] or water droplets. All reaction and photolysis rates are treated according to JPL recommendations [14]. In the specific case of N_2O_5 hydrolysis on sulphate aerosols, the ULAQ model includes an explicit dependence from aerosol acidity (which in turn depends on temperature and pressure); gaseous HNO_3 is released from this reaction, both in the stratosphere and troposphere. An evaluation example of the ULAQ model tropospheric chemistry is given in Brunner *et al.* [32]. The model includes the major components of stratospheric and tropospheric aerosols (sulphate, carbonaceous, soil dust, sea salt, PSCs). Nitrate aerosols (*i.e.*, ammonium nitrate) are not explicitly considered in the ULAQ aerosol module. This is not expected to be a significant sink term for HNO_3 from aircraft NO_x, since NH_3 originates from ground sources, is highly reactive and easily removed by wet and dry deposition and not present in significant concentrations at cruise altitudes. The updated radiative transfer module operating on-line in the

ULAQ-CCM, is a two-stream delta-Eddington approximation model for the calculation of photolysis rates in the ultra-violet (UV) to visible (VIS) wavelengths and for solar heating rates and radiative forcing in UV-VIS and solar near-infrared (NIR) bands. In addition, a companion broadband, k-distribution longwave radiative module is used to compute radiative transfer and heating rates in the planetary infrared spectrum [33].

2.2. Oslo CTM3

The Oslo CTM3 [34] is a three dimensional off-line CTM. Its predecessor, Oslo CTM2, will be described in the next section. CTM3 includes improvements in the following parameterizations: a new version of transport scheme parameterization, wet scavenging processes, lightning parameterization and calculations of photodissociation rates. These improvements are all described by Søvde *et al.* [34]. The model domain spans 60 layers between the surface and 0.1 hPa, with a horizontal resolution of T42 (approximately 2.8 × 2.8 degrees). CTM3 is driven by 3-hour forecasts generated by the Integrated Forecast System of the European Centre for Medium-Range Weather Forecasts (ECMWF), cycle 36r1. Chemical species are advected using the improved second order moments scheme [34,35], where horizontal winds are from ECMWF and vertical winds are calculated from the continuity equation.

Convective transport of tracers is based on the convective upward flux from the ECMWF model, and this allows entrainment or detrainment of tracers to occur when the upward flux changes. The entrainment and detrainment processes are improved by balancing the detrainment rates from the meteorological data with the upward flux. This allows the mixing with ambient air to increase or decrease as the plume rises. Turbulent mixing in the boundary layer is treated according to the Holtslag K-profile scheme [36]. Oslo CTM3 comprises several chemical and aerosol modules, namely the tropospheric chemistry, which accounts for the most important parts of the O_3-NO_x-hydrocarbon chemistry cycle [37]; the tropospheric sulphur cycle [38], and also comprehensive stratospheric chemistry, as documented by Søvde *et al.* [39]. CTM3 also includes tropospheric sea salt aerosols and nitrate aerosols. The N_2O_5 hydrolysis on the surface of sulphate aerosols uses a constant uptake coefficient of 0.1, which is within JPL recommendations, but currently the model does not include weighting factors due to H_2SO_4. Gaseous HNO_3 is released from this reaction, both in the stratosphere and troposphere. The quasi steady state approximation (QSSA) chemistry solver [40] is used for the chemistry calculations.

2.3. Oslo CTM2

The CTM3 predecessor, the Oslo CTM2 [39], has been widely used in previous studies [10,37,38,41]. The model resolution is the same as for CTM3, as are the meteorological data used to drive the model. The advection of chemical species in CTM2 is calculated by the older second-order moment method [42], and while convective transport is similar to CTM3, it does not use detrainment rates from meteorological data, thus resulting in a more effective transport to higher altitudes [34]. As in CTM3, the Holtslag K-profile scheme [36] is used for turbulent mixing. The chemistry schemes in Oslo CTM2 are the same as Oslo CTM3 and this includes tropospheric and stratospheric chemistry and the tropospheric sulphur scheme. However, in this study, due to computational constraints, CTM2 does not include tropospheric nitrate aerosols and sea salt aerosols. The hydrolysis of N_2O_5 on sulphate aerosols is treated as in CTM3. Emissions in CTM2 are the same as in CTM3, except for NO_x emissions from lightning. In CTM2 lightning NO_x emissions are coupled on-line to the convection in the model using the Price *et al.* [43] parameterization, along with their seasonal variation described by Berntsen and Isaksen [44]. In our study, we have updated the vertical lightning NO_x distribution to follow Søvde *et al.* [34], using vertical profiles from Ott *et al.* [45] instead of Pickering *et al.* [46], and scaling the profiles to extend from the surface and up to the top of convection. This vertical distribution allows no lightning emissions above the top of a convective plume, and is the same distribution as that used in Oslo CTM3. Søvde *et al.* [24] showed that compared with the old vertical distribution, the new vertical distribution of lightning NO_x emissions reduces NO_x in the UTLS and this resulted in a larger effect of aircraft emissions in this region.

2.4. Emission Inventories

In REACT4C, aircraft emissions inventories have been produced using FAST [47,48,49]. FAST is an emissions model approved for use by the International Civil Aviation Organization (ICAO)'s Committee on Aviation Environmental Protection (CAEP) Modelling and Database Group (MDG) [50]. The base case aircraft emissions for the year 2006 were calculated using flight data from the CAEP Round 8 MDG work programme. The flight data used were made up of radar data from North American and European airspace, which accounts for ~80% of global civil aviation traffic; and the Official Airline Guide (OAG) schedule data for the rest of the world. Great circle trajectories were assumed for all routes, with corrections applied to the distance and fuel consumption. The corrections were based on an empirical formula derived from the CAEP

Round 8 Goals modelling work, which estimated the relationship between great circle assumptions and actual distance and fuel burn from global civil aviation [51]. In FAST, the 2006 global fleet were represented by 42 generic aircraft types with the fuel flow estimated using the PIANO aircraft performance model [52]. These fuel flow data were used by FAST as the basis for calculating NO_x emissions that relate sea-level emissions from NO_x certification data to altitude [47]. The relationships between fuel, soot emissions and altitude were derived from the Aero2k project [53]. The FAST gridding utility then produced annual emissions inventory of fuel use, NO_x and soot (see Table 1) on a 1 × 1 degree horizontal spacing and 2000 ft vertical spacing, which corresponds to flight intervals. Table 1 shows FAST results compared with another ICAO approved model, AEDT [54]. This highlights the possible emissions variation that could arise from different inventories and therefore, how this may also influence modelling results from other studies.

Table 1. Total aviation emissions considered in this study from FAST, compared with another emissions inventory (International Civil Aviation Organization, ICAO) approved model, AEDT.

Field	FAST	AEDT
FUEL (Tg/year)	178	188
NO_x (Tg-NO_2/year)	2.33	2.67
EI-NO_x (g-NO_2/kg-fuel)	13.1	14.1
SOOT (Tg-BC/year)	4.07×10^{-3}	6.81×10^{-3}
EI-BC (g-BC/kg-fuel)	0.0229	0.0362
EI-SO_2 (g-SO_2/kg-fuel)	0.8	0.85

2.5. Numerical Experiment Setup

Three simulations were performed: NO-AIRCRAFT (NA, simulation with no aircraft emissions), base case AIRCRAFT EMISSION (AE: simulation with only aircraft gaseous emissions) and base case AIRCRAFT EMISSION* (AE*: simulation with both gas and particle emissions from aircraft). It should be noted that our perturbation study may also cover non-linear effects of chemistry, which depend on the background of NO_x and hydrocarbons. Non-linear effects are generally small for small NO_x changes [3]. While the AE-NA perturbation may comprise some non-linear effects, our main focus is the impact of including particle emissions. As already noted, these provide surfaces for heterogeneous chemical reactions, converting NO_x to HNO_3. As the NO_x emissions are the same, the reduction in NO_x impact from aircraft and the accompanying non-linear effects should be small. By doing a larger set of small perturbation runs, the presence of non-linearity in chemistry may be revealed, but to further quantify these effects a tagging approach [22] is probably the best tool, which is beyond the scope of our study.

Each ULAQ-CTM experiment was run for 15 years (1996–2010), where the first five years were considered spin-up, thus, allowing the aircraft perturbation to reach a satisfactory steady-state. The CTM3 and CTM2 were spun-up without aircraft emissions for five years, using the ULAQ-CTM monthly mean background SAD (*i.e.*, produced from the experiment NA) for heterogeneous reactions on aerosols, instead of their usual background SAD based on observations. After spin-up, these CTMs were then run for an additional six years for each of the NA, AE and AE* experiments using ULAQ-CTM monthly SAD values from NA, AE and AE* cases, respectively. All annual mean values refer to an average over the modelling years 2001–2010 (ULAQ) or 2001–2006 (CTM2 and CTM3), unless specified otherwise. Surface mixing ratios of long-lived species (e.g., CH_4 and N_2O) as well as surface emissions of ozone precursors (e.g., NO_x, CO, VOCs) and aircraft emissions are kept fixed at year 2006 values. Only the meteorology is time-dependent, ensuring inter-annual variability. For further details see Søvde *et al.* [24]. The length of these time-integrations has allowed all models to reach a satisfactory steady-state condition.

3. RESULTS AND DISCUSSION

Results for the numerical simulations described in Section 2 are discussed in the following subsections. Section 3.1 will evaluate aerosol results from the ULAQ model, while the aerosol perturbations and resulting impact on ozone photochemistry due to NO_x/aerosol coupled emissions from global aviation will be discussed in Section 3.2 and Section 3.3, respectively. Section 3.4 will provide the radiative forcing estimates due to ozone and CH_4 changes, and also from other species that may play a role in the UTLS region (*i.e.*, aerosols, ice particles and stratospheric water vapour).

3.1. Evaluation of Aerosol Model Results

Sulphate aerosols dominate the aerosol population and mass distribution in the mid-upper troposphere, *i.e.*, the region directly perturbed by aviation emissions. The particle size distribution is determined by homogeneous and heterogeneous nucleation, condensation/evaporation, water vapour growth, coagulation, dry and wet deposition, sedimentation and large-scale transport [25]. The main removal processes of tropospheric aerosol particles as parameterized in the ULAQ-CTM are: washout, which is calculated as a function of the precipitation rate (wet deposition); scavenging by direct and turbulent deposition to the surface (dry deposition) and loss caused by gravitational settling (sedimentation). Aerosol particles are distributed in size bins and each size category is transported separately. Mechanically generated particles (*i.e.*, soil dust and sea salt) and carbonaceous aerosols (*i.e.*, black and organic carbon) are treated with size-binned emissions [55]. Carbonaceous aerosols are treated separately for the organic, soluble fraction (OC), which is the purely scattering component; and the hydrophobic black carbon fraction (BC), for which an ageing time of 1 day is assumed [56] (although this time may vary significantly, depending upon co-emitted species). The BC fraction has a single scattering albedo that is significantly lower than unity (about 0.4–0.5 on average) and thus is very efficient in absorbing the incoming solar radiation [57,58].

The ULAQ-CTM aerosol predictions have been evaluated as follows: (a) SAGE-II data are used to evaluate the calculated extinction vertical profiles at 0.55 μm and 1.020 μm (Figure 1). (b) Vertical profiles of BC mass mixing ratios are compared with recent aircraft measurements at tropical, mid-latitude and at high latitude locations over North America (Figure 2). (c) A comparison between measured and calculated total optical depth is made for the North Atlantic region. In this case, remote sensing measurements from satellites (AVHRR, MODIS and TOMS retrievals) are used (Figure 3). (d) The calculated aerosol optical depth at 0.55 μm, integrated between 5.5 km and 20 km altitude, is compared to SAGE-II derived values (Figure 4).

Aerosol extinction profiles at $\lambda = 0.55$ μm, for altitudes between 5 km and 20 km and 45° N latitude (Figure 1, upper panels) have been measured by the SAGE II satellite instrument. The use of these data has provided information on the global distribution of several aerosol properties [59]. The ratios for aerosol extinction profiles at $\lambda = 1.020$ μm have also been derived (Figure 1, bottom panels). The model performs quite well with respect to the observed vertical extinction profiles. This is an indication of a realistic representation of emissions, wet and dry removal, microphysical processes for sulphate aerosols, particle transport and the suitability of the refractive indices. In particular, the calculated extinction ratios showed that a realistic aerosol size distribution has been applied in the model.

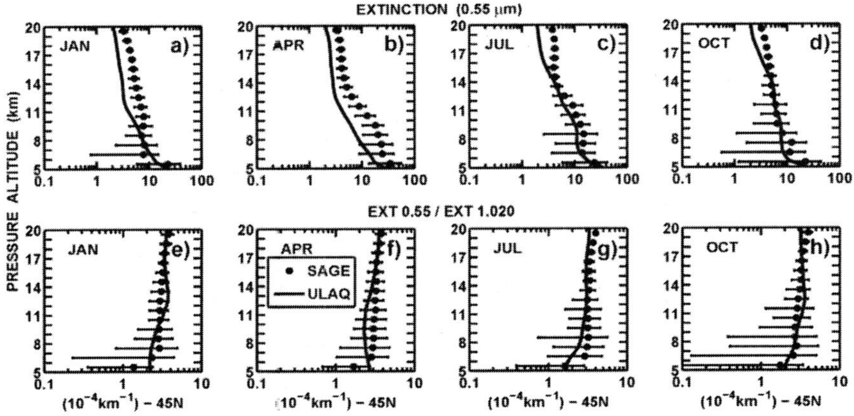

Figure 1. Extinction profiles at $\lambda = 0.55$ µm (upper panels) and extinction ratio between wavelength channels at $\lambda = 0.55$ µm and $\lambda = 1.020$ µm (bottom panels): ULAQ calculations (solid line) and SAGE-II derived values (asterisks) are shown at 45N for January, April, July and October, from left to right panels, respectively (average 1996–2000).

Next, to get a qualitative sense of how the ULAQ-CTM performs in the mid-upper troposphere, we consider the BC model profiles in the vicinity of recent aircraft measurements (see Figure 2). The measurements were made with Single Particle Soot Absorption Photometers (SP2s) onboard of NOAA research aircrafts at tropical, mid-latitudes and high-latitudes over North America [60,61]. We also present a comparison of model calculated total aerosol mass mixing ratios (sulphate, carbonaceous and mechanically generated particles) to airborne observations during the AVE-Houston campaigns (Figure 2e). Further details of the campaigns are provided in Table 2. The lower-latitude campaign observations indicated polluted boundary layers with BC concentrations decreasing by 1–3 orders of magnitude between the surface and the mid-upper troposphere. Some of the large data values can be explained by the sampling of highly polluted conditions, due to heavy biomass burning. The ULAQ model used climatological biomass burning emission scenarios and this did not include specific fire conditions. Nevertheless, it did show consistent mid-tropospheric mean BC levels in the mid-latitudes and in the tropics. The springtime Arctic campaign observed maximum BC above the surface, which may occur from background "Arctic haze" pollution that could have originated from lower latitudes, and is transported to the Arctic by meridionally lofting along isentropic surfaces [56]. Results from the ULAQ model reflects these conditions, although with smaller mixing ratios values.

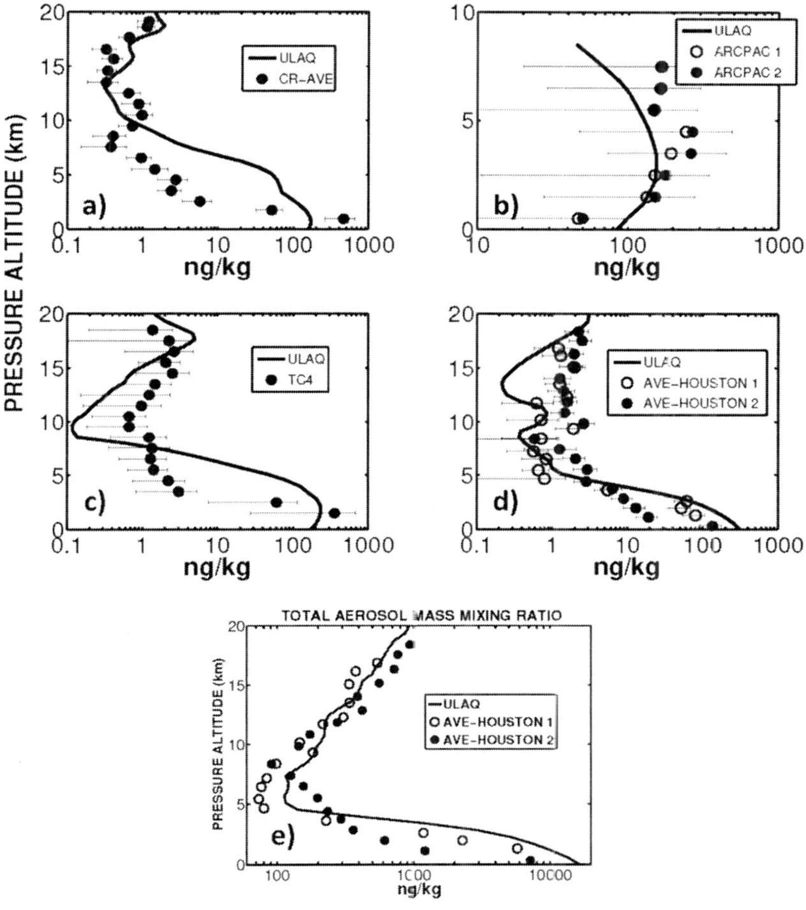

Figure 2. Panels (**a**–**d**): black carbon aerosol vertical profiles calculated in the ULAQ model and compared to measurements from aircraft campaigns [60] (see Table 2). Measurements were made onboard of NOAA research aircrafts at tropical latitudes (left panels (a) and (c)), high latitudes (upper right panel (b)) and mid latitudes (bottom right panel (b)), over Central and North America. Panel (**e**): comparison of calculated and observed total aerosol mass mixing ratio (Houston campaigns).

In the ULAQ model, aerosol mass loading is converted to aerosol optical depth (or thickness, AOT) separately for each aerosol type, depending on their size bin properties. This is done through suitable mass-to-extinction conversion coefficients obtained using the calculated size distribution in a Mie scattering program [62], along with appropriate values of the refractive indices. Comparisons between the measured and calculated optical depths are conducted on a regional basis. The selected regions are small enough to accommodate distinct differences in sources and transport but large enough to minimize the effects from differences in spatial resolution. The regional choices included one ocean region and two land regions relative to the North Atlantic sector. In this comparison, remote sensing measurements from satellites are used (AVHRR, TOMS and MODIS retrievals), for which the aerosol properties are vertically integrated and type-combined. Thus, meaningful evaluations of particular aerosol types are largely limited to regions or seasons where these aerosol types dominate [63]. The large spread in independent satellite retrievals is not surprising due to the many assumptions made in the retrieval codes with one of the most critical aspects being the specification of surface albedo. AVHRR data are normally considered of very high quality over the ocean (where surface albedo is well known). In this case, the ULAQ model did a good job, except during summer months where an under-prediction of approximately 50% was found (Figure 3). Aerosol emissions used in the ULAQ model are from the international assessment project AeroCom [64] (*i.e.*, Aerosol Comparisons between Observations and Models), and these include anthropogenic sources, biomass burning and non-explosive volcanoes.

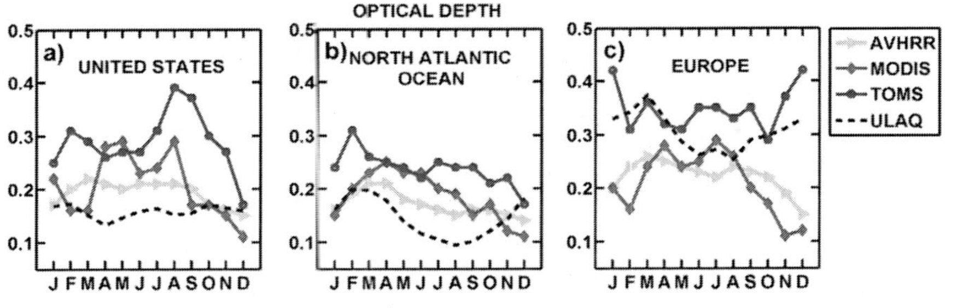

Figure 3. Annual cycle of aerosol optical depth at λ=0.55 μm over United States (left panel (**a**)), North Atlantic (mid panel (**b**)) and Europe (right panel (**c**)). AVHRR, MODIS and TOMS satellite retrievals are used for comparison with the ULAQ model calculations.

The aerosol optical depth in the mid-upper troposphere is shown in Figure 4. Aerosol mass is converted to aerosol extinction separately for each aerosol type and each size bin, at all altitude layers. Observations of AOT at λ = 0.55 μm are derived by integrating SAGE-II extinction measurements between 5.5 and 20 km altitude and then compared with ULAQ results, for which the same method is also used. The latitudinal variation of optical depth is presented in Figure 4, with each panel referring to a different season. A realistic reproduction of SAGE-II measurements was observed in the ULAQ calculations, especially during winter and spring months. Moreover, the model performs well in reproducing the maximum optical depth at high northern latitudes during springtime.

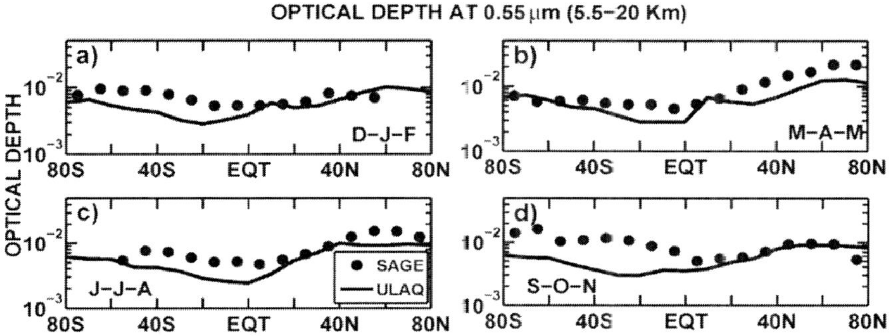

Figure 4. ULAQ model calculations (solid line) and SAGE-II derived values (circles) of aerosol optical depth at λ = 0.55 μm, integrated between 5.5 km and 20 km altitude, for winter (upper left panel (**a**)), spring (upper right panel (**b**)), summer (bottom left panel (**c**)) and fall months (bottom right panel (**d**)) (average 1996–2000).

Table 2. Details of BC mass mixing ratio measurements, made with Single Particle Soot Absorption Photometers onboard of NOAA research aircraft [60].

Field Campaign	Aircraft Platform	Investigator Group	Dates	Latitude Range	Longitude Range	Altitude Range (km)
AVE Houston	NASA WB-57F	NOAA	10–12 November 2004	29–38° N	88–98° W	0–18.7
CR-AVE	NASA WB-57F	NOAA	6–9 February 2006	1° S–10° N	79–85° W	0–19.2
TC4	NASA WB-57F	NOAA	3–9 August 2007	2–12° N	80–92° W	0–18.6
ARCPAC	NOAA WP-3D	NOAA	12–21 April 2008	65–75° N	126–165° W	0–7.4

3.2. Aviation Impact on Aerosols

Sulphur emissions from aircraft engines may form new H_2SO_4 aerosols in aircraft wakes that are then transported globally as quasi-passive tracers [10,65]. The emissions also increase the size of pre-existing sulphuric acid aerosols, from the additional condensation of H_2SO_4 formed through OH oxidation of aircraft emitted SO_2 [9,66]. The largest increase of the aerosol SAD comes from the newly formed aerosols, having typical small particle sizes of 5–20 nm. Zonal and annual averages of the sulphate aerosol SAD changes calculated in the ULAQ-CTM are presented in Figure 5a (assuming a 10% fraction of emitted sulphur will form ultrafine H_2SO_4 particles in aircraft wakes): they represent an increase of approximately 10% in the main aircraft emission region at the Northern Hemisphere (NH) mid-latitudes. The calculated zonal mean aircraft SAD shows a maximum increase of approximately 0.4 $\mu m^2/cm^3$ at current cruise altitudes and poleward of 40N. This change is roughly consistent with that calculated by Kärcher and Meilinger [17] under comparable conditions. Assuming an average background SAD of 2.5 $\mu m^2/cm^3$ [67], a 10% plume particle formation fraction and a 25% additional aircraft contribution to SAD from gas phase SO_2 emissions, the Kärcher and Meilinger [17] calculation gives an aircraft SAD change of 0.37 $\mu m^2/cm^3$. The ULAQ model predicted a SAD increase above the aircraft source region that was qualitatively consistent with the results of the fuel tracer experiment described by Danilin *et al.* [65]. A negative change was found at low altitudes, due to small changes in OH affecting sulphate production. As noted in Section 2, the monthly mean SAD produced from the ULAQ-CTM simulations are used in Oslo CTM3 and CTM2. A comparison of the calculated background SAD values in the ULAQ-CTM with SAGE-II derived values is presented in Figure 5b–d. The calculated values are well inside the uncertainty bars of the observations at all latitude bands. It is interesting to note how the model reproduces the SAD maximum in the high-latitude upper troposphere and lowermost stratosphere below 12 km altitude. This maximum results from the high-latitude stratospheric downwelling and from persistent cold temperatures during winter-spring months. Low temperatures decrease the weight percentage of sulphuric acid in supercooled H_2O-H_2SO_4 aerosols, thus increasing the available surface area density for equivalent sulphate mass available.

Formation of upper tropospheric ice particles may be favored by aircraft emitted BC particles (*i.e.*, “soot-cirrus”). A simple parameterization is adopted from climatological frequencies of relative humidity, where a mean value and a standard deviation are calculated for the spatial distribution of water vapour mixing ratios. Local ice super-saturation conditions ($RH_{ICE} > 100\%$) are a result of turbulent adiabatic ascent [19] and can be found at temperatures below 233K and below the tropopause. A normal-distribution probability for $RH_{ICE} > 100\%$ can be calculated using the water vapour mixing ratio distribution. This represents the probability that an ice particle can be formed via heterogeneous freezing on a preexisting population of ice condensation nuclei (P_{HET}), typically mineral dust transported from the surface or freshly emitted non-hydrophobic aviation BC particles. Using the assumption of a 0.1% fraction for non-hydrophobic emitted BC (N_{BC}) available as ice nuclei [68], the number of aviation soot-cirrus particles (N_{HET}) is calculated using the ULAQ-CTM microphysical scheme for polar stratospheric cloud ice particles formation [25]. Here, N_{BC} is used as the population of available condensation nuclei and P_{HET} as the probability that at any model grid point $RH_{ICE} > 1$.

Upper tropospheric ice particles may also form by means of homogeneous freezing of supercooled aerosols [19]. This process takes place for higher ice supersaturation ratios, with respect to heterogeneous freezing, typically $RH_{ICE} > 1.4$ [68]. The numerical scheme adopted in the ULAQ model is the one developed in Kärcher and Lohmann [19], and this assumes that basic physical processes determine the number of ice crystals (N_{HOM}) formed during an adiabatic ascent, including temperature and updraft speed. The latter is calculated as a function of the turbulent kinetic energy, following the method outlined in Lohmann and Kärcher [69]. In Figure 6, we show soot and soot-cirrus number density changes due to aircraft emissions. BC particles at cruise altitudes behave as a passive tracer, to a first approximation, so that the steady-state zonal mean BC accumulation is largely governed by horizontal mixing and high-latitude downwelling, thus resembling the sulphate aerosol SAD change of Figure 5. The latitude-longitude BC map near cruise altitudes (Figure 6c) shows how the changes in soot number density follow the geographical pattern of aviation emissions, with coupling to westerly winds horizontal transport. This was expected due to the passive tracer behaviour of the soot number density. Cirrus ice number density changes are largest in the Northern Hemisphere mid-high latitudes below the tropopause, *i.e.*, where ice super-saturation conditions tend to occur. The calculated maximum changes of soot and soot-cirrus optical thickness at $\lambda = 0.55$ μm are 1.2×10^{-5} and 1.2×10^{-3}, respectively.

3.3. Ozone Photochemistry

Based on the numerical experiments NA, AE and AE*, we evaluate the impact of gaseous aircraft emissions (AE-NA) and the impact of aerosol aircraft emissions (AE*-AE) using the models described in Section 2. In Figure 7, the AE-NA annual-zonal mean differences of NO_y, NO_x ($NO+NO_2$) and O_3 are presented. The structure of the perturbations is similar in the models, giving a maximum increase of approximately 3–5 ppbv for O_3. It should be mentioned that other studies indicate 10%–25% reduced impact on O_3 when including aircraft plume parameterizations [7]. However, the possible impact on OH and CH_4 is uncertain. We use monthly averaged emissions in the models, treating them as effective emissions. Possible reductions due to plume effects are thus not accounted for.

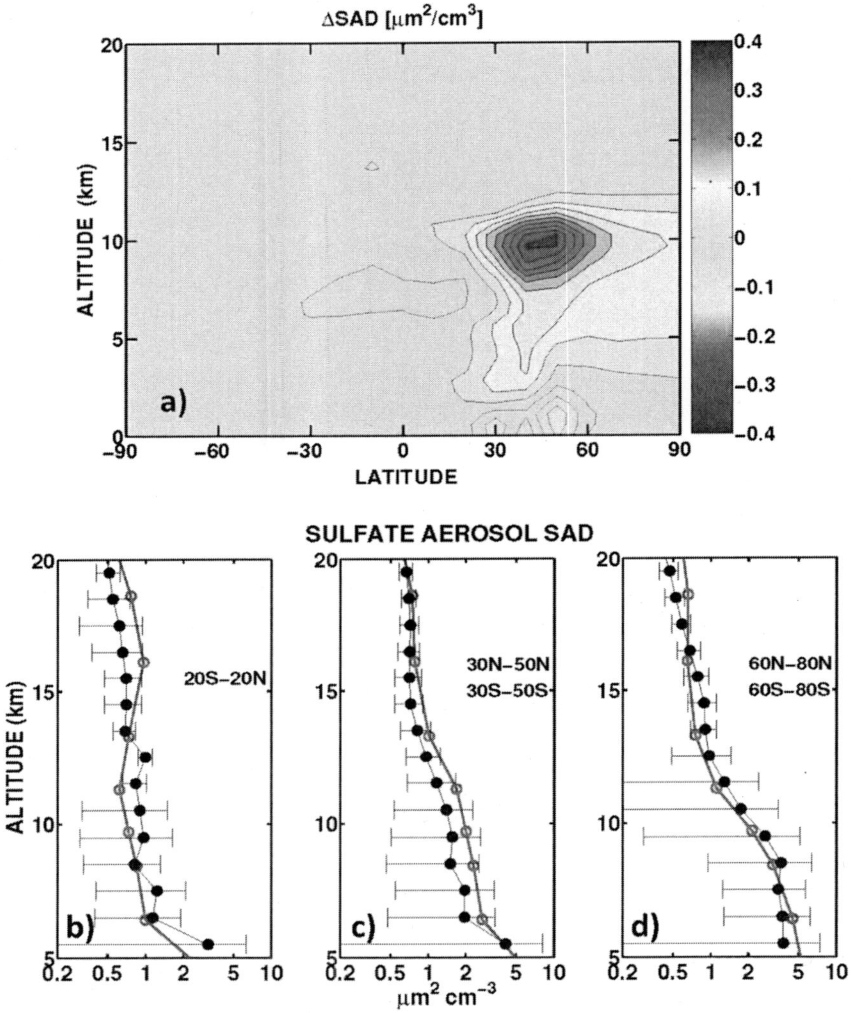

Figure 5. Top panel (**a**): zonally and annually averaged perturbations of sulphate aerosols surface area density, due to aviation emissions and calculated in the ULAQ model. Bottom panels (**b–d**): ULAQ model calculations (red line) and SAGE-II derived values (black line with circles and uncertainty bars) of sulphate aerosol SAD, shown as vertical profiles at different latitude bands: tropics (left panel (b)), mid-latitudes (mid panel (c)) and high-latitudes (right panel (d)) (average 1999–2002). Units are $\mu m^2/cm^3$.

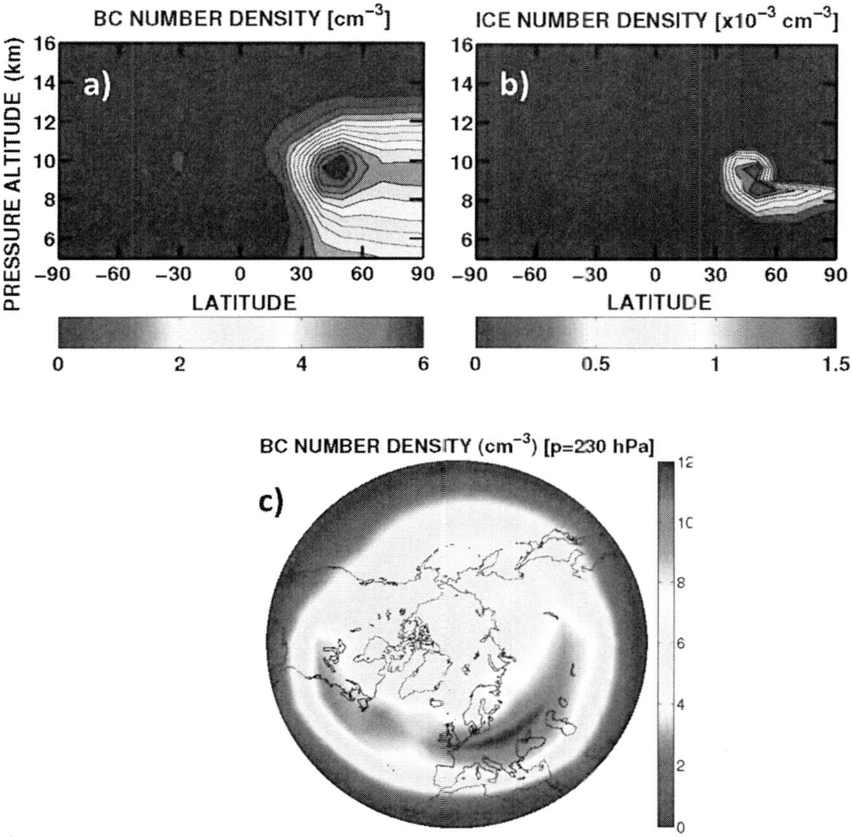

Figure 6. Top panels: zonally and annually averaged aircraft perturbations of soot number density (left panel (**a**)) and soot cirrus number density (right panel (**b**)), due to aviation emissions and calculated in the ULAQ model. Bottom panel: as in panel (a), but for the lat/lon map at p = 230 hPa (cm^{-3}).

The ULAQ-CTM produces larger NO_x and NO_y perturbations, indicating less effective removal processes from the source region, *i.e.*, less horizontal mixing with the tropical latitudes, coupled to a less effective HNO_3 rainout loss in the extra-tropics. Also, the maximum changes of NO_y and O_3 in ULAQ-CTM is found at about 40 N, whereas CTM3 and CTM2 have their maxima at higher latitudes. This points to a less effective horizontal mixing in the ULAQ model. A somewhat stronger stratospheric Brewer-Dobson circulation in the latter model is another explanation for the larger NO_x perturbation in the tropical lower stratosphere and the larger O_3 change in the Southern Hemisphere UTLS. As explained in Section 2, the meteorological fields in the models differ, and hence, may have a significant impact on large-scale species transport. Examples of the effects of different large-scale transport in global models on the calculated distribution of tracers from aircraft are discussed in Rogers *et al.* [70]. A complete discussion of the background NO_x chemistry in the models, along with the treatment of anthropogenic and natural emissions of NO_x and other tropospheric ozone precursors (CO and VOC),*i.e.*, fossil fuel, biomass burning and lightning are discussed in Søvde *et al.* [24].

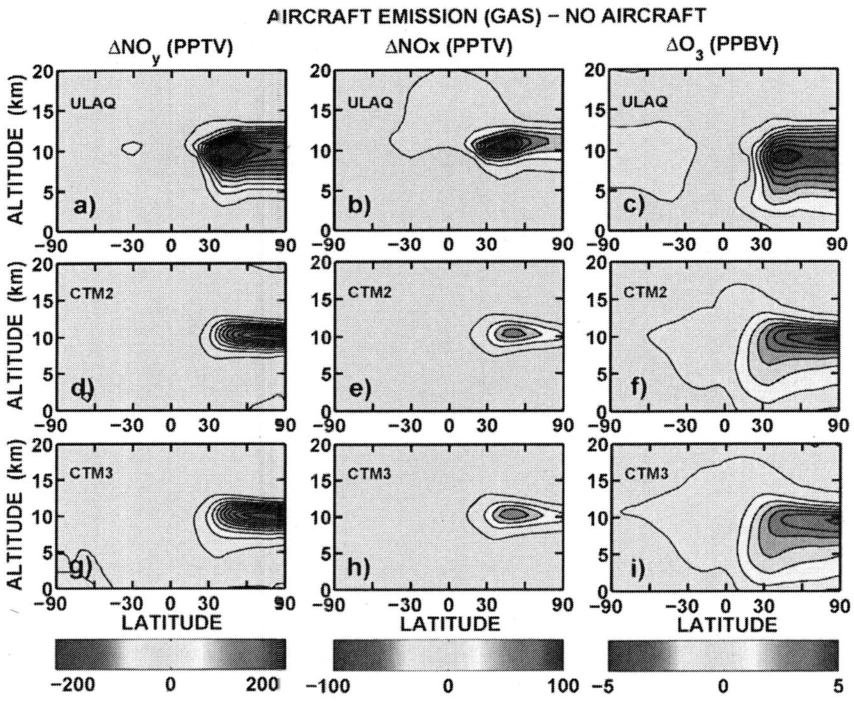

Figure 7. Zonally and annually averaged changes produced by gaseous aircraft emissions (AE-NA) of NO_y (left panels), NO_x (mid panels) and O_3 (right panels), for ULAQ, CTM2 and CTM3 models, from upper to lower panels, respectively. Black contours are shown every 25 pptv for NO_x and NO_y, and 0.5 ppbv for O_3.

The increase of upper tropospheric NO_x produced by aircraft emissions has the indirect effect of enhancing OH, mainly via reaction $NO + HO_2 \rightarrow NO_2 + OH$. Model calculated OH changes are presented in Figure 8 and for all three models they are consistently localized in the NH region of highest NO_x emissions, with maxima in the order of 0.20–0.25 ×

10^6molec/cm^3. This OH increase plays an important role in atmospheric chemistry by reducing the CH_4 lifetime and then inducing a negative RF (see Section 3.4.3). The CH_4 lifetime is primarily driven by tropical OH values and Figure 8illustrates reasonable behavior from the models when compared with climatological values [71] showing a reasonable behavior. The calculated lifetime ranges between 8.4 years from the ULAQ model, 10 years from CTM2 and 8.6 years from CTM3. The lower CH_4 lifetime in Oslo CTM3 with respect to CTM2 can be explained by the overestimation of OH in the upper troposphere. The reason for this is that CTM3 has higher NO_x in that region compared with CTM2 and that the photolysis is somewhat more effective. Differences in NO_x between CTM2 and CTM3 originates mainly from different wet scavenging of species. These higher tropical OH values in CTM3 will be investigated in future studies.

The aerosol SAD changes may effectively impact the NO_x photochemistry via heterogeneous chemical reactions, involving nitrogen reservoir species such as N_2O_5 and $BrONO_2$ producing HNO_3. An overall loss mechanism of both background and aircraft-emitted NO_x is from the wet removal of HNO_3 by rain. A reduction in atmospheric NO_x may result in lower O_3 production, and increased reactive halogens results in increased O_3 loss in the lower stratosphere through well-known Cl-Br catalytic cycles. The aerosol-reduced NO_x may in fact bind less halogen species in reservoir species such as $BrONO_2$, and may, therefore, increase O_3 loss. The coupled NO_x/aerosol aviation emissions (AE*) could produce a large-scale steady-state NO_x change that differs significantly from the NO_x only AE case. The inclusion of aviation aerosols in the three models for experiment AE* showed a reduction of up to 0.6 ppbv O_3 annual average mixing ratio (*i.e.*, AE*-AE), as shown in Figure 9. This is approximately 10%–15% of the aviation perturbation produced by NO_x emissions alone.

As noted above, the O_3 decrease is mostly due to reduced production via reduced NO_x, and in part from an increased loss from more active halogens, which is caused by more effective heterogeneous chemistry in the extra-tropical lowermost stratosphere. Inter-model differences of the altitude-latitude location of the maximum changes are again largely due to the coupling of less pronounced horizontal mixing and stronger UTLS downwelling in the high-latitude branch of Brewer Dobson circulation in ULAQ-CTM.

Figure 10 presents the latitude-longitude maps of the aviation-induced change of the ozone column, which in turn, determines the net radiative forcing (see for example Gauss *et al.* [72], for the changes in tropospheric ozone as one of the major drivers of climate change in the past decades). The perturbation produced by aviation NO_x emissions alone (AE-NA) reaches maximum values close to 1.2 DU in all three models. As expected, it is located at the northern mid-latitudes, east of the regions with the largest emissions from aviation. The fact that the O_3 changes shown in Figure 7 have peak values over the polar region is not contradictory with the maximum column changes found at the Northern mid-latitudes, because here the vertical distribution of the ozone mixing ratio perturbation is much wider. Consistent with Figure 7, the ULAQ-CTM features a stronger confinement in the extra-tropics of the O_3 column changes. The aerosol induced perturbation (AE*-AE) is of the order of −0.15 DU in all three models. Part of the inter-model differences may be attributable to the transport inconsistency of SAD changes imported into CTM2 and CTM3 from the aerosol calculation in the ULAQ model. In addition, the latter model includes hydrolysis of N_2O_5 and $BrONO_2$ as heterogeneous chemical reactions that may efficiently convert NO_x into HNO_3 on the surface of sulphate aerosols, whereas CTM2 and CTM3 include only the hydrolysis of N_2O_5.

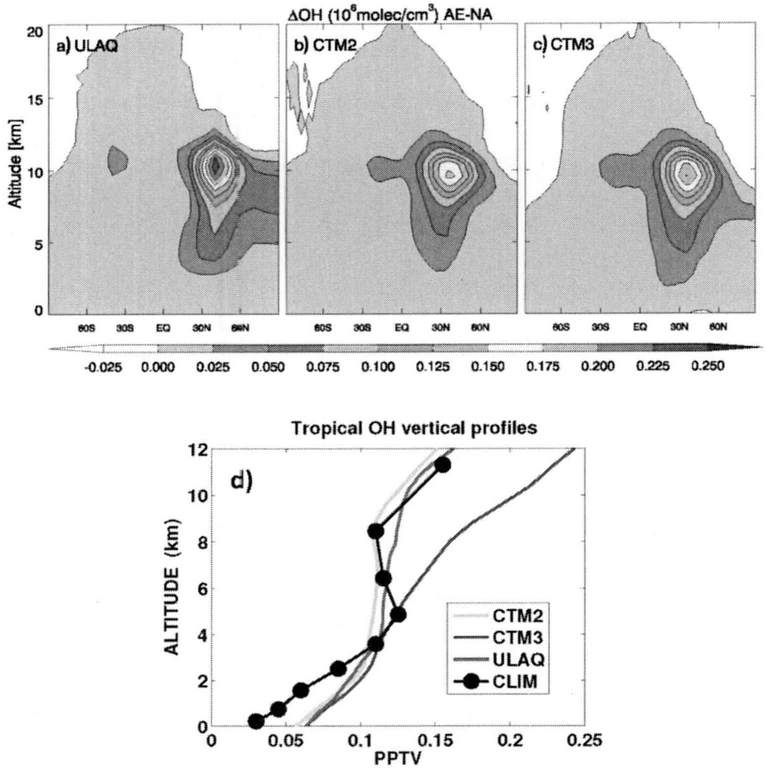

Figure 8. Top panels: zonally and annually averaged OH changes produced by aircraft emissions (AE-NA), for ULAQ (left panel (**a**)), CTM2 (mid panel (**b**)) and CTM3 (right panel (**c**)), respectively. Units are 10^6 molec/cm^3. Bottom panel (**d**): annually averaged vertical tropical profiles of OH mixing ratio (30S–30N) (pptv) from the three models, compared with climatological values [71].

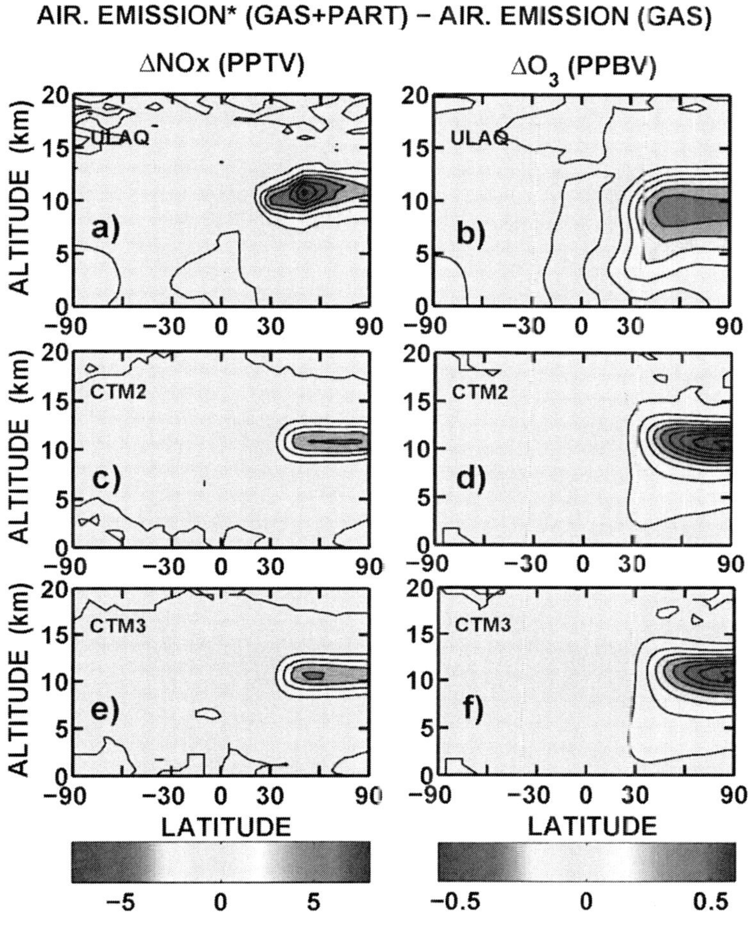

Figure 9. Zonally and annually averaged changes produced by aircraft aerosol emissions (AE*-AE) of NO_x (left panels **(a,c,e)**) and O_3 (right panels **(b,d,f)**), for ULAQ, CTM2 and CTM3 models, from upper to lower panels, respectively. Black contours are shown every 2.5 pptv for NO_x, and 0.05 ppbv for O_3.

3.4. Radiative Forcing

The radiative impact at the tropopause due to O_3 short- and long-term response, as well as to CH_4 lifetime changes are calculated with two independent radiative transfer models (ULAQ and Oslo) using the 3D monthly averages of ozone mixing ratio calculated in the CTMs. The ULAQ radiative model is also used to evaluate the radiative impact due to other radiatively-active species involved (*i.e.*, aerosols, cirrus ice particles and stratospheric H_2O). A short description of the radiative transfer models is presented in the following subsections, followed by a discussion of their main results that are relevant for the present study.

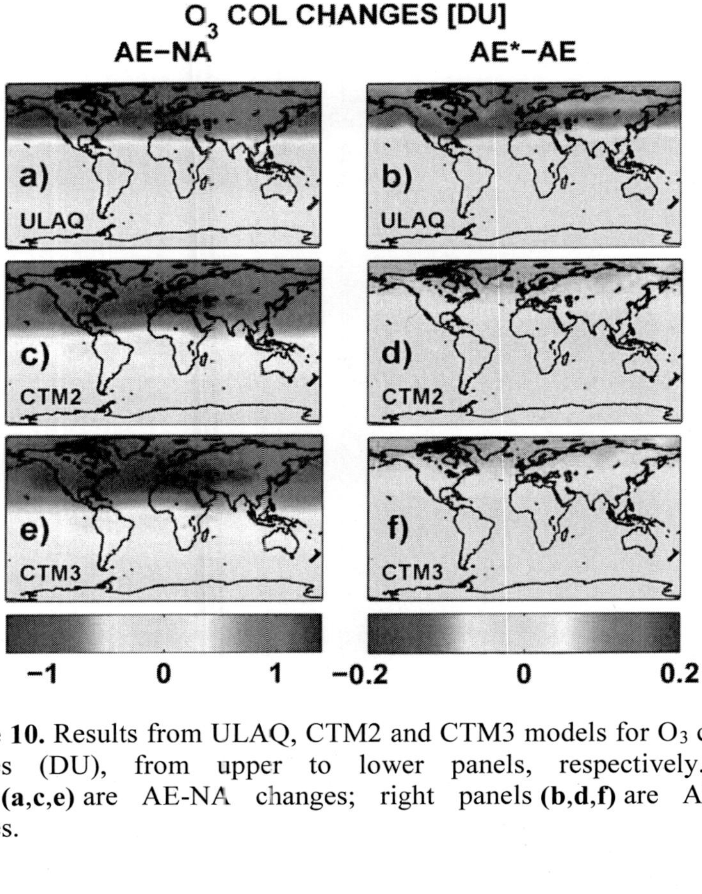

Figure 10. Results from ULAQ, CTM2 and CTM3 models for O_3 column changes (DU), from upper to lower panels, respectively. Left panels **(a,c,e)** are AE-NA changes; right panels **(b,d,f)** are AE*-AE changes.

3.4.1. ULAQ Radiative Transfer Model

The updated ULAQ radiative transfer module accounts for solar and planetary infrared radiation, both in clear and all sky conditions. Solar radiation is split among 150 bins in the UV and visible range, and 100 bins in the NIR, covering the solar spectrum from Lyman-alpha up to 7 μm for chemical species photolysis rate calculation, solar heating rates and radiative forcings. The solar radiation code considers sun-earth distance changes, solar cycle, sphericity of the atmosphere and its refraction. Optically active species at solar wavelengths are O_3, O_2, NO_2, SO_2, H_2O, CO_2 and aerosol particles. Multiple scattering is represented through a two-stream delta-Eddington approximation model [73]. The planetary infrared code uses a k-distribution formulation over 10 spectral bands [33] to compute heating rates and radiative forcings, including absorption/emission by CO_2, O_3, H_2O, N_2O, CH_4 and aerosols. Stratospheric temperature adjustment is performed in the Fixed Dynamic Heating (FDH) approximation [74]. Mie optical parameters for aerosols are computed using wavelength-dependent refractive indices from the OPAC database and the aerosol size distributions from the ULAQ model. The ULAQ model's photolysis rates, surface and top-of-atmosphere radiative fluxes have been validated in the framework of SPARC-CCMVal and AeroCom inter-comparison campaigns [75,76].

3.4.2. Oslo Radiative Transfer Model

The Oslo radiative transfer model [77] applies a long-wave broad band radiative transfer code [78] and the multi-stream DISORT code [79] for short-wave radiation. Following the standard definition of radiative forcing [80], stratospheric temperature adjustment is included in the simulations. Meteorological data from the European Centre for Medium range Weather Forecast (ECMWF) were used for simulations on a monthly mean time scale. For these calculations, O_3 (mixing ratio) from ULAQ-CTM model was re-gridded to the Oslo CTM resolution, both horizontally and vertically.

3.4.3. Radiative Forcing Results

The global annual average RFs calculated off-line with the radiative transfer models (ULAQ, Oslo), for the three numerical simulations (NA, AE, AE*) and all three models (ULAQ, CTM2 and CTM3) are presented in Table 3 and Table 4. On average, the three models calculate an ozone column difference due to the aircraft induced increase in aerosol SAD of −5%, which is consistent with the −7% calculated by Unger [21]. The ULAQ estimate is closer to this independent value, because it includes the hydrolysis of $BrONO_2$ in addition to that of N_2O_5, in contrast to CTM2 and CTM3. The tables give RF values for clear and total sky conditions, with stratospheric temperature adjustment applied for the longwave forcing in the case of total sky conditions. A good consistency between the radiative transfer models is clear from the instantaneous longwave forcing in Table 3: with a 0.5% difference found between the two radiative models. A 15% difference for SW and 5%–10% for total sky

LW RFs (Table 4) are due to slight differences in surface albedo and cloud cover adopted in the two models. Results in Table 4 also indicate the importance of stratospheric temperature adjustment, whereby the aircraft induced O_3 increase in the lower stratosphere, warming this region both in the UV and in the planetary infrared 9.6 μm band. This produced an indirect negative forcing from well-mixed greenhouse gases (e.g., CO_2, CH_4, N_2O) [81].

The results of additional ULAQ-CTM simulations (*i.e.*, model version labeled ULAQ_FLX) have been included in both tables. In this case, the CH_4 prediction is made using a flux boundary condition instead of a fixed mixing ratio at the surface (as in ULAQ, CTM2 and CTM3). Gridded fluxes of CH_4 are used at the surface for anthropogenic and natural sources [82,83]. Due to the long CH_4 lifetime, the ULAQ_FLX numerical simulations are extended up to 50 years, to allow the model to reach equilibrium between the predicted CH_4 field and O_3 (via OH and NO_x). This means that the O_3 RF from ULAQ_FLX includes both the short-term ozone response and its long-term response (*i.e.*, the so-called primary mode O_3 RF) [84].

Latitude-longitude maps of the short-term O_3 RF are presented in Figure 11, in terms of average values over the three models (ULAQ, CTM2 and CTM3) and the two off-line radiative calculations (ULAQ, Oslo). Globally, the aviation aerosol effect on O_3 photochemistry is quantified as −4% of the aviation NO_x effect, in terms of radiative forcing. As expected, the geographical patterns of the calculated radiative forcing are closely linked to those of the ozone column changes in Figure 10.

A summary of the radiative forcing due to the aviation NO_x perturbation is presented in Table 5. We note that the short-term O_3 RF is close to, or slightly above, the mean value reported by Holmes *et al.* [5]. Further discussion on short-term O_3RF from our models can be found in Søvde *et al.* [24]. Also listed in Table 5 is CH_4 radiative forcing. The latter is calculated as a function of the lifetime percentage reduction ($\Delta\tau$-CH_4) due to the OH increase produced by the aircraft-induced NO_xincrease (via $NO + HO_2 \rightarrow NO_2 + OH$), with the following formula: $CH_4\text{-RF}(mW/m^2) = \chi\text{-}CH_4(ppbv) \times 0.37 \times \Delta\tau\text{-}CH_4(\%)/100 \times 1.4$ [77], where τ is the CH_4 lifetime. A magnification factor 1.4 is used to take into account that the CH_4 prediction in the models is usually made using a fixed mixing ratio boundary condition at the surface, so that the CH_4 adjustment to the upper tropospheric OH perturbation is underestimated. As explained in IPCC [85], the factor 1.4 may be used as a best estimate to adjust the calculated CH_4 lifetime change to the missing feedback of the upper tropospheric OH perturbation on the calculated lower tropospheric CH_4 mixing ratios. An exception to this is for the ULAQ_FLX experiment, where the CH_4adjustment to the changing OH field is accounted for by the use of a flux boundary condition at the surface.

Incidentally, we note that the ULAQ_FLX to ULAQ ratio of the CH_4 lifetime change is 1.39 for both AE-NA and AE*-NA, which is very close to the 1.4 factor suggested in IPCC [85].

Table 3. Summary of O_3 RF mean values for clear sky conditions. Differences are calculated between NA (experiment with no aircraft emission), AE (experiment with only gaseous emission [NO_x+H_2O]) and AE* (experiment with gaseous and particle emission [NO_x+H_2O+SO_4]). RF values are calculated with Oslo and ULAQ radiative code at ECMWF and National Center for Environmental Prediction (NCEP) tropopause level, respectively. LW, SW, NET are longwave, shortwave and net RFs, respectively. Only LW instantaneous values are considered here.

Model	EXP	O_3 col [DU]	Oslo LW-inst [mW/m²]	Oslo SW [mW/m²]	Oslo NET-inst [mW/m²]	ULAQ LW-inst [mW/m²]	ULAQ SW [mW/m²]	ULAQ NET-inst [mW/m²]
ULAQ	AE-NA	0.463	20.65	3.43	24.08	20.48	2.84	23.52
	AE*-NA	0.430	19.73	3.27	23.00	19.53	2.72	22.25
	AE*-AE	−0.033	−0.92	−0.16	−1.08	−0.95	−0.12	−1.07
CTM2	AE-NA	0.576	25.88	3.[illegible]3	29.02	25.97	2.56	28.53
	AE*-NA	0.549	25.02	3.09	28.11	25.14	2.47	27.61
	AE*-AE	−0.026	−0.86	−0.04	−0.91	−0.83	−0.09	−0.92
CTM3	AE-NA	0.632	27.46	3.49	30.95	27.56	2.84	30.40
	AE*-NA	0.606	26.63	3.42	30.05	26.79	2.78	29.57
	AE*-AE	−0.026	−0.83	−0.07	−0.90	−0.77	−0.06	−0.83
ULAQ_FLX	AE-NA	0.264	14.12	3.52	17.64	14.10	2.93	17.03
	AE*-NA	0.251	13.66	3.37	17.03	13.64	2.84	16.48
	AE*-AE	−0.013	−0.46	−0.15	−0.61	−0.46	−0.09	−0.55

Table 4. As in Table 3, but for total sky conditions. LW-RF and NET-RF with stratospheric temperature adjustment are also included here.

model	EXP	O_3 col [DU]	Oslo LW-inst [mW/m²]	Oslo LW-adj [mW/m²]	Oslo SW [mW/m²]	Oslo NET-adj [mW/m²]	ULAQ LW-inst [mW/m²]	ULAQ LW-adj [mW/m²]	ULAQ SW [mW/m²]	ULAQ NET-adj [mW/m²]
ULAQ	AE-NA	0.463	14.82	12.59	5.26	17.85	16.43	13.64	4.52	18.16
	AE*-NA	0.430	14.16	12.03	5.02	17.05	15.72	13.03	4.32	17.35
	AE*-AE	−0.033	−0.66	−0.56	−0.24	−0.80	−0.71	−0.61	−0.20	−0.81
CTM2	AE-NA	0.576	18.70	16.19	4.7[illegible]	20.94	21.65	17.99	4.01	22.00
	AE*-NA	0.549	18.06	15.50	4.6[illegible]	20.13	20.95	17.26	3.89	21.15
	AE*-AE	−0.026	−0.64	−0.69	−0.12	−0.81	−0.70	−0.73	−0.12	−0.85
CTM3	AE-NA	0.632	19.77	17.11	5.27	22.38	22.80	19.16	4.44	23.60
	AE*-NA	0.606	19.18	16.59	5.14	21.63	22.25	18.52	4.35	22.87
	AE*-AE	−0.026	−0.59	−0.62	−0.13	−0.75	−0.64	−0.64	−0.09	−0.73
ULAQ_FLX	AE-NA	0.264	10.09	7.25	4.73	11.98	11.05	8.12	4.14	12.26
	AE*-NA	0.251	9.77	7.06	4.54	11.60	10.73	7.90	3.95	11.85
	AE*-AE	−0.013	−0.32	−0.19	−0.19	−0.38	−0.32	−0.22	−0.19	−0.41

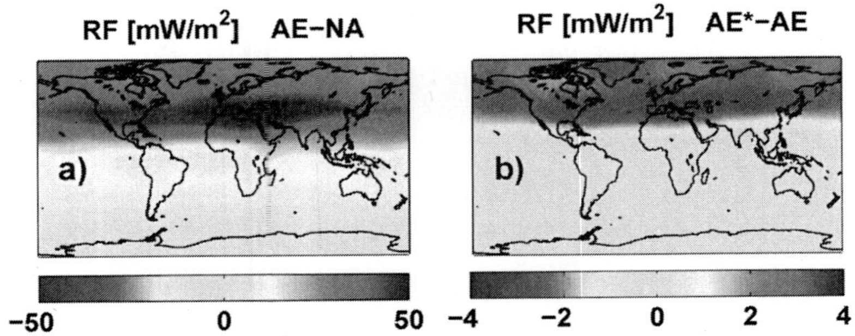

Figure 11. O_3 RF (adjusted) (mW/m^2) from an average of ULAQ, CTM2 and CTM3 model results and both radiative transfer codes. Left panel **(a)** is for AE-NA (globally averaged value: 20.82 mW/m^2); right panel **(b)** is for AE*-AE (globally averaged value: −0.79 mW/m^2).

The uncertainty in the calculated CH4 RF is estimated to be approximately ±10% [86], which would not significantly change our result. In the NO_x-related RF, we also include the effect of CH_4 on stratospheric water vapour. This is estimated to give an additional RF of 15% of the CH_4 RF [87], although the uncertainty is large, about 70% [86]. The CH_4 RF equation is a simplification of the detailed formula reported by IPCC [86]; those formulas are the basis for calculating the specific forcing, which is linked to the unperturbed CH_4 concentrations so that a higher CH_4 concentration yields a lower specific forcing. Our approach is often used when assessing impacts of small CH4 perturbations, and we find that our CH_4 RF differs by less than 1% compared to the full equations using CH_4 and N_2O. Scaling up our lifetime changes linearly to match 1 Tg(N)/year, we get about 1.1%–1.5%, which is in the lower range of Holmes *et al.* [5]. Gottschaldt *et al.* [23] finds a similar value of 1.5%, which is in agreement with our higher values. Correspondingly, our CH_4 RF is also in the lower range of Holmes *et al.* [5].

Ozone changes produced by tropospheric NO_x emissions are not only associated with a short-term response of the $NO_x$and HO_x perturbations (see Table 3 and Table 4). The long-term response to CH_4 changes driven by the OH perturbation has to be considered as well. As discussed above, CTMs are normally run using a fixed mixing ratio boundary condition at the surface and this does not allow lower tropospheric CH_4 to adjust to the upper tropospheric OH perturbation produced by aviation NO_x emissions. In this case, a parametric formula, such as the one adopted here, *i.e.*, PMO-RF = 0.5 × CH_4-RF [86], has to be used in order to estimate the long-term "primary mode" effect. The O_3 RF reported in Table 5 for ULAQ_FLX includes both the short-and long-term responses of ozone.

The difference between ULAQ and ULAQ_FLX represents in a first approximation the long-term O_3 response in ULAQ_FLX, which is approximately 77% of the ULAQ_FLX CH_4-RF. This is a larger value with respect to the 50% used in the parametric formula above, simply because the calculated difference between ULAQ and ULAQ_FLX in the short-term O_3 RF includes not only the PMO RF, but also the feedback in ULAQ_FLX of the changing CH_4 on stratospheric water vapour and finally on HO_x and NO_x catalytic cycles for ozone destruction.

Table 5. Summary of NO_x-related RF mean value terms: O_3, CH_4 and PMO (primary mode ozone, *i.e.*, long-term ozone response). Ozone RF values are obtained as a mean from Oslo and ULAQ radiative code calculations. Differences are calculated between NA (experiment with no aircraft emission), AE (experiment with only gaseous emission [NO_x+H_2O]) and AE* (experiment with gaseous and particle emission [NO_x+H_2O+SO_4]). See text for details on the calculations of CH_4 RF, PMO RF and the stratospheric H_2O RF induced by CH_4 changes. Note that for ULAQ_FLX, O_3-RF includes both short-term and long-term (primary mode) effects, whereas for the other models, only short-term O_3 RF is included (long-term effects are then included in "PMO RF").

Species	EXP	ULAQ	CTM2	CTM3	ULAQ_FLX
	AE-NA	0.463	0.576	0.632	0.264
ΔO_3 col [DU]	AE*-NA	0.430	0.549	0.606	0.251
	AE*-AE	−0.033	−0.026	−0.026	−0.013
	AE-NA	18.00	21.47	22.99	12.12
O_3 RF [mW/m^2]	AE*-NA	17.20	20.64	22.25	11.72
	AE*-AE	−0.80	−0.83	−0.74	−0.40
	AE-NA	−0.82	−1.24	−1.13	−1.14
$\Delta\tau$-CH_4 [mo]	AE*-NA	−0.76	−1.21	−1.09	−1.06
	AE*-AE	−0.06	−0.03	+0.04	+0.08
	AE-NA	−0.81	−1.03	−1.09	−1.17
$\Delta\tau$-CH_4 [%]	AE*-NA	−0.75	−1.01	−1.05	−1.09
	AE*-AE	−0.06	−0.02	+0.04	+0.08
	AE-NA	−7.30	−9.33	−9.88	−7.60
CH_4 RF [mW/m^2]	AE*-NA	−6.86	−9.14	−9.57	−7.08
	AE*-AE	−0.44	−0.18	+0.32	+0.52
	AE-NA	−1.10	−1.40	−1.48	−1.14
Stratospheric H_2O RF [mW/m^2] from CH_4 changes	AE*-NA	−1.03	−1.37	−1.43	−1.06
	AE*-AE	−0.07	−0.03	+0.05	+0.08
	AE-NA	−3.65	−4.66	−4.94	–
PMO RF [mW/m^2]	AE*-NA	−3.43	−4.57	−4.78	–
	AE*-AE	−0.22	−0.09	+0.16	–
	AE-NA	5.96	6.08	6.69	3.38
TOTAL RF from NO_x [mW/m^2]	AE*-NA	5.88	5.56	6.47	3.58
	AE*-AE	−0.08	−0.52	−0.22	+0.20

A summary of the RF from aviation aerosol is presented in Table 6. The calculated normalized direct forcings for sulphate and BC (−140 and 1700 W/g, respectively) are comparable with those reported in IPCC [85] (*i.e.*, −125 to −214 and 1100–3000 W/g, respectively). The BC RF (+0.8 mW/m^2, globally) is due to direct soot emission from the aircraft (with an effective radius of 0.14 μm). The sulphate RF (−3.5 mW/m^2, globally) is due to aircraft emissions of SO_2 (EI = 0.8 g-SO_2/kg-fuel), which is oxidized by OH to SO_4 with additional condensation in the accumulation mode of the sulphate aerosol size distribution (with a calculated effective radius of 0.16 μm). The RF signs of BC (positive) and SO_4 (negative) are due to the dominant solar radiation absorption and scattering, respectively. The indirect RF of aviation soot through formation of upper tropospheric cirrus-ice particles via heterogeneous freezing (*i.e.*, the soot-cirrus RF) is obtained on the basis of a particle size distribution with a calculated effective radius of 5 μm (+4.9 mW/m^2, globally). The largest uncertainty on this soot-cirrus RF comes from the BC fraction assumed to act as ice nuclei (0.1% in our case). Values in the range 0.1%–1.0% (or even smaller) are suggested by Hendricks *et al.* [68]. For soot-cirrus, the longwave contribution dominates over the shortwave forcing, due to the size of these particles, whereas for BC and SO_4 the solar forcing dominates. Table 7 shows that the indirect BC results are consistent with Gettelman and Chen [20], where an aviation impact <10 mW/m^2 was observed when a fixed nucleating efficiency of 0.1% for BC was applied. Furthermore, the RFs per unit BC emission from both studies are within 2.5% of each other. However, it should be noted that the treatment of background ice nucleation, together with the uncertainties in the ice nucleation potential of "pre-activated" aircraft soot, could produce large uncertainties surrounding the indirect effect of aircraft soot, resulting in either negative or positive RF, as shown by Zhou and Penner [88] and also summarized in Table 7.

Table 8 presents an overall summary of the aviation RFs calculated in the present study, excluding CO_2, contrails and contrail-cirrus, which are not explicitly considered in the present study. The contribution of aviation water vapour accumulating above the tropopause is also considered, and the calculated value (0.6 mW/m^2) is consistent with that obtained and discussed in Wilcox *et al.* [89] (0.9 ± 0.5 mW/m^2). According to our calculations, aircraft emitted aerosols tend to decrease the O_3 production and its positive RF (by −0.27±0.2 mW/m^2), but on the other hand the net contribution of direct and indirect soot forcings (positive) plus the direct sulphate forcing (negative) (+2.7 mW/m^2) tends to increase the overall aviation RF. The small and negative SO4 forcing (−0.2 mW/m^2) in the AE-NA case is due to the OH increase produced by aviation NO_x emissions, which favours the SO_2 oxidation into sulphate. According to our calculations, the net NO_x-induced RF appears to be the dominant term (excluding CO_2 and contrails), although it might be somewhat overestimated, because the effect of NOx emissions on nitrate aerosols is not included in this study. However, the existence of such an effect (from aviation emissions) and its magnitude are not well established. An increasing contribution to solar radiation scattering would add a negative term to the NO_x-related RF, as demonstrated by Unger *et al.* [90]. However, there are not enough studies whereby nitrates are included in CTM simulations and therefore, a reliable RF estimate that is comparable to the rest of the aviation RFs reported in the study is not available. In addition, a rather marginal effect of aircraft emissions on nitrate aerosol could be expected when taking into account that the formation of nitrate aerosols ($NH4^+NO3^-$) requires a significant presence of ammonium, which may be the case only in the lower troposphere. Holmes *et al.* [5] have shown a net NO_x effect amounting to 4.5 ± 4.5 mW/m^2 for 1 Tg(N)/year aircraft emissions. Our results are consistent with that estimate.

Table 6. Summary of aerosol Radiative Forcing mean values: direct contributions from sulphate and soot (BC) and indirect BC forcing (soot-cirrus) (AE*-NA): ULAQ-CTM values and ULAQ radiative code. Here $\Delta\tau$ is the particle optical thickness perturbation at $\lambda = 0.55$ μm. The final column shows the normalized RF (*i.e.*, absolute value of the net-RF divided by the globally averaged Δ-load).

Species	Δ-load TOT [μg/m²]	Δ-load STRAT [μg/m²]	Δτ [λ = 0.55 μm]	SW [mW/m²]	LW-inst [mW/m²]	LW-adj [mW/m²]	NET-adj [mW/m²]	NRF [W/g]
SO_4	25.0	6.6	2.4E-04	−5.2	1.8	1.7	−3.5	140
BC	0.46	0.10	3.0E-06	0.85	0.01	−0.08	0.78	1700
Soot Cirrus	320	0	1.0E-04	−1.7	6.6	6.6	4.9	15

Table 7. RF from other soot-cirrus studies. Relevant experiments not directly comparable to this study are also included for illustrative purposes.

	This study	Gettelman and Chen [20]	Zhou and Penner [88]
Model	ULAQ-CTM	CAM5/MAM	CAM5.2/IMPACT
Experimental setup	Fixed nucleating efficiency (0.1%), with aviation NO_x and SO_4, no contrails.	Fixed nucleating efficiency (0.1%), no aviation NO_x/SO_4/contrails. [Also with SO_4 and contrails]	[Varying nucleating efficiencies to include pre-activated soot and sensitivity to background. No aviation NO_x, with contrails]
Aircraft BC emission (Tg-BC/year)	0.00407	0.00681	0.00681
Soot-cirrus TOA RF (mW/m²)	4.9	8.0 [−21 ± 11]	[−350 to 90]
RF per unit BC emission (Wm^{-2}/Tg-BC/year)	1.20	1.17	

Table 8. Summary of Radiative Forcing (RF) mean values relevant for aviation emissions, excluding CO_2, contrails and contrail-cirrus, which are not explicitly considered in the present study. RF values from NO_x are obtained as a mean from the CTMs (ULAQ, CTM2 and CTM3) and Oslo and ULAQ radiative code calculations. The uncertainty interval is calculated as the range of the net NO_x-related RF from the three models.

species	RF [mW/m²] AE-NA	RF [mW/m²] AE*-NA	RF [mW/m²] AE*-AE
TOTAL RF from NO_x	6.2 ± 0.4	6.0 ± 0.4	−0.27 ± 0.2
Direct SO_4 RF	−0.2	−3.5	−3.3
Direct BC RF	0.0	0.8	0.8
Indirect BC RF (soot cirrus)	0.0	4.9	4.9
Stratospheric H_2O	0.6	0.6	0.0
TOTAL RF	6.6	8.8	2.2

4. CONCLUSIONS

In this work, we have estimated the effect of aircraft gaseous and particulate sulphur emissions in three CTMs, the ULAQ-CTM, Oslo CTM3 and Oslo CTM2. The surface area densities of sulphate aerosols were calculated on-line by the ULAQ-CTM, and were included as monthly means in the Oslo CTMs because of their lack of aerosol modules capable of handling aircraft aerosols. Aerosol mass prediction with the ULAQ-CTM has been widely validated through the participation in international inter-comparison projects, but was also validated here. The model performed well in reproducing observed aerosol mass concentrations, with discrepancies in a few South Atlantic observational sites, which could be attributed to the inefficient parameterization of removal processes. The processing of emission sources towards an altitudinal distribution of aerosol mass involves sulphate chemistry, transport and removal processes. These mechanisms have a large potential for errors, so they have to be controlled by the optical depth validation and extinction simulated values. Optical thickness has been validated by means of SAGE-II measurements and AVHRR, TOMS and MODIS retrievals; with the ULAQ model reproducing the seasonal variation for both oceanic and land regions of the North Atlantic. Moreover, the ULAQ model performed well when compared with SAGE-II data. While analysing the contribution to extinction from different aerosol types, sulphate aerosols gave the largest contribution at the tropopause height. We found that the inclusion of aircraft sulphur and aerosols emissions produced a slightly smaller impact on O_3 than when only aircraft NO_x emissions were considered. According to our calculations, this changed the net O_3-RF (*i.e.*, short- and long-term O_3 RF) by −0.63 mW/m²(*i.e.*, −3.8%). A change in OH will also set up a small increase in CH_4 and its RF, which we found to be +0.36 mW/m² (*i.e.*, +3.5%) and this includes also the CH_4 feedback on stratospheric water vapour. The calculated aerosol-driven net change of the NO_x-related RF produced by aircraft emissions of sulphate aerosols accounts then for −0.27 mW/m^2 (*i.e.*, −4.4%)

The long-term response to CH_4 changes due to aviation NO_x emissions were also considered in this study. In the ULAQ_FLX experiment, using a CH_4 surface flux boundary condition, we find a long-term impact on O_3 amounting to 77% of the CH_4 RF. This is near the upper limit of the IPCC recommended range of values, when taking into account a 55% uncertainty for the PMO. Our larger number originates from the feedback of the changing CH_4 on stratospheric water vapour and finally on HO_x and NO_x catalytic cycles for stratospheric ozone destruction.

The RF from direct SO_4 and BC were found to be −3.3 mW/m^2 and +0.8 mW/m^2, respectively, while the indirect BC result (soot-cirrus) was estimated to be +4.9 mW/m^2. The direct effects were comparable to other studies. As for soot-cirrus, there are still large uncertainties surrounding the estimate, not just in terms of its magnitude but also whether it has a warming or cooling effect. This study shows that aerosols should be included in model simulations, with the largest aerosol impacts seen in the indirect effects of BC on clouds and not their effects on NO_x chemistry (excluding the effect of nitrates). However, further work is necessary to quantify these indirect effects and their sensitivities to parameters such as background sulphate concentrations and the interaction between aviation aerosols and contrails-cirrus.

ACKNOWLEDGMENTS

The authors acknowledge funding of the European Commission REACT4C project, under Grant No. ACP8-GA-2009-233772. The authors acknowledge use of SAGE-II data for aerosol products evaluation. One of the authors (NDL) would like to thank Dr. Joshua Schwarz for kindly providing the dataset of BC mass mixing ratio measurements, made with Single Particle Soot Absorption Photometers onboard of NOAA research aircraft.

AUTHOR CONTRIBUTIONS

Giovanni Pitari: Overall coordination and responsibility for the ULAQ chemistry-transport model. Daniela Iachetti: Responsibility for the ULAQ-CTM modules for aircraft emissions and tracer transport. Glauco Di Genova: Responsibility for the ULAQ radiative transfer model. Natalia

De Luca: Evaluation of aerosol results with observations. Ole Amund Søvde: Responsibility for the Oslo chemistry-transport models. Øivind Hodnebrog: Responsibility for the Oslo radiative transfer model calculations. David S. Lee: Responsibility for the MMU models for aviation emissions. Ling Lim: Set up of aviation emission inventories for CTM use.

CONFLICTS OF INTEREST

The authors declare no conflict of interest.

REFERENCES

1 Sausen, R.; Isaksen, I.; Grewe, V.; Hauglustaine, D.; Lee, D.S.; Myhre, G.; Köhler, M.O.; Pitari, G.; Schumann, U.; Stordal, F.; Zerefos, C. Aviation radiative forcing in 2000: An update on IPCC (1999). Meteorol. Z. 2005, 14, 555–561.

2 Lee, D.S.; Pitari, G.; Grewe, V.; Gierens, K.; Penner, J.; Petzold, A.; Prather, M.; Schumann, U.; Bais, A.; Berntsen, T.; et al. Transport impacts on atmosphere and climate: Aviation. Atmos. Environ. 2010, 44, 4678–4734.

3 Köhler, M.O.; Rädel, G.; Dessens, O.; Shine, K.P.; Rogers, H.L.; Wild, O.; Pyle, J.A. Impact of perturbations to nitrogen oxide emissions from global aviation. J. Geophys. Res. 2008, 113, D11305.

4 Hoor, P.; Borken-Kleefeld, J.; Caro, D.; Dessens, O.; Endresen, O.; Gauss, M.; Grewe, V.; Hauglustaine, D.; Isaksen, I.S.A.; Jöckel, P.; et al. The impact of traffic emissions on atmospheric ozone and OH: results from QUANTIFY. Atmos. Chem. Phys. 2009, 9, 3113–3136.

5 Holmes, C.D.; Tang, Q.; Prather, M.J. Uncertainties in climate assessment for the case of aviation NO. P. Natl. Acad. Sci. USA 2011, 108, 10997–11002.

6 Kraabøl, A.G.; Berntsen, T.K.; Sundet, J.K.; Stordal, F. Impacts of NOx emissions from subsonic aircraft in a global three-dimensional chemistry transport model including plume processes. J. Geophys. Res. 2002, 107.

7 Cariolle, D.; Caro, D.; Paoli, R.; Hauglustaine, D.A.; Cuénot, B.; Cozic, A.; Paugam, R. Parameterization of plume chemistry into large-scale

atmospheric models: Application to aircraft NOx emissions. J. Geophys. Res. 2009, 114, D19302.

8 Weisenstein, D.; Ko, M.; Dyominov, I.; Pitari, G.; Ricciardulli, L.; Visconti, G.; Bekki, S. The effects of sulfur emissions from HSCT aircraft: a 2-D model intercomparison. J. Geophys. Res. 1998, 103, 1527–1547.

9 Pitari, G.; Mancini, E.; Bregman, A. Climate forcing of subsonic aviation: indirect role of sulfate particles via heterogeneous chemistry. Geophys. Res. Lett. 2002, 29.

10 Søvde, O.A.; Gauss, M.; Isaksen, I.S.A.; Pitari, G.; Marizy, C. Aircraft pollution—A futuristic view. Atmos. Chem Phys.2007, 7, 3621–3632.

11 Hendricks, J.; Kärcher, B.; Lohmann, U.; Ponater, M. Do aircraft black carbon emissions affect cirrus clouds on the global scale? Geophys. Res. Lett. 2005, 32, L12814.

12 Burkhardt, U.; Kärcher, B. Global radiative forcing from contrail cirrus. Nat. Clim. Chang. 2011, 1, 54–58.

13 Meilinger, S.K.; Kärcher, B.; Peter, T. Suppression of chlorine activation on aviation-produced volatile particles.Atmos. Chem. Phys. 2002, 2, 307–312.

14 Sander, S.P.; Friedl, R.R.; Barker, J.R.; Golden, D.M.; Kurylo, M.J.; Wine, P.H.; Abbatt, J.P.D.; Burkholder, J.B.; Kolb, C.E.; Moortgat, G.K.; et al. Chemical Kinetics and Photochemical Data for Use in Atmospheric Studies, Evaluation No. 17; Tech. Rep. 10–06; California Institute of Technology: Pasadena, CA, USA, 2011.

15 Pueschel, R.F.; Hallett, J.; Strawa, A.W.; Howard, S.D.; Ferry, G.V.; Foster, T.; Arnott, W.P. Aerosol and cloud particles in tropical cirrus anvil: importance to radiation balance. J. Aerosol. Sci. 1997, 28, 1123–1136.

16 Curtius, J.; Sierau, B.; Arnold, F.; Baumann, R.; Busen, R.; Schulte, P.; Schumann, U. First direct sulfuric acid detection in the exhaust plume of a jet aircraft in flight. Geophys. Res. Lett. 1998, 25, 923–926.

17 Kärcher, B.; Meilinger, S.K. Perturbation of the aerosol layer by aviation-produced aerosols: A parameterization of plume processes. Geophys. Res. Lett. 1998, 25, 4465–4468.

18 Fahey, D.W.; Keim, E.R.; Woodbridge, E.L.; Gao, R.S.; Boering, K.A.; Daube, B.C.; Wofsy, S.C.; Lohmann, R.P.; Hinsta, E.J.; Lowenstein, M.; et al. In situ observations in aircraft exhaust plumes in the lower stratosphere at midlatitudes. J. Geophys. Res. 1995, 100, 3065–3074.

19 Kärcher, B.; Lohmann, U. A parameterization of cirrus cloud formation: homogeneous freezing of supercooled aerosols. J. Geophys. Res. 2002, 107.

20 Gettelman, A.; Chen, C. The climate impact of aviation aerosols. Geophys. Res. Lett. 2013, 40, 1–5.

21 Unger, N. Global climate impact of civil aviation for standard and desulfurized jet fuel. Geophys. Res. Lett. 2011, 38, L20803.

22 Grewe, V.; Tsati, E.; Hoor, P. On the attribution of contributions of atmospheric trace gases to emissions in atmospheric model applications. Geosci. Model Dev. 2010, 3, 487–499.

23 Gottschaldt, K.; Voigt, C.; Jöckel, P.; Righi, M.; Deckert, R.; Dietmüller, S. Global sensitivity of aviation NOx effects to the HNO3-forming channel of the HO2 + NO reaction. Atmos. Chem. Phys. 2013, 13, 3003–3025.

24 Søvde, O.A.; Matthes, S.; Skowron, A.; Iachetti, D.; Lim, L.; Hodnebrog, Ø.; Di Genova, G.; Pitari, G.; Lee, D.S.; Myhre, G.; Isaksen, I.S.A. Aircraft emission mitigation by changing route altitude: A multi-model estimate of aircraft NOxemission impact on O3 photochemistry. Atmos. Environ. 2014.

25 Pitari, G.; Mancini, E.; Rizi, V.; Shindell, D.T. Impact of future climate and emission changes on stratospheric aerosols and ozone. J. Atmos. Sci. 2002, 59, 414–440.

26 Eyring, V.; Butchart, N.; Waugh, D.W.; Akiyoshi, H.; Austin, J.; Bekki, S.; Bodeker, G.E.; Boville, B.A.; Bruhl, C.; Chipperfield, M.P.; et al. Assessment of temperature, trace species, and ozone in chemistry-climate model simulation of the recent past. J. Geophys. Res. 2006, 99, D22308.

27 Morgenstern, O.; Giorgetta, M.A.; Shibata, K.; Eyring, V.; Waugh, D.W.; Shepherd, T.G.; Akiyoshi, H.; Austin, J.; Baumgärtner, A.; Bekki, S.; et al. A review of CCMVal-2 models and simulations. J. Geophys. Res. 2010, 115, D00M02.

28 Pitari, G.; Aquila, V.; Kravitz, B.; Robock, A.; Watanabe, S.; Cionni, I.; De Luca, N.; Di Genova, G.; Mancini, E.; Tilmes, S. Stratospheric Ozone Response to Sulfate Geoengineering: Results from the Geoengineering Model Intercomparison Project (GeoMIP). J. Geophys. Res. 2014, 119, 2629–2653.

29 Minschwaner, K.; Salawitch, R.J.; McElroy, M.B. Absorption of Solar Radiation by O2: Implications for O3 and Lifetimes of N2O, CFCl3, and CF2Cl2. J. Geophys. Res. 1993, 98, 543–561.

30 Rayner, N.A.; Parker, D.E.; Horton, E.B.; Folland, C.K.; Alexander, L.V.; Rowell, D.P.; Kent, E.C.; Kaplan, A. Global analyses of sea surface temperature, sea ice, and night marine air temperature since the late nineteenth century. J. Geophys. Res. 2003, 108.

31 Pitari, G.; Rizi, V.; Ricciardulli, L.; Visconti, G. High-Speed civil transport impact: The role of Sulfate, Nitric acid Trihydrate and ice aerosols studied with a two-dimensional model including aerosol physics. J. Geophys. Res. 1993, 98.

32 Brunner, D.; Staehelin, J.; Rogers, H.L.; Köhler, M.O.; Pyle, J.A.; Hauglustaine, D.; Jourdain, L.; Berntsen, T.K.; Gauss, M.; Isaksen, I.S.A.; Meijer, E.; et al. An evaluation of the performance of chemistry transport models—Part 2: Detailed comparison with two selected campaigns. Atmos. Chem. Phys. 2005, 5, 107–129.

33 Chou, M.-D.; Suarez, M.J.; Liang, X.-Z. A Thermal Infrared Radiation Parameterization for Atmospheric Studies; NASA Goddard Space Flight Center: Greenbelt, MD, USA, 2001; p. 55.

34 Søvde, O.A.; Prather, M.J.; Isaksen, I.S.A.; Berntsen, T.K.; Stordal, F.; Zhu, X.; Holmes, C.D.; Hsu, J. The chemical transport model Oslo CTM3. Geosci. Model Dev. 2012, 5, 1441–1469.

35 Prather, M.J.; Zhu, X.; Strahan, S.E.; Steenrod, S.D.; Rodriguez, J.M. Quantifying errors in trace species transport modeling. Proc. Natl. Acad. Sci. USA 2008, 105, 19617–19621.

36 Holtslag, A.A.M.; DeBruijn, E.I.F.; Pan, H.L. A high resolution air mass transformation model for short-range weather forecasting. Mon. Wea. Rev. 1990, 118, 1561–1575.

37 Berntsen, T.; Isaksen, I.S.A. A global 3-D chemical transport model for the troposphere, Model description and CO and Ozone results. J. Geophys. Res. 1997, 102, 21239–21280.

38 Berglen, T.F.; Berntsen, T.K.; Isaksen, I.S.A.; Sundet, J.K. A global model of the coupled sulfur/oxidant chemistry in the troposphere: The sulfur cycle. J. Geophys. Res. 2004, 109, D19310.

39 Søvde, O.A.; Gauss, M.; Smyshlyaev, S.P.; Isaksen, I.S.A. Evaluation of the chemical transport model Oslo CTM2 with focus on Arctic winter ozone depletion. J. Geophys. Res. 2008, 113, D09304.

40 Hesstvedt, E.; Hov, O.; Isaksen, I.S A. Quasi steady-state approximation in air pollution modelling: Comparison of two numerical schemes for oxidant prediction. Int. J. Chem. Kin. 1978, X, 971–994.

41 Søvde, O.A.; Hoyle, C.R.; Myhre, G.; Isaksen, I.S.A. The HNO3 forming branch of the HO2 + NO reaction: preindustrial-to-present trends in atmospheric species and radiative forcings. Atmos. Chem. Phys. 2011, 11, 8929–8943.

42 Prather, M.J. Numerical advection by conservation of second-order moments. J. Geophys. Res. 1986, 91, 6671–6681.

43 Price, C.; Penner, J.; Prather, M. NOx from lightning 1. Global distribution based on lightning physics. J. Geophys. Res.1997, 102, 5929–5941.

44 Berntsen, T.K.; Isaksen, I.S.A. Effects of lightning and convection on changes in upper tropospheric ozone due to NOxemissions from aircraft. Tellus 1999, 51, 766–788.

45 Ott, L.E.; Pickering, K.E.; Stenchikov, G.L.; Allen, D.J.; De Caria, A.J.; Ridley, B.; Lin, R.-F.; Lang, S.; Tao, W.K. Production of lightning NOx and its vertical distribution calculated from three-dimensional cloud-scale chemical transport model simulations. J. Geophys. Res. 2010, 115, D04301.

46 Pickering, K.E.; Wang, Y.S.; Tao, W.K.; Price, C.; Muller, J.F. Vertical distributions of lightning NOx for use in regional and global chemical transport models. J. Geophys. Res. 1998, 103, 31203–31216.

47 Lee, D.S.; Owen, B.; Fichter, C.; Lim, L.L.; Dimitriu, D. Allocation of International Aviation Emissions from Scheduled Air Traffic - Present Day and Historical (Report 2 of 3); Final Report to the UK Department for Environment, Food and Rural Affairs(DEFRA) Global Atmosphere Division; Manchester Metropolitan University: Manchester, UK, 2005.

48 Lee, D.S.; Fahey, D.W.; Forster, P.M.; Newton, P.J.; Wit, R.C.N.; Lim, L.L.; Owen, B.; Sausen, R. Aviation and global climate change in the 21st century. Atmos. Environ. 2009, 43, 3520–3537.

49 Owen, B.; Lee, D.S.; Lim, L.L. Flying into the future: Aviation emissions scenarios to 2050. Environ. Sci. Technol. 2010,44, 2255–2260.

50 ICAO. International Civil Aviation Organization. 2013. Available online: http://www.icao.int/environmental-protection/Pages/modelling-and-databases.aspx (accessed on 29 May 2015).

51 ICAO/CAEP. International Civil Aviation Organization (ICAO)/Committee on Aviation Environmental Protection (CAEP), Agenda Item 4: Modeling and Databases Task Force (MODTF) goals assessment results. In Proceedings of the ICAO/CAEP Working Paper, Steering Group Meeting, Salvador, Brazil, 22–26 June 2009.

52 Simos, D. PIANO User's Guide Version 5.0; Lissys Limited: Loughborough, UK, 2008; Available online: www.piano.aero (accessed on 29 May 2015).

53 Eyers, C.J.; Norman, P.; Middel, J.; Plohr, M.; Michot, S.; Atkinson, K.; Christou, R.A. AERO2k Global Aviation Emissions Inventories for 2002 and 2025. QINETIQ/04/01113. QINETIQ Ltd.: Hampshire, UK, December 2004. Available online: http://aero-net.info/fileadmin/aeronet_files/links/documents/AERO2K_Global_Aviation_Emissions_Inventories_for_2002_and_2025.pdf (accessed on 29 May 2015).

54 Barrett, S.; Prather, M.; Penner, J.; Selkirk, H.; Balasubramanian, S.; Dopelheuer, A.; Fleming, G.; Gupta, M.; Halthore, R.; Hileman, J.; et al. Guidance on the use of AEDT Gridded Aircraft Emissions in Atmospheric Models, Version 2.0; Technical Note LAE-2010-008-N. US Federal Aviation Administration: Washington, DC, USA, August 2010. Available online: http://lae.mit.edu/uploads/LAE_report_series/2010/LAE-2010-008-N.pdf (accessed on 29 May 2015).

55 Penner, J.E.; Hegg, D.; Andreae, M.; Leaitch, D.; Pitari, G.; Annegarn, H.; Murphy, D.; Nganga, J.; Barrie, L.; Feichter, H. Aerosols and Indirect Cloud Effects, IPCC Third assessment report. In IPCC Climate Change 2001; Cambridge University Press: Cambridge, UK, 2001; pp. 289–348.

56 Koch, D.; Schulz, M.; Kinne, S.; Schuster, G.; Bond, T.C.; Clarke, A.; Fahey, D.W.; Freitag, S.; McNaughton, C.; Spackman, J.R.; et al. Evaluation of black carbon estimations in global aerosol models. Atmos. Chem. Phys. 2009, 9, 9001–9026.

57 Bond, T.C.; Doherty, S.J.; Fahey, D.W.; Forster, P.M.; Berntsen, T.; DeAngelo, B.J.; Flanner, M.G.; Ghan, S.; Kärcher, B.; Koch, D.; et al. Bounding the role of black carbon in the climate system: A scientific assessment. J. Geophys. Res. 2013,118, 5380–5552.

58 Flanner, M.G. Arctic climate sensitivity to local black carbon. J. Geophys. Res. 2013, 118, 1840–1851.

59 Yue, G.K.; Poole, L.R.; Wang, P.H.; Chiou, E.W. Stratospheric aerosol acidity, density, and refractive index deduced from SAGE-II and NMC temperature data. J. Geophys. Res. 1994, 99, 3727–3738.

60 Schwarz, J.P.; Gao, R.S.; Fahey, D.W.; Thomson, D.S.; Watts, L.A.; Wilson, J.C.; Reeves, J.M.; Darbeheshti, M.; Baumgardner, D.G.; Kok, G.L.; et al. Single-particle measurements of mid latitude black carbon

and light-scattering aerosols from the boundary layer to the lower stratosphere. J. Geophys. Res. 2006, 111, D16207.

61 Koch, D.; Schulz, M.; Kinne, S.; Schuster, G.; Bond, T.C.; Clarke, A.; Fahey, D.W.; Freitag, S.; McNaughton, C.; Spackman, J.R.; et al. Corrigendum to "Evaluation of black carbon estimations in global aerosol models" published in Atmos. Chem. Phys., 9, 9001–9026, 2009. Atmos. Chem. Phys. 2010, 10, 79–81.

62 Mishchenko, M.I.; Travis, L.D.; Lacis, A.A. Scattering, Absorption, and Emission of Light by Small Particles; Cambridge University Press: Cambridge, UK, 2002.

63 Kinne, S.; Lohmann, U.; Ghan, S.; Easter, R.; Chin, M.; Ginoux, P.; Takemura, T.; Tegen, I.; Koch, D.; Herzog, M.; et al. Monthly averages of aerosol properties: A global comparison among models, satellite data, and AERONET ground data. J. Geophys. Res. 2003, 108.

64 Kinne, S.; Schulz, M.; Textor, C.; Guibert, S.; Balkanski, Y.; Bauer, S.; Berntsen, T.; Berglen, T.; Boucher, O.; Chin, M.; et al. An AeroCom initial assessment—Optical properties in aerosol component modules of global models. Atmos. Chem. Phys. 2006, 6, 1815–1834.

65 Danilin, M.Y.; Fahey, D.W.; Schumann, U.; Prather, M.J.; Penner, J.E.; Ko, M.K.W.; Weisenstein, D.K.; Jackman, C.H.; Pitari, G.; Koeler, I.; et al. Aviation fuel tracer simulation: Model intercomparison and implications. Geophys. Res. Lett.1998, 25, 3947–3950.

66 Pitari, G.; Iachetti, D.; Mancini, E.; Montanaro, V.; De Luca, N.; Marizy, C.; Dessens, O.; Rogers, H.; Pyle, J.; Grewe, V.;et al. Radiative forcing from particle emissions by future supersonic aircraft. Atmos. Chem. Phys. 2008, 8, 4069–4084.

67 Thomason, L.; Poole, L.; Deshler, T. A global climatology of stratospheric aerosol surface area density deduced from stratospheric aerosol and gas experiment II measurements: 1984–1994. J. Geophys. Res. 1997, 102, 8967–8976.

68 Hendricks, J.; Kärcher, B.; Lohmann, U. Effects of ice nuclei on cirrus clouds in a global climate model. J. Geophys. Res.2011, 116, D18206.

69 Lohmann, U.; Kärcher, B. First interactive simulations of cirrus a clouds formed by homogeneous freezing in the ECHAM GCM. J. Geophys. Res. 2002, 107.

70 Rogers, H.; Teyssedre, H.; Pitari, G.; Grewe, V.; Van Velthoven, P.; Sundet, J. Model intercomparison of the transport of aircraft-like emissions from sub- and supersonic aircraft. Meteorol. Z. 2002, 11, 151–159.

71 Spivakovsky, C.M.; Logan, J.A.; Montzka, S.A.; Balkanski, Y.J.; Foreman-Fowler, M.; Jones, D.B.J.; Horowitz, L.W.; Fusco, A.C.; Brenninkmeijer, C.A.M.; Prather, M.J.; et al. Three-dimensional climatological distribution of tropospheric OH: Update and evaluation. J. Geophys. Res. 2000, 105, 8931–8980.

72 Gauss, M.; Myhre, G.; Isaksen, I.S A.; Grewe, V.; Pitari, G.; Wild, O.; Collins, W.J.; Dentener, F.J.; Ellingsen, K.; Gohar, L.K.; et al. Radiative forcing since preindustrial times due to ozone change in the troposphere and the lower stratosphere. Atmos. Chem. Phys. 2006, 6, 575–599.

73 Toon, O.B.; McKay, C.P.; Ackerman, T.P.; Santhanam, K. Rapid calculation of radiative heating rates and photodissociation rates in inhomogeneous multiple scattering atmospheres. J. Geophys. Res. 1989, 94, 16287–16301.

74 Ramanathan, V.; Dickinson, R.E. The role of stratospheric ozone in the zonal and seasonal radiative energy balance of the earth-troposphere system. J. Atmos. Sci. 1979, 36, 1084–1104.

75 Chipperfield, M.P.; Liang, Q.; Abraham, L.; Bekki, S.; Braesicke, P.; Dhomse, S.; Di Genova, G.; Fleming, E.L.; Hardiman, S.; Iachetti, D.; et al. Multi-model estimates of atmospheric lifetimes of long-lived Ozone-Depleting Substances: Present and future. J. Geophys. Res. 2014, 119, 2555–2573.

76 Randles, C.A.; Kinne, S.; Myhre, G.; Schulz, M.; Stier, P.; Fischer, J.; Doppler, L.; Highwood, E.; Ryder, C.; Harris, B.; et al. Intercomparison of shortwave radiative transfer schemes in global aerosol modeling: Results from the AeroCom Radiative Transfer Code Experiment. Atmos. Chem. Phys. 2013, 13, 2347–2379.

77 Myhre, G.; Shine, K.; Readel, G.; Gauss, M.; Isaksen, I.; Tang, Q.; Prather, M.; Williams, J.; van Velthoven, P.; Dessens, O.; et al. Radiative forcing due to changes in ozone and methane caused by the transport sector. Atmos. Environ. 2011,45, 387–394.

78 Myhre, G.; Stordal, F. Role of spatial and temporal variations in the computation of radiative forcing and GWP. J. Geophys. Res. 1997, 102, 11181–11200.

79 Stamnes, K.; Tsay, S.C.; Wiscombe, W.; Jayaweera, K. Numerically stable algorithm for discrete-ordinate-method radiative-transfer in multiple-scattering and emitting layered media. Appl. Opt. 1988, 27, 2502–2509.

80 Forster, P.; Ramaswamy, V.; Artaxo, P.; Berntsen, T.; Betts, R.; Fahey, D.; Haywood, J.; Lean, J.; Lowe, D.; Myhre, G.; et al. Changes in

Atmospheric Constituents and in Radiative Forcing. In Climate Change 2007: The Physical Science Basis. Contribution of Working Group I to the Fourth Assessment Report of the Intergovernmental Panel on Climate Change; Cambridge University Press: Cambridge, UK; New York, NY, USA, 2007; pp. 130–234.

81 IPCC/TEAP. Special report on safeguarding the ozone layer and the global climate system, 2005: Issues related to hydrofluorocarbons and perfluorocarbons. In Intergovernmental Panel on Climate Change; Metz, B., Kuijpers, L., Solomon, S., Andersen, S.O., Davidson, O., Pons, J., de Jager, D., Kestin, T., Manning, M., Meyer, L., Eds.; Cambridge University Press: Cambridge, UK, 2005; pp. 1–488.

82 IPCC. Climate change 2013. Chapter 6; In The Physical Science Basis; Stocker, T.F., Qin, D., Plattner, G.-K., Tignor, M.M.B., Allen, S.K., Boschung, J., Nauels, A., Xia, Y., Bex, V., Midgley, P.M., Eds.; Cambridge University Press: Cambridge, UK, 2013; pp. 465–570.

83 Lamarque, J.F.; Bond, T.C.; Eyring, V.; Granier, C.; Heil, A.; Klimont, Z.; Lee, D.; Liousse, C.; Mieville, A.; Owen, B.; et al. Historical (1850–2000) gridded anthropogenic and biomass burning emissions of reactive gases and aerosols: methodology and application. Atmos. Chem. Phys. 2010, 10, 7017–7039.

84 Wild, O.; Prather, M.J.; Akimoto, H. Indirect long-term radiative cooling from NOx emissions. Geophys. Res. Lett. 2001,28, 1719–1722.

85 IPCC. Aviation and the global atmosphere. In Intergovernmental Panel on Climate Change; Penner, J.E., Lister, D.H., Griggs, D.J., Dokken, D.J., McFarland, M., Eds.; Cambridge University Press: Cambridge, UK, 1999; pp. 1–373.

86 Myhre, G.; Shindell, D.; Bréon, F.-M.; Collins, W.; Fuglestvedt, J.; Huang, J.; Koch, D.; Lamarque, J.-F.; Lee, D.; Mendoza, B.; et al. Anthropogenic and natural radiative forcing. In Climate Change 2013: The Physical Science Basis. Contribution of Working Group I to the Fifth Assessment Report of the Intergovernmental Panel on Climate Change; Cambridge University Press: Cambridge, UK; New York, NY, USA, 2013; pp. 659–740.

87 Myhre, G.; Nielsen, J.S.; Gulstad, L.; Shine, K.P.; Rognerud, B.; Isaksen, I.S.A. Radiative forcing due to stratospheric water vapour from CH4 oxidation. Geophys. Res. Lett. 2007, 34, L01807.

88 Zhou, C.; Penner, J.E. Aircraft soot indirect effect on large-scale cirrus clouds: Is the indirect forcing by aircraft soot positive or negative? J. Geophys. Res. 2014.

89 Wilcox, L.J.; Shine, K.P.; Hoskins, B.J. Radiative forcing due to aviation water vapour emissions. Atmos. Environ. 2012,63, 1–13.

90 Unger, N.; Zhao, Y.; Dang, H. Mid-21st century chemical forcing of climate by the civil aviation section. Geophys. Res. Lett. 2013, 40, 641–645.

CHAPTER 9

A Polarized Atmospheric Radiative Transfer Model for Calculations of Spectra of the Stokes Parameters of Shortwave Radiation Based on the Line-by-Line and Monte Carlo Methods

Boris Fomin [1,*] and Victoria Falaleeva [2]

[1] Central Aerological Observatory, 3 Pervomayskaya Street, Dolgoprudny, Moscow Region 141700, Russia

[2] Moscow Institute of Physics and Technology, 9 Institutskiy Alley, Dolgoprudny, Moscow Region 141700, Russia

ABSTRACT

This paper presents a new version of radiative transfer model called the Fast Line-by-Line Model (FLBLM), which is based on the Line-by-Line (LbL) and Monte Carlo (MC) methods and rigorously treats particulate and molecular scattering alongside absorption. The advantage of this model consists in the use of the line-by-line model that allows for the computing of high-resolution spectra quite quickly. We have developed the model by taking into account the polarization state of light and carried out some validations by comparison against benchmark results. FLBLM calculates the Stokes parameters spectra of shortwave radiation in vertically inhomogeneous atmospheres. This update makes the model applicable for the assessment of cloud and aerosol influence on radiances as measured by the SW high-resolution polarization spectrometers. In sample results we demonstrate that the high-resolution spectra of the Stokes parameters contain more detailed information about clouds and aerosols than the medium- and low-resolution spectra wherein lines are not resolved. The presented model is rapid enough for many practical applications (e.g., validations) and might be useful especially for the remote sensing. FLBLM is suitable for development of the reliable technique for retrieval of optical and microphysical properties of clouds and aerosols from high-resolution satellites data.

KEYWORDS

solar radiation; aerosol; polarization; remote sensing; line-by-line model

1. INTRODUCTION

Clouds and aerosols play an important role in the atmospheric radiative transfer but "remain the dominant uncertainty in radiative forcing" (IPCC, 2007) [1]. Monitoring their global distribution and properties are thus the challenging objective in satellite remote sensing. The current satellite instruments, such as MODIS (King *et al.*, 2003) [2], provide images of the integral properties of the cloud or aerosol layers, e.g., optical thicknesses in certain spectral bands. However, it is hard to affirm whether researchers have enough information about their structures. To solve the problem of such uncertainties in radiative forcing, suitable satellite instruments and a reliable technique for the retrieval of the cloud and aerosol structures still needs to be improved or developed.

Applications of the cross-nadir SW polarimetry and spectroscopy seem very promising, because the outgoing solar radiation is polarized and contains information about the atmosphere and its surface. However, the quality of the retrieval information depends on the spectral resolution and number of channels (bands) in measurements. As is known, the half-widths of the air spectral lines are proportional to pressure and change from ~0.1 cm^{-1} in low troposphere to ~0.01 cm^{-1} in low stratosphere. Therefore, the resolution of 0.01 cm^{-1} is necessary to obtain the maximum amount of information encoded in the spectra of the reflected solar radiation.

NASA and CNES plan to launch the Aerosol Polarimetry Sensor (APS-NG, 2018 year) and Multi-viewing Multi-channel Multi-polarization Imagers (3MI, 2020–2036 years) (Bizzarri, 2011) [3] for measurements of a triplet of the I, Q and U Stokes parameters of outgoing radiation in the shortwave spectral region from ~300 to ~2,200 nm (~4,500–33,000 cm^{-1}). There are ~10 channels in APS-NG and 3MI with bandwidths of ~100 cm^{-1} (10–40 nm), which is similar in number to POLDER, MODIS and other satellite instruments that have been already launched. Unfortunately, this resolution (bandwidth) is much greater than ~0.01 cm^{-1} discussed above. On the other hand, TANSO-FTS (GOSAT) and some other instruments (Bizzarri, 2011) [3] measure spectra with resolution ~0.2–0.5 cm^{-1}. This encourages the development of polarization spectrometers with a resolution of ~0.1 cm^{-1} and better might be feasible. However, real usefulness of such instruments needs preliminary investigations by means of simulations of the satellite experiments and it forced us to develop a suitable model.

We present a "forward" model for a remote-sensing technique based on the underlying principle that radiance at different wavenumber points possesses different penetration depths inside the scattering media due to high variability of the molecular absorption. This principle makes it possible to get information about the media structure using the multi-channel sensors. Such a technique has been proposed by Chang and Li (2002) [4]: to derive the vertical profile of droplet effective radius using the MODIS data. Then this technique has been successfully applied for the retrieval of cloud-top and cloud-bottom heights within the oxygen A-band (~0.761 μm) using the Medium-Resolution Imaging Spectrometer (Fisher and Grassl, 1991) [5] and SCIAMACHY (Kokhanovsky, 2006) [6]. Furthermore, it is useful to note that the spectra of outgoing radiation contain 'rich information on the composition and structure of clouds' in the φ-band H_2O (at 1.38 μm) (Liou *et al.*, 1996; Mishchenko *et al.*, 2000) [7,8].

The technique needs rigorous radiative transfer models, which can take into account, in detail, the molecular absorption along with the particulate scattering. In principle, any vector code can do this by using the precalculated molecular spectra as the codes involved in recent validations (Kokhanovsky, 2010) [9]. It takes only to run the code in each spectral point. However, for MODIS and other instruments mentioned above it takes about ~100/0.01 = 10^4 times to run the code (bandwidth/resolution). It should be noted that simulation of the TANSO-FTS and analogous FT spectrometers needs not ~0.5 cm^{-1}/0.01 = 50 monochromatic calculations in each channel as it seems, but ~10 $cm^{-1}/0.01$ = 1,000 due to rather long wings of the channel response functions shown in Figure 1.

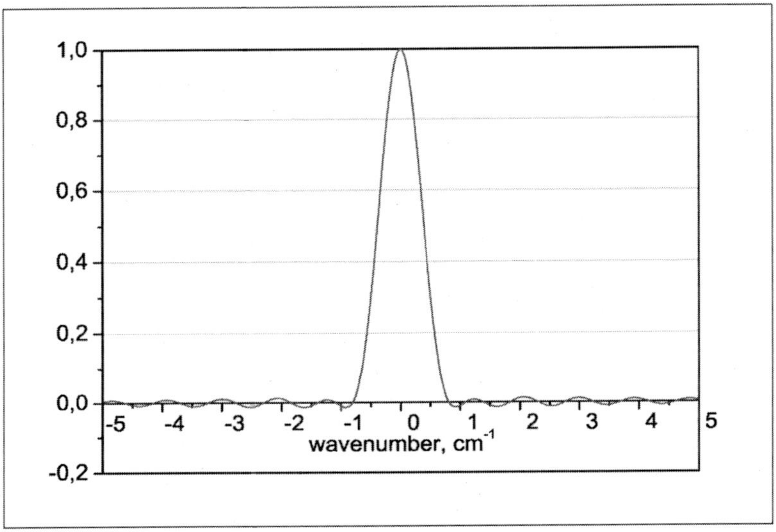

Figure 1. Response function of the Fourier Transform Spectrometer for spectral resolution equal to 0.5 cm^{-1}.

The vector and scalar codes are thus usually used together with the approximate k-distribution method, which however can only be applied to the medium- and low-resolution measurements. For example, Mishchenko *et al.* (2000) [8] used the k-distributions with 1 cm^{-1} resolution. However, this popular method is restricted in terms of accuracy and possesses other disadvantages (Fomin and Correa, 2005) [10]. For developing k-distribution researchers usually use the correlated-k technique based on the "sorting procedure", which leads to losses of any information about overlapping of the absorbing species. It makes difficult the accurate simultaneous treatment of several absorbers: optical active gases and aerosol or cloud particles in our case (Tarasova and Fomin, 2007) [11]. Therefore, calculations using the k-distribution method need validations by means of the benchmark Line-by-Line (LbL) models.

This was an additional reason for us to develop this polarized LbL radiative transfer model, which is unrestricted in spectral resolution but rapid enough for many practical applications (e.g., validations). It is a modification of author's scalar Monte Carlo model FLBLM (Fomin and

Mazin, 1998) [12], which successfully passed intercomparison with other codes (Halthore *et al.*, 2005), (Forster *et al.*, 2011) and (Oreopoulos *et al.*, 2012) [1,2,3,4,5,6,7,8,9,10,11,12,13,14,15]. For our vector model we have borrowed some algorithms from 3DMCpol (Cornet *et al.*, 2010) [16] and other MC codes discussed in Section 2.3. But in contrast to 3DMCpol, which is used to assess the 3-D radiative effects on polarized reflectance of clouds, FLBLM is suitable to assess the high-resolution spectral effects of the 1-D clouds and aerosols. Thus in this paper we consider the most important details of the rapid but accurate treatments of the molecular spectra in the MC models. Also we provide an example of high-resolution retrieval and perform some analysis on how the errors in the retrieval values depend on spectral resolution.

2. DESCRIPTION OF THE MODEL

The presented radiative transfer model follows the standard MC, LbL and other techniques. In this section we describe mostly those details of the applied algorithms, which essentially increase the computation speed. Therefore, we consider: (a) precalculations of the parameters of electromagnetic scattering on cloud and aerosol particles (spherical polydispersions in this study); (b) computations of the molecular spectra; (c) the applied MC technique.

2.1. Computation of the Parameters of Electromagnetic Scattering on Spherical Polydispersions

The applied algorithms are based on the Mie theory and the geometric optics approach (GOA) along with the Fraunhofer diffraction (for details see (Fomin and Mazin, 1998) [12]). The last one is used when direct computations using the Mie theory become too slow in cases wherein the scattering particle size is much greater than the wavelength. In this case, we apply an effective 'inverse' algorithm in order to calculate scattering parameters at any angle (not interval). It makes it easy to combine calculations based on the Mie theory (for "small" particles), which defines scattering parameters at determined angles, with calculations based on GOA (for "large" particles). Thus the angular resolution in our calculations is practically unrestricted, as demonstrated in Figure 2. This figure shows a part of the phase function around rainbows calculated using GOA (solid line) and the Mie theory (dashed line). For this illustration we used a "large" water drop (diameter $D = 1$ mm, wavelength $\lambda = 0.412$ µm).

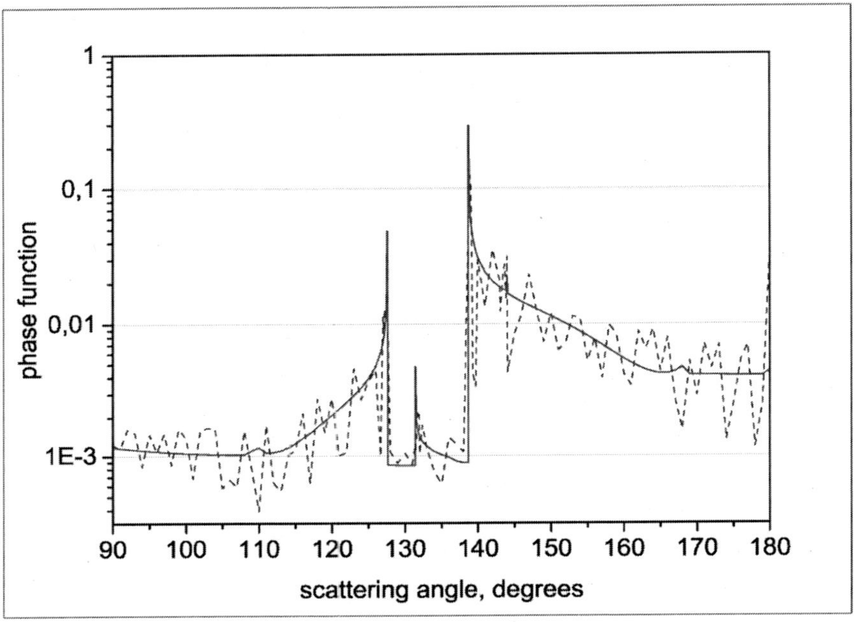

Figure 2. Part of the phase function around rainbows calculated using the geometric optics approach (GOA) (solid line) and the Mie theory (dashed line) for the water drop (diameter $D = 1$ mm, wavelength $\lambda = 0.412$ µm).

There is a problem with the GOA approach that needs to be discussed. It is divergence of the light intensities at the angles associated with rainbows, where the ratio of the solid angles of the scattering and initial rays becomes infinite. However, the integration of intensity over the scattering angle gives ray energy that is limited. So inside the neighborhoods of singular points (~0.1°) the average values are used, which are calculated analytically as suggested in (Fomin and Mazin, 1998) [12]. Here it is useful to note that the use of a phase matrix for the radiative transfer calculations generally leads to such integration and this algorithm can be considered as "exact" for the most of practical applications.

For the spherical polydispersions all the scattering parameters of a single particle must be integrated over radius a using the given size distribution function $n(a)$. We apply the Mie algorithms when a $< (\lambda/2\pi)X$ and after that the GOA algorithms. The X parameter depends on the accuracy of the calculations and usually lies in the interval (100, 1,000). The special numerical technique for this integration is also described in (Fomin and Mazin, 1998) [12]. Usually, it takes only several minutes to obtain optical properties of a typical cloud or aerosol model in the whole spectral region (from microwave to UV) using an ordinary computer (e.g., notebook).

Generally, such calculations are made in advance for the entire spectral range and the optical properties of clouds and aerosols are stored in the database. However, the remote sensing problems often require performing calculations in a narrow spectral region (e.g., near infrared) and only a few spectral points but for many polydispersions. In this case calculation is performed in seconds and if necessary, this block can be included in the model, which provides its flexibility.

2.2. Computation of the Molecular Spectra

The monochromatic volume absorption coefficient, $K(\nu)$, at any j-th wavenumber point ν_j usually can be calculated by formula

$$K(\nu_j) = \left[f_1(\nu_j) + f_2(\nu_j) ... + f_i(\nu_j) ... \right] + \text{continual absorption}, \quad (1)$$

where $f_i(\nu)$ is the absorption profile of the i-th spectral line. As was shown above it takes to consider at least $\sim 10^3$–10^4 points ν_j in each spectral channel of the satellite instruments, whereas a number of functions $f_i(\nu)$ may reach tens of thousands and more. Moreover the absorption profile $f_i(\nu)$ depends on temperature, pressure, gas volume mixing ratio and constants associated with the i-th spectral transition. So the problem of efficient LbL algorithms is of extreme importance. In FLBLM we use the algorithm suggested in (Fomin, 1995) [17], which is one of the most effective for the Summation (1) as has been tested by Kuntz and Hopfner (1998) [18]. In these algorithms we use a set of the uniform wavenumber (frequency) grids with doubling wavenumber steps. It allows the possibility for calculating each profile $f_i(\nu)$ independently of others: in detail at the line center and more rarely in the far wings as shown in Figure 3.

Usually it takes less than 100 wavenumber points to approximate each $f_i(\nu)$ with the required accuracy (~1%). When all the lines are taken into account, we apply the simple (parabolic) interpolation procedure to obtain the absorption spectrum at the required $\sim 10^3$–10^4 equidistant wavenumber points. It should be stressed that the procedure is performed at once to the all spectral lines that takes negligible computer times. This algorithm thus allows for increasing the calculation speed up to ~$(10^3–10^4)/100$ = 10–100 times and the spectra calculations usually require less computer time than further solving of the radiative transfer equations. Finally it should be mentioned that the HITRAN-2008 spectral database (Rothman *et al.*, 2009) [19] is used for the line profile calculation as well as the MT_CKD-2.5 water vapor and other continuum models from the Line-by-Line Radiative Transfer Model (LBLRTM) (Clough, 2005) [20].

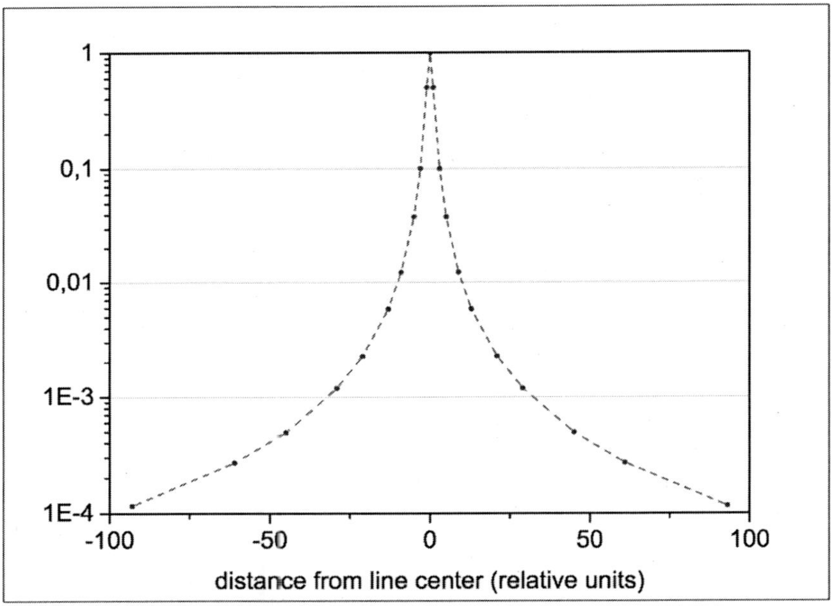

Figure 3. Illustration of the effective interpolation technique using the nonuniform wavenumber grid for the line profile calculation.

In this version, photon's trajectories are obtained by MC for scattering but non absorbing medium. The selective gas absorption is then taken into account with the help of the photon's "weight":

$$W_\nu\left(M^*\right)=\exp\left(-u_\nu\left(M^*\right)\right), \tag{2}$$

where M* is the trajectory's point and μ_ν(M*) is the photon's optical path. Due to the above effective LbL technique the consideration of the gaseous absorption usually needs less computer time than the MC calculations so that it is not necessary to apply the K-distribution technique.

2.3. Applied MC Technique

The scalar version of the model is a usual forward MC (Marchuk *et al.*, 1980) [21], (Evans and Marshak, 2005) [22] combined with the LbL technique. It has been developed as long ago as in the 1990s (Fomin, *et al.*, 1993) [23]. We will now describe the polarized version, which can calculate the Stokes vector $S = (I, Q, U, V)$ (Chandrasekhar, 1960) [24]. The applied technique is based on the "local estimation" MC and follows the technique that has been described in detail in a paper by Cornet *et al.* (2010) [16]. There are two differences in the vector approach compared to the scalar one. The first difference is related to the fact that the scattering azimuth angle of a photon now is not a uniformly distributed random value between 0° and 180° degrees but depends on the initial polarization state and the scattering zenith angle. To overcome this difficulty we applied an algorithm suggested by Cornet *et al.* (2010) [16]. The second difficulty is related to the fact that the scattering plane rotates after each scattering so that it takes to perform a pair of rotations between the meridian and this planes. Here we follow method described in (Sushkevich, 2005) [0].

In FLBLM these monochromatic MC algorithms for the scattering but non absorbing media were combined with the above LbL algorithms to take into account the particulate and gaseous absorption by means of the photon "weights" (see (Marchuk *et al.*, 1980), (Evans and Marshak, 2005)) [21,22]. The presented model is based on the "local estimation" MC because it allows reducing the statistical errors in calculated radiances and consequently this method is more effective for applications in the remote sensing problems.

To perform computations for any spectral regions (e.g., the instrument channels) we use two methods. In the first method, photons are uniformly distributed (usually randomly) over the given spectral region. This method has been used for a long time in the scalar version and was found very effective for calculations in the wide spectral regions due to statistical errors in each wavenumber points, which have a tendency to annihilate each other. Thus the final errors mostly depend on the total number of photons. The second method (Romanov *et al.*, 1991) [26] wherein the same group of the photon's trajectories (in nonabsorbing media) is used at each wavenumber point is more suitable for the very high-resolution computations: e.g., spectra themselves. Both methods are demonstrated in Figure 4, wherein a spectrum of the normalized *Q*-parameter at the zenith angle = 50° is shown.

In our model the cloud and aerosol optical properties are constant in the spectral range no wider than 10 cm^{-1} (calculation is performed in such intervals). For the next interval these optical properties are recalculated. Here we use the fact that the optical properties of aerosol and cloud vary much more smoothly compared to selective gas absorption. It should be stressed that the first method is more rapid in practical calculations and we will discuss it further in Section 4.

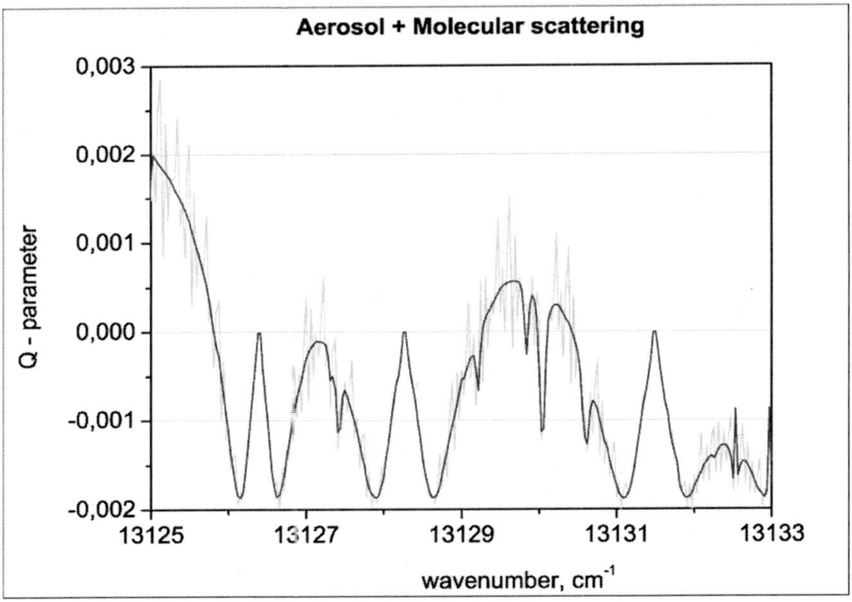

Figure 4. Spectra of the normalized *Q*-parameter at the zenith angle = 50° calculated by the first (light line) and second methods (dark line) (see text).

For these calculations we used the aerosol model from the recent intercomparison with benchmark results (Kokhanovsky*et al.*, 2010) [9] (see Section 3). The molecular scattering is also taken into account at λ ~ 0.76 μm (to illustrate situation in the A-band O_2), which is comparable with the aerosol scattering at the given angle. Also we used the standard "mid-latitude summer" atmospheric model with the homogeneous aerosol layers between 0.0 and 1.0 km above the underlying surface. The spectral region is a part of the A-band O_2, which is rather popular in the remote sensing.

3. VALIDATIONS AGAINST BENCHMARK RESULTS

The update of the model (solution of the vector radiative transfer equations) was carefully validated by means of the benchmark computations of SCIATRAN from a recent intercomparison of the vector radiative transfer codes (Kokhanovsky*et al.*, 2010) [9]. So we use the same cloud and aerosol models that have been applied in (Kokhanovsky *et al.*, 2010) [9]. In both models the radii a of particles follow the lognormal distributions

$$n(a) = \frac{1}{\sqrt{2\pi}sa}\exp\left(-\ln^2(a/a_0)/2s^2\right) \tag{3}$$

where $a_0 = 0.3$ µm and $a_0 = 5$ µm, $s = 0.92$ and $s = 0.4$ for the aerosol and cloud models, respectively. It was also assumed that all these aerosol and cloud particles are nonabsorbing but refractive and their refractive indexes are 1.385 and 1.339, respectively.

Unfortunately, the "local estimation" MC mentioned above still requires a lot of computer time in a case of strong forward scattering of large particles. To overcome this difficulty we applied a usual trick: cutting-off a strong forward (diffraction) peak to smooth the phase matrix elements and then appropriate correction (decreasing) the scattering coefficients. It means disregarding scattering within small angles (~1°). Both the aerosol and cloud phase functions are shown in Figure 5(a,b).

These figures show the transformed phase functions after smoothing of the phase matrix elements and their normalization. Finally, it should be noted that in the benchmark calculations the optical thicknesses of the aerosol and cloud layers were assumed to be equal to 0.3262 and 5.0, respectively. In the case of application in our calculations of the transformed (smoothed) matrix elements the aerosol and cloud optical thicknesses have been decreased to 0.3262 × 0.95216 = 0.31088 and 5.0 × 0.68252 = 3.4126 values. The factors of 0.95216 and 0.68252 have been obtained during normalization of the phase matrix elements just after the above correction.

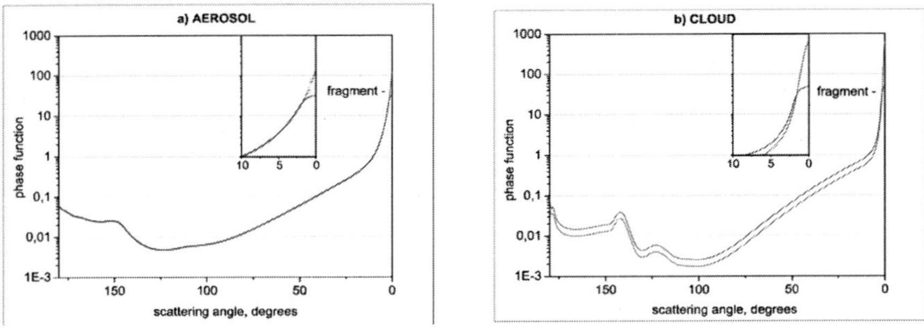

Figure 5. Phase functions of the aerosol (**a**) and cloud (**b**) models used in this work: initial (from (Kokhanovsky *et al.*, 2010) [9]) and after cutting-off a strong forward peak.

Results show rather good agreement with the benchmark computations; some of them are presented below. Calculations have been performed for the homogeneous nonabsorbing atmosphere and the black underlying surface. The molecular scattering has been excluded. In these calculations the solar zenith angle (SZA) and the relative azimuth between line of sight and the initial solar beam are equal to 60° and 180°, respectively. For comparison with the benchmark calculations the Stokes parameters have been normalized by factor (π/(cos(60°) × solar irradiance)). Figure 6 demonstrates an example comparison of Q-parameter in transmitted light for aerosol case (initial matrix) for different relative azimuths. Figure 7shows the advantage of transformed matrix over the initial one for cloud model. For both these figures we used 10^5 photons.

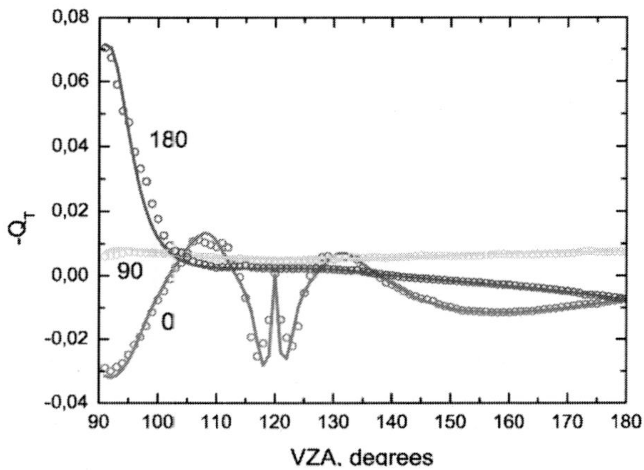

Figure 6. The normalized-Q-parameter of the Stokes vector (line—SCIATRAN, circles—Fast Line-by-Line Model (FLBLM)) in transmitted light for aerosol case and initial matrix. The number of photons is 10^5. Relative azimuths are 0°, 90°, and 180°. Azimuths counter clock-wise.

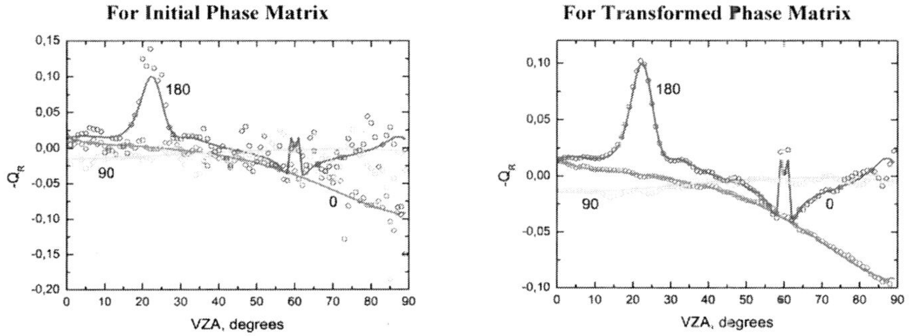

Figure 7. The normalized -Q-parameter of the Stokes vector (line—SCIATRAN, circles—FLBLM) in reflected light for cloud case. The number of photons is 10^5. Relative azimuths are 0°, 90°, and 180°. Azimuths counter clock-wise.

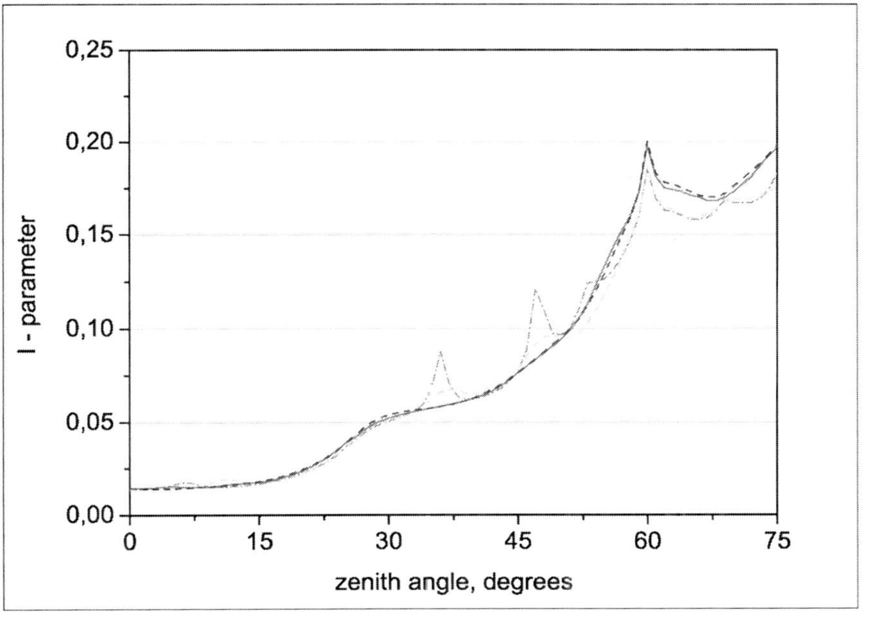

Figure 8. The normalized intensity calculated for the aerosol model. Optical thickness = 0.3262, SZA = 60°, relative azimuth = 180°. Red solid line—benchmark calculations, blue dashed line—calculations using 10^5 photons with the transformed scattering matrix, gray dashed-dotted and green dashed lines—calculations using 10^4 photons with the initial and transformed scattering matrixes, respectively.

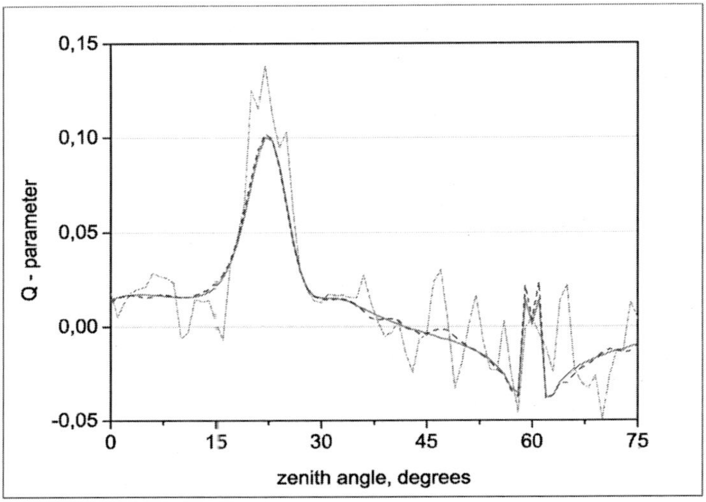

Figure 9. The normalized -*Q*-parameter calculated for the cloud model. Optical thickness = 5.0, SZA = 60°, relative azimuth = 180°. Red solid line—benchmark calculations, blue dashed and gray dashed-dotted lines—calculations using 10^5photons with the transformed and initial scattering matrixes, respectively.

Besides the calculations with modified phase matrices, one of the most important things to vary is the number of photons. Some results of computations of solar radiation reflected from the above aerosol and cloud layers are shown in Figure 8 andFigure 9, respectively. These figures also demonstrate usefulness of the above cutting-off of the forward peak in the scattering matrix elements. We have performed several calculations using 10^4 and 10^5 photons for the initial and transformed scattering matrices (see Figure 8). As one can see, this trick helps to decrease number of photons in 10–100 times. It gives a hope that ~10^4–10^5 photons will be enough to reach accuracy needed for the most practical applications. (There have been used 10^8 photons in the benchmark calculations (Kokhanovsky *et al.*, 2010) [9]).

4. APPLICATION TO REMOTE SENSING

4.1. Retrieval of the Structure of Clouds and Aerosols

A measured high-resolution spectrum represents *N* experimental values that make it possible to evaluate *J* retrieval values. It should be stressed that $J << N$ due to experimental errors and a fact that the "inverse" problem is ill-conditioned in many practical situations. Also $J << N$ because of the statistical errors in the N measurements as well as several points contain identical information about clouds and aerosols.

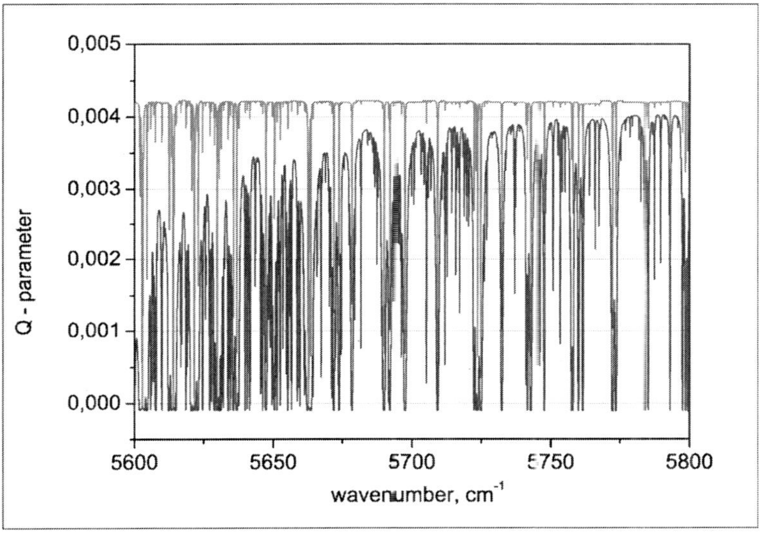

Figure 10. Differences in a spectrum of the reflected solar radiation caused by change in the aerosol structure. The same that Figure 8 (the single aerosol layer between 0–1 km, optical thickness = 0.3262, Mid-Latitude Atmospheric model)but in the spectral region of the Ω—band H_2O (1.87 μm). Red line – for the above homogeneous aerosol layer and blue line—for a layer of the same optical thickness but with linear dependence on altitude of the volume scattering coefficient from maximal at the surface to zero at 1 km.

From high-resolution spectra of Stokes parameters measured in the molecular absorption bands of the reflected solar radiation, it should be possible to obtain several parameters describing the cloud and aerosol structures. This statement is confirmed by Figure 10, which demonstrates clearly distinguishable differences in a spectrum of the reflected solar radiation caused by changes in the aerosol structure only. The reason of the difference between homogeneous aerosol layer and a layer with extinction coefficient increasing from the top to the bottom is obvious. Indeed, the first layers near the top are more reflective leading to a weaker absorption of the photons by the atmospheric molecules. In the same way, the differences in signals between the channels are linked to the absorption and could give information at different levels.

An important question arises in such kind of satellite experiments: how do errors in the retrieval values depend on the spectral resolution and the Signal to Noise Ratio (SNR)? To make this question clear and indirectly to evaluate applicability of our model, we considered a simple but realistic numerical experiment, which simulates investigation of aerosol structures in the boundary layer. In this experiment it is supposed that the solar and satellite zenith angles = 0.0° and the aerosol optical thickness is small enough to use the "single scattering" approximation. The spectral region and atmospheric model are the same as in Figure 10. The aerosol volume scattering coefficient ρ is defined by expression

$$\rho(z)=\sum_{j=1}^{J} x_j \delta(z-z_j) \tag{4}$$

where z is altitude, the nodes z_j are equidistant points from 0.0 to 1.0 km and x_j are the retrieval values. Thus in our numerical experiment the signal y_i (normalized intensity I, Q or U component) in each spectral point is equal to

$$y_i=\sum_{j=1}^{J} B_i^j x_j+\varepsilon_i \tag{5}$$

where ε_i is the Gaussian (normal) distributed errors with variance $\sigma^2{}_i \sim (y_i{*}SNR)^2$. The $B^j{}_i$ coefficients are

$$B_i^j=\exp(-2\tau_i^j) \tag{6}$$

where $\tau^j{}_i$ is the molecular absorption optical depth at the z_j altitude and the i-th spectral point. It should be stressed that these coefficients are realistic because have been calculated by the LbL method for the given atmospheric model. We have been performed statistical analysis using the theory of maximum likelihood estimation (Rodgers, 2000) [27] that gives a system of J linear equations for J unknown values x_j:

$$\sum_{j=1}^{J} x_j \left(\sum_{i}^{N} B_i^j B_i^l\right) = \sum_{i}^{N} y_i B_i^l, (l = 1, 2, \ldots, J). \tag{7}$$

where N is a number of the spectral points. For the selected spectral region 5,600–5,800 cm^{-1} in our numerical experiments we have tested $N = 6{,}400$, 800 and 200 (spectral resolution = 1/32, 1/4, and 1 cm^{-1}) and $J = 3, 4\ldots$ Statistical errors of the x_j-values, which are related to experimental errors ε_i, can be evaluated using the Fisher information matrix (Rodgers, 2000) [27] or by means of the series of L repetitions of the experiment. We chose the second method as more universal, wherein we have selected $L = 1{,}000$ repetitions. In each l-th repetition $y_i^{(l)}$ have been obtained by Formula (5), where initial values $<x_j> = 0.1 \times 2/[J(J + 1)] \times (J - j + 1)$ (it corresponds to the linear aerosol profile with optical thickness = 0.1). Errors $\varepsilon_i^{(l)}$ have been obtained using a program that independently generates the normally distributed quantities (the diagonal covariance matrix of y_i is supposed). Thus in each l-th "retrieval" ($l = 1, 2, \ldots, L$) we have obtained $x_j^{(l)}$ values for the statistical data processing. The result is shown in Table 1, where relative errors in percent η_j for the x_j values are calculated by expression

$$\eta_j = \frac{100}{<x_j>}\left[\frac{1}{(L-1)}\sum_{l=1}^{L}(x_j^{(l)} - <x_j>)^2\right]^{0.5} \tag{8}$$

Table 1. Relative errors (%) in a case of four retrieval values in the numerical experiment that simulates investigation of aerosol structures in the boundary layer (see text). X—the retrieved value and N—the number of spectral measurements.

	$X_1 = 0.04$			$X_2 = 0.03$			$X_3 = 0.02$			$X_4 = 0.01$		
SNR	**N = 6,400**	**800**	**200**	**N = 6,400**	**800**	**200**	**N = 6,400**	**800**	**200**	**N = 6,400**	**800**	**200**
1,000	1.7	5	10	5	16	30	7	21	40	5	16	30
333	5	15	30	16	47	91	21	64	121	16	48	89
100	17	52	102	53	156	305	71	213	402	53	160	297

These preliminary results confirm that retrieval of the aerosol structure by the discussed method is possible even in the thin boundary layer.

However, it needs to take into account spectra consisting from thousands of the spectral points and SNR must be greater than ~300 (as in TANSO-FTS). Also it can be expected that number of the independent parameters (degree of freedom) hardly exceeds $J = 3–4$ for real instruments.

As a whole errors η_j, as was expected, are inversely proportional to a factor $SNR*(N)^{0.5}$. So it may seem that it is the same to use different spectral resolutions and spectral intervals if only the number of points N is the same. In Table 2, we have tested the effect of the spectral resolution on the relative error by keeping the number of spectral points to N = 200 and repeated our experiment in the spectral regions 6,200–6,206.25 cm^{-1} and 6,200–6,400 cm^{-1} with spectral resolutions 1/32 cm^{-1} and 1 cm^{-1}, respectively.

Table 2. The same that Table 1 but for spectral resolutions are 1/32 and 1.0 cm^{-1} and spectral regions are 6,200–6,206.5 and 6,200–6,400, respectively. The number of the spectral points $N = 200$.

	X_1=0.04		X_2=0.03		X_3=0.02		X_4=0.01	
SNR	1/32	1.0	1/32	1.0.	1/32	1.0.	1/32	1.0
1,000	7.9	10	20	30	21	40	14	30
333	24	30	59	91	63	121	41	89
100	79	102	196	305	211	402	137	297

As one can see errors in the high-resolution retrieval are essentially less than in other case and, consequently, the high-resolution measurements are preferable: evidently because they can better use more informative parts of a spectrum. Also it should be mentioned that measurements of the Q and U components of the Stokes vector allow better investigation of aerosols and clouds because they often are the main polarizing factor. Thus the high-resolution polarized measurements seem rather promising.

4.2. Time Considerations for High Resolution Spectra Calculations

To investigate applicability of the model to remote sensing as a “forward” model we have performed some numerical experiments with an imagined hybrid of POLDER and TANSO-FTS. The result was a simulation of the measurements by FTS with the spectral resolution equals to 0.5 cm^{-1} (see Figure 1) in a band 13,000–13,200 cm^{-1}. In this section we present some computer time estimations for calculation of simulated spectra measurements. It took ~10 hours for the calculations performed by the second method (see Section 2.3) using an ordinary computer (CPU 3.20 GHz) with a spectral resolution equal to 1/32 = 0.03125 cm^{-1} in the spectral region 13,000–13,200 cm^{-1} (the POLDER's channel 768.35–757.45 nm) for a triplet of the relative azimuth angles (0, 90 and 180 degrees) and ten of the zenith angles (0, 10, 20 … 90 degrees). The number of photons was 2,000,000 in each 8.0 cm^{-1} spectral subinterval, where the scattering properties are supposed constant. But the number of photons can be essentially decreased if we apply the more rapid first method. It took only 20,000 instead of 2,000,000 photons and ~10 minutes of the computer time. Figure 11 demonstrates the initial and convoluted spectra obtained by the both methods for the aerosol layer between 0 and 1 km.

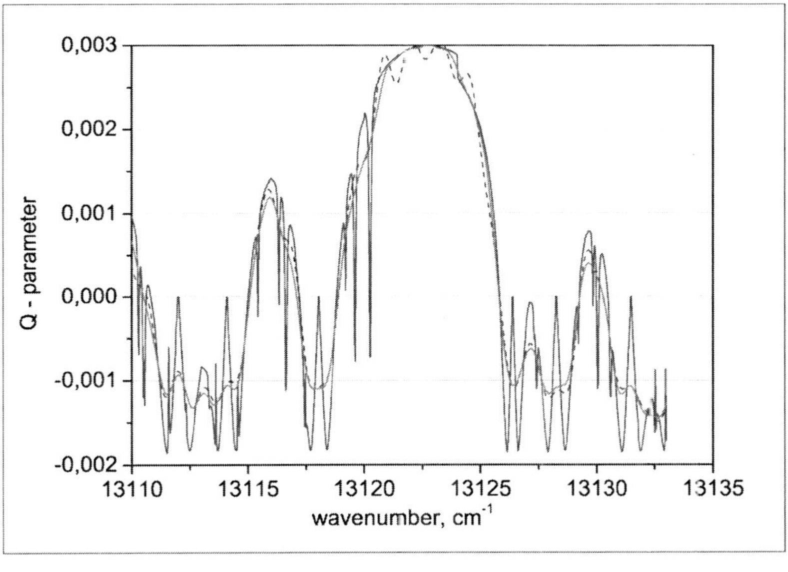

Figure 11. Initial (blue line) and convoluted with the response function (Figure 1) (red and gray lines) spectra of the normalized *Q*-parameter in the standard "Mid-latitude Summer Atmosphere". The homogeneous aerosol layer is between 0–1 km. Solid blue line—initial spectrum; red solid line and dashed gray lines—convoluted spectra obtained by the second method using 2 × 10^6 and first method using 2 × 10^4 photons, respectively.

Nevertheless the visible statistical errors are less than the spectrum distortion because of the convolution. It should be noted that these errors become practically invisible for ~100,000 photons and it takes ~1 hour of the computer time. Due to the fact that the statistical errors mostly depend on the total number of photons (not spectral interval) the computer time is inversely proportional to the spectral resolution. Thus calculations for POLDER need ~100 times less computer time than for TANSO-FTS (several minutes only).

5. CONCLUSIONS

The update of the presented model has been successfully validated by means of the recent intercomparison (Kokhanovsky*et al.*, 2010) [9]. Being installed on a typical PC, FLBLM simulates high-resolution measurements (with resolution ~0.5 cm^{-1}) in a ~100 cm^{-1} spectral region during ~10–100 minutes (depending on the required accuracy). It takes only several minutes to obtain results for the medium resolution measurements (e.g., by POLDER, 3MI, *etc.*).

Our results confirm that the high-resolution spectrum of the Stokes parameters of the reflected solar radiation contain much more useful information about structures of the scattering layers than the medium-resolution one. The presented model thus seems applicable for the remote sensing as a simulator and for validations.

ACKNOWLEDGMENTS

This work was supported by grants from RFBR (11-01-00021 and 09-01-00071).

REFERENCES

1 IPCC. Summary for Policymakers. Available online: http:// www.ipcc.ch/pdf/assessment-report/ar4/wg1/ar4-wg1-spm.pdf (accessed on 15 December 2007).

2 King, M.D.; Menzel, P.W.; Kaufman, Y.J.; Tanre, D.; Gao, B.C.; Platnick, S.; Ackerman, S.A.; Remer, L.A.; Pincus, R.; Hubanks, P.A. Cloud and aerosol properties, precipitable water, profiles of temperature and water vapor from MODIS. *IEEE Trans. Geosci. Remote Sens.* **2003**, *41*, 442–458.

3 Bizzari, B. The space-based global observing system in 2011 (GOS-2011). Available online: ftp://ftp.wmo.int/ Documents/PublicWeb/sat/DossierGOS (accessed on 1 September 2011).

4 Chang, F.L.; Li, Z. Estimating the vertical variation of cloud droplet effective radius using multispectral near-infrared satellite measurements. *J. Geophys. Res.* **2002**, *107*, 1–12.

5 Fisher, J.; Grassl, H. Detection of cloud top-height from backscattered radiances within the oxygen A-band. Part 1: Theoretical study. *J. Appl. Meteorol.* **1991**, *30*, 1245–1259.

6 Kokhanovsky, A.A. Reference. In *Cloud Optics*; Springer: Berlin, Germany, 2006.

7 Liou, K.N.; Goody, R.M.; West, R. CIRRUS: A Low Cost Cloud/Climate Mission. Office of Space Science NASA: Washington, DC, USA, 1996.

8 Mishchenko, M.I.; Hovenier, J.W.; Travis, L.D. Reference. In *Light Scattering by Nonspherical Particles*; Academic Press: San Diego, CA, USA, 2000.

9 Kokhanovsky, A.A.; Budak, V.P.; Cornet, C.; Duan, M.; Emde, C.; Katsev, I.L.; Klyukov, D.A.; Korkin, S.V.; C-Labonnote, L.; Mayer, B.; Min, Q.; Nakajima, T.; Ota, Y.; Prikhach, A.S.; Rozanov, V.V.; Yokota, T.; Zege, E.P. Benchmark results in vector atmospheric radiative transfer. *J. Quant. Spectrosc. Rad. Transfer.* **2010**, *111*, 1931–1946.

10 Fomin, B.A.; Correa, M.P. A k-distribution technique for radiative transfer simulation in inhomogeneous atmosphere: 2. FKDM, fast k-distribution model for the shortwave. *J. Geophys. Res.* **2005**.

11 Tarasova, T.A.; Fomin, B.A. The use of new parameterizations for gaseous absorption in the CLIRAD-SW solar radiation code for models. *J. Atmos. Ocean. Technol.* **2007**, *24*, 1157–1163.

12 Fomin, B.A.; Mazin, I.P. Model for an investigation of radiative transfer in cloudy atmosphere. *Atmos. Res.* **1998**.

13 Halthore, R.N.; Crisp, D.; Schwartz, S.E.; Anderson, G.P.; Berk, A.; Bonnel, B.; Boucher, O.; Chang, F.-L.; Chou, M.-D.; Clothiax, E.E.; Dubuisson, P.; Fomin, B.; Fouquart, Y.; Freidenreich, S.; Gautier, C.; Kato, S.; Lazlo, I.; Li, Z.; Plana-Fattori, A.; Ramaswamy, V.; Ricchiazzi, P.; Shiren, Y.; Trischenko, A.; Wiscombe, W. Intercomparison of shortwave radiative transfer codes and measurements. *J. Geophys. Res* **2005**.

14 Forster, P.; Fomichev, V.; Rozanov, E.; Cagnazzo, C.; Jonsson, A.; Langematz, U.; Fomin, B.; Iacono, M.; Mayer, B.; Mlawer, E.; Myhre, G.; Portmann, R.; Akiyoshi, H.; Falaleeva, V.; Gillett, N.; Karpechko, A.; Li, J.; Lemennais, P.; Morgenstern, O.; Oberländer, S.; Sigmond, M.; Shibata, K. Evaluation of radiation scheme performance within chemistry climate models. *J. Geophys. Res.* **2011**.

15 Oreopulos, L.; Mlawer, E.; Delamere, J.; Shippert, T.; Cole, J.; Fomin, B.; Iacono, M.; Jin, Z.; Manners, J.; Raisanen, P.; Rose, F.; Zhang, Y.; Wilson, M.; Rossow, W. The continual intercomparison of radiation codes: Results from phase I. *J. Geophys.Res.* **2012**.

16 Cornet, C.; C-Labonnote, L.; Szcap, F. Three-dimensional polarized Monte Carlo atmospheric radiative transfer model (3DMCPOL): 3D effects on polarized visible reflectances of a cirrus cloud. *J. Quant. Spectrosc. Rad. Transfer.* **2010**, *111*, 174–186.

17 Fomin, B.A. Effective interpolation technique for line-by-line calculations of radiation absorption in gases. *J. Quant. Spectrosc. Rad. Transfer.* **1995**, *53*, 663–669.

18 Kuntz, M.; Hopfner, M. Efficient line-by-line calculation of absorbing coefficients. *J. Quant. Spectrosc. Rad. Transfer.* **1999**, *110*, 97–104.

19 Rothman, L.S.; Gordon, I. E.; Barbe, A.; Benner, D.C.; Bernath, P.F.; Birk, M.; Boudon, V.; Brown, L.R.; Campargue, A.; Champion, J.-P.; Chance, K.; Coudert, L.H.; Dana, V.; Devi, V.M.; Fally, S.; Flaud, J.-M.; Gamache, R.R.; Goldman, A.; Jacquemart, D.; Kleiner, I.; Lacome, N.; Lafferty, W.J.; Mandin, J.-Y.; Massie, S.T.; Mikhailenko, S.N.; Miller, C.E.; Moazzen-Ahmadi, N.; Naumenko, O.V.; Nikitin, A.V.; Orphal, J.; Perevalov, V.I.; Perrin, A.; Predoi-Cross, A.; Rinsland, C.P.; Rotger, M.; Šimečková, M.; Smith, M.A.H.; Sung, K.; Tashkun, S.A.; Tennyson, J.; Toth, R.A.; Vandaele, A.C.; Vander Auwera, J. The HITRAN 2008 molecular spectroscopic database. *J. Quant. Spectrosc. Rad. Transfer.* **2009**,*110*, 533–572.

20 Clough, S.A.; Shepard, M.W.; Mlawer, E.J.; Delamere, J.S.; Jacono, M.J.; Cady-Pereira, K.; Boukabara, S.; Brown, P.D. Atmospheric radiative transfer modeling: A summary of the AER codes. *J. Quant. Spectrosc. Rad. Transfer.* **2005**, *91*, 233–244.

21 Marchuk, G.I.; Mikhailov, G.A.; Nazaraliev, M.A.; Darbijan, R.A.; Kargin, B.A.; Elepov, B.S. Reference. In *The Monte Carlo Method in Atmospheric Optics*; Springer: New York, NY, USA, 1980.

22 Evans, K.F.; Marshak, A. Numerical Methods. In *3D Radiative Transfer in Cloudy Atmospheres*; Springer: Berlin, Germany, 2005.

23 Fomin, B.A.; Rublev, A.N.; Trotsenko, A.N. Line-by-line Procedures to Compute Radiative Transfer Parameters in Scattering Atmosphere. In Proceeding of IRS'92: Current Problems in Atmospheric Radiation. A.DEEPAK Publishing: Hampton, VA, USA, 1993.

24 Chandrasekhar, S. Reference. In *Radiative Transfer*; Dover: New York, NY, USA, 1960.

25 Sushkevich, T.A. Reference. In *Mathematical Models of Radiation Transfer*; BINOM: Moscow, Russia, 2005. (in Russian)

26 Romanov, S.V.; Trotsenko, A.N.; Fomin, B.A. Reference. In *Application of the Numerical Methods to Describe the Transfer of Solar Radiation in The Scattering Atmosphere with Accurate Treatment of The Selective Gas Absorption*; Preprint Institute of Atomic Energy: Moscow, Russia, 1991. (in Russian)

27 Rodgers, C.D. Reference. In *Inverse Method for Atmospheric Sounding*; World Scientific: River Edge, NJ, USA, 2000.

CHAPTER 10

Characteristics of Low-Molecular Weight Carboxylic Acids in PM2.5 and PM10Ambient Aerosols From Tanzania

Stelyus L. Mkoma[1], Gisele O. da Rocha[2], José D.S. da Silva [2] and Jailson B. de Andrade[2]

[1] Faculty of Science, Sokoine University of Agriculture, Tanzania

[2] Instituto de Química, Universidade Federal da Bahia, Brazil

1. INTRODUCTION

Carboxylic acids together with water-soluble inorganic ions are an important group of water-soluble organic compounds in the atmospheric aerosols (Jacobson et al., 2000, Bourotte et al., 2007). They are highlighted because they account for substantial portion of atmospheric aerosols, and potentially control chemical and physical properties of the particles. Consequently, they may have direct and indirect effects on the earth's radiation balance by scattering incoming solar radiation, which counteracts the global warming (IPCC, 2007). More attention has been paid to carboxylic acids due to their potential to modify the hygroscopic properties of atmospheric particles, including cloud condensation nuclei activity and hence to change global radiation balance (Kerminen, 2001; Peng et al., 2001). Major water-soluble inorganic ions are associated with atmospheric visibility degradation, adverse human health effects, and acidity of precipitation (Dockery & Pope, 1996; IPCC, 2007).

Among the organic acids, low molecular weight carboxylic acids such as acetic, oxalic and malonic are generally most abundant in the atmospheric aerosols. Carboxylic acids in variable concentrations have been reported in various environments including rural and urban atmosphere (Kawamura &Sakaguchi, 1999; Kerminen et al., 2000; Nicolas et al., 2009) and have different source origin, including biomass burning, fossil fuel combustion (Kawamura 1987; Narukawa et al., 1999), sea spray, traffic and industrial emissions and photochemical oxidation of precursors from anthropogenic and biogenic origin (Kawamura &Sakaguchi, 1999; Limbeck&Puxbaum, 1999; Kumar et al., 2001; Chakraborty& Gupta, 2010). Other sources for carboxylic acids in the marine atmosphere include in-cloud and heterogeneous formations (Warneck, 2003).

Chemical composition of PM2.5 and even that of PM10 aerosols is important to gain insights into sources and of their toxicity and to evaluate effectiveness of abatement strategies for relevant emission sectors. Particulate matter (PM) with aerodynamic diameter less than 2.5 μm (PM2.5) exhibited stronger relation with health than those with aerodynamic diameter less than 10 μm (PM10), but other studies have reported a strong potential of PM10 to human health (Salma et al., 2002; Kappos et al., 2004). Most studies on low molecular weight carboxylic acids and their related compounds (Limbeck et al., 2001; Limon-Sanchez et al., 2002; Kawamura &Yasui, 2005) and major ions (Harrison et al., 2004; Karthikeyan&Balasubramanian, 2006; Mariani et al., 2007; Kundu et al., 2010; Mkoma et al., 2010) have extensively been reported.

In Africa similar aerosols measurements especially of organic components are missing. Therefore, a full scenario of air quality is far from being revealed because some pollutants including carboxylic acids have not been measured. The knowledge of elucidating chemical composition, levels, and source profiles of aerosols in the Tanzania atmosphere remains a challenge and is needed for both scientific and policy reasons. The continuous changes in socioeconomic and political environments in Tanzania result in changes in development, particularly in transport, industry, energy, and construction sectors. This chapter reports for the first time in Tanzania, composition of low molecular weight carboxylic acids in PM2.5 and PM10 aerosol samples collected from a rural background atmosphere in Morogoro. An insight of characteristics of water-soluble inorganic ions is also discussed in this chapter.

2. EXPERIMENTAL

2.1. Aerosol Sampling Site

Aerosol samples were collected at a rural site in Morogoro (300,000 inhabitants) between 26 April and 10 May 2011. This site is located at about 200 km west of the Indian Ocean and the city of Dar es Salaam, a business capital in Tanzania(Fig. 1). The samples were collected at Solomon Mahlangu Campus of Sokoine University of Agriculture (06°47'41"S, 37°37'44"E, altitude 504ma.s.l.). This site is located about 5 km from Morogoro central area and major road systems and possible aerosol sources include biomass burning, agriculture, livestock and soil dust. Approximately 70% of this area is covered by vegetation and about 15% with pasture field. Conversely, tropical savannah is the most important land cover in large part of the sampling site.

2.2. Aerosol Collection

Two samplers were used in parallel to collect aerosol particles: a "Gent" PM2.5 and PM10 filter holder each with two quartz fibre filters (Whatman QM-A) in series. Quartz fibre filters can adsorb volatile organic compounds (VOCs) causing positive artifacts when measuring PM and particulate OC. On the other hand, semi-volatile organic compounds (SVOCs) in aerosols may partially evaporate during sampling resulting in negative artifacts (Turpin et al., 2000; Mader et al., 2003; Hitzenberger et al., 2004). The quartz fibre filters were pre-fired at 550 °C during 24 h before use. Samplers operated at a flow rate of 17 L/min and were mounted on grass survey at SMC synoptic station approximately 2.7 m above ground level. The sampling was carried out approximately in 24 h intervals and exchange of filters during sampling periods was done at 7:30 am. A total of 11 sets of actual filed samples and 2 blanks were collected for each sampler and used in this chapter. After sampling the exposed filters were folded in half face to face, placed in polyethylene plastic bags and kept frozen at -4 °C during storage and transported cool to the laboratory of research and development in chemistry (LPQ) at the Institute of chemistry, Federal University of Bahia (UFBA). The samples were stored in a freezer at −20°C prior to analysis. All the procedures were strictly quality-controlled to avoid any possible contamination of the samples.

During the sampling period meteorological data were collected from the site. The daily winds were predominantly south-easterly with an average speed of 6.8 m/s. Average temperature was 26.8 °C and average relative humidity was 73%. The recorded maximum temperature and relative humidity were

29.8 °C and 79.5%, while minimum values were 23.7 °C and 63.5%, respectively. During the campaigns 5 days hand rainfall of a total 19.9 mm.

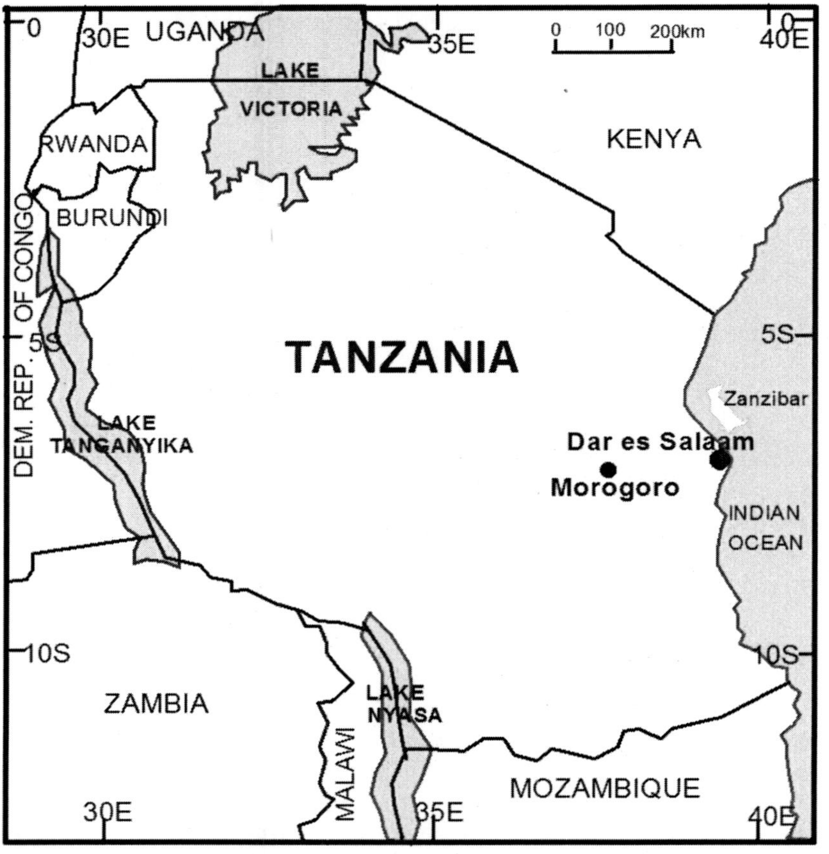

Figure 1. Location of the sampling site in Morogoro, Tanzania

2.3. Aerosol Analyses

For particulate mass measurements, the filter samples were weighed before and after sampling with an analytical microbalance balance Mettler Toledo MX5 (reading precision 1 μg). Before weighing, the filters were conditioned in a chamber equipped with hydro-thermometer clock at a temperature of 20 °C and relative humidity of 40% for 48 h and the weightings were done under these conditions.

For determination of carboxylic acids and water-soluble ions one-half of 12.88 cm^2 portions punched from of each PTFE filter was extracted using 5 ml Milli-Q ultrapure water (resistivity of 18.2 MΩcm, Barnstead International, USA) in a shaker tubes Model AT56 (Fanem Ltd, Sao Paulo, Brazil) for 5 minutes, followed by filtering through Polytetrafluoroethylene (PTFE) filter (0.45 μm pore size, Sartorius Stedim, Germany). The concentrations of aqueous extracts were determined by Dionex ion chromatography ICS 1100 and ICS 2100 for acids/anions and cations respectively which was equipped with an auto sampler (Dionex ICS Series AS-DV). An analytical column AS16 (3 x 50 mm) with AG16 guard column (3 x 50 mm) and CSRS-300 I (2 mm) suppressor in ion-exchange mode was used to determine carboxylates (monocarboxylates: formateand acetate; dicarboxylates: oxalate, malonate, succinate, and maleate; ketocarboxylate: pyruvate) and water-soluble anions (chloride Cl^-, nitrate NO_3^- and sulphate SO_4^{2-}). The eluent gradient programme was sweeping from 6.0 to 8.0 mmol/L KOH in 35 minutes under flow rate of 0.38 μL/min, except for acetic acid which was determined in another run, reducing injection time to avoid overlap of peaks. For determination of water-soluble cations (NH_4^+, Na^+, K^+, Mg^{2+} and Ca^{2+}) an analytical column CS16 and Guard column CG16 (both 3 x 50 mm) and CSRS-I (2 mm) suppressor in a chemical mode were used. An eluent of 17.5 mmol/L H_2SO_4 was used at flow rate of 0.35 μL/min. The injection volume was 25 μL for all detection. Peak identification was confirmed based on a match of ion chromatograph retention times and standard samples. Limit of detection determined as mean equal to 3 times standard deviation of the field blank value corresponded to a range of 0.008 to 0.017 ng/L for carboxylates, 0.008 to 0.023 ng/L for anions and 0.021 to 0.083 ng/L for cations. Limits of quantification were between 0.026 and 0.058 ng/L for carboxylates, 0.028 and 0.078 ng/L for anions and 0.063 and 0.252 ng/L for cations.

3. RESULTS AND DISCUSSION

3.1. Concentrations Of Pm Mass

Mean PM mass concentrations and associated standard deviations and ranges as derived from the two low-volume samplers are shown in Table 1. The results showed that mean mass concentration of PM2.5 and PM10 aerosols during the campaign were 13±3.5 μg/m^3 and 16±2.3 μg/m^3, respectively. The percentages of PM2.5 mass in PM10 size fraction (Fig. 2) found to range from 44–99% with a mean of 83±29%. These results indicate that most of PM mass was in PM2.5 size fraction. High PM2.5/PM10 ratios for PM mass indicate that there is small contribution from soil dust, which is known to be mostly associated with PM10 aerosols. Currently in Tanzania, the ambient air quality standard limit values for inhalable particulate matter are 60 to 90 μg/m^3 for PM10 (TBS, 2006). The mean concentrations for PM10 mass at our site in Morogoro were below these average limit values. In addition, the current data sets were in line with levels reported in our previous studies (Mkoma et al., 2009a,b; Mkoma et al., 2010). Nevertheless, when compared PM mass data from our rural site in Tanzania are in line with few available other data sets for rural sites in Southern Africa (Nyanganyura et al., 2007). They are also comparable to or lower to other sites in Europe and Asia (Van Dingenen et al., 2004; Gu et al., 2010; Maenhaut et al., 2011; Ram &Sarin, 2011).

3.2. Concentrations Of Carboxylates Ions

Table 1 present mean total concentrations and range of carboxylates (TCAs) which were 23.7±6.5 ng/m^3 (range: 13.3-36.5 ng/m^3) in PM2.5 and 36.4±12 ng/m^3 (range: 10.7-58.2 ng/m^3) in PM10 aerosols. Oxalate and malonate were most abundant carboxylates in PM2.5 accounting for 32.5% and 31.85% of total carboxylates, respectively, whereas in PM10 acetate was most abundant accounted for 62.5% of total carboxylates followed by oxalate which accounted for 32.6% of total carboxylates. Other studies have also reported oxalates to be most abundant carboxylate in aerosol samples (Mochida et al., 2003; Warneck, 2003). Pyruvate was also found in substantial amount and formate the least abundant counting on average 3% of total carboxylates in each of the aerosol fractions. Succinate and malonate were below detection limit in PM2.5 and PM10 aerosols, respectively. The total carboxylates accounted for 0.18% to total PM2.5 mass and 0.22% to PM10 mass. In comparison with other studies, the mean concentrations of all measured carboxylates in Tanzania were lower to those reported in urban and rural sites around the world (Souza et al., 1999; Kerminen et al., 2000; Yao et al., 2003;Kawamura &Yasui, 2005).

3.3. Water-Soluble Inorganic Ions And Ratios

Chemical characteristics of water-soluble inorganic ions and their relative abundances in PM2.5 and PM10 aerosols are also shown in Table 1. In both aerosol fractions, water-soluble Mg^{2+} was the most important cation and SO_4^{2-} the main anionic species. On average Mg^{2+} accounted for 44.4% of total water-soluble ions in PM2.5 and 24.7% in PM10 whereas SO_4^{2-} accounted for 22.8% and 35.2% of total ions in PM2.5 and PM10, respectively. High levels of crustal element Mg^{2+} together with Ca^{2+} are essentially attributable to soil/mineral dust dispersal. As to reasonable NH_4^+ levels (8% of total ions) in PM2.5, this may be due to presence of ammonia gas from biomass burning especially during smoldering combustion (Andreae&Merlet, 2001) and from agricultural activities in particular cattle raising (Street et al., 2003; Stone et al., 2010). Water-soluble K^+, a good indicator for biomass burning, was second most abundant cation in PM2.5 accounted for 10.6% of total water-soluble ions.

For SO_4^{2-} the higher levels could be attributed to its efficient formation by in-cloud processing of SO_2 (Yao et al., 2003) and from secondary formation processes (Allen et al., 2004). As to low NO_3^-levels, this is likely due to the fact that the site is rural with little or no traffic and undoubtedly there are less anthropogenic emissions of precursor gas NO_x. Also as to low concentrations of Na^+ whichis mainly derived from sea-salt, this is presumably due to long distance (about 200 km) from the Indian Oceanto our sampling site. The observed levels for water-soluble ions are comparable with those reported in our previous work in Morogoro (Mkoma et al., 2009a; Mkoma et al., 2010). It appears that the levels of SO_4^{2-}, NO_3^-, and NH_4^+ in PM10 fractions are substantially lower in Tanzania than at European rural sites (Putaud et al., 2004) and Asia (Aggarwal& Kawamura, 2009;Pavuluri et al., 2011).

Table 1. Mean concentrations, ranges and relative abundances (%) of carboxylates and water-soluble inorganic ions in PM2.5 and PM10 aerosols from Morogoro.

Species	PM2.5					PM10				
	Mean	SD	Min.	Max.	Rel. Ab.	Mean	SD	Min.	Max.	Rel. Ab.
PM mass ($\mu g/m^3$)	13.3	3.5	8.2	19.5	-	16.2	2.3	12.5	20.5	-
Carboxylates ions (ng/m^3)										
Formate, FA	0.71	0.30	0.38	1.27	3.00	1.2	0.6	0.5	2.6	3.4
Acetate, Ac	5.4	2.2	0.4	7.8	22.9	22.7	3.3	16.3	27.5	62.5
Oxalate, Oxa	7.7	2.7	4.7	13.1	32.5	11.8	7.2	4.5	31.0	32.6
Malonate, Mal	7.5	3.9	4.8	18.3	31.8	-	-	-	-	-
Succinate, Suc	-	-	-	-	-	1.6	2.4	0.2	6.5	4.5
Pyruvate, Pyr	2.4	0.7	1.5	3.8	9.9	2.2	0.7	0.9	3.3	6.1
Total carboxylate	23.7	6.5	13.3	36.5	-	36.4	12.0	10.7	58.2	-
Water-soluble ions (ng/m^3)										
NH_4^+	37.8	11.7	21.0	66.5	8.4	26.2	12.5	10.7	54.0	4.1
NO_3^-	4.5	1.6	1.2	8.0	1.0	25.1	12.4	10.5	52.0	3.9
SO_4^{2-}	102	27	69.1	160	22.8	237	125	38.8	487	36.6
Cl^-	2.3	0.4	1.6	3.1	0.5	9.6	5.7	3.3	20.2	1.5
Na^+	15.6	2.5	13.2	21.5	3.5	98.5	42.0	41.1	174.8	15.2
K^+	47.5	24.1	22.0	97.0	10.6	37.6	28.2	6.3	38.5	5.8
Mg^{2+}	199	33.7	155	265	44.4	158	69	16.4	228	24.5
Ca^{2+}	39.3	8.5	28.1	58.8	8.8	56.2	42.9	1.0	107	8.7
Total ions	448	88	348	585	-	646	214	256	1108	-

Rel. Ab. = Relative abundances

To determine the impact of marine sources on chemical composition of aerosol particles in PM2.5 and PM10 fractions, sea-salt ratios were calculated for each inorganic ion using Na as a reference element, assuming all Na to be of marine origin. The ratios for Cl^-/Na^+, SO_4^{2-}/Na^+, K^+/Na^+, Mg^{2+}/Na^+ and Ca^{2+}/Na^+ in PM2.5 were 0.15 (0.25), 6.48 (1.81), 2.95 (0.04), 12.97 (0.04), 2.44 (0.12), respectively. The corresponding values in PM10 were 0.10 (0.25), 2.51 (1.81), 0.43 (0.04), 1.68 (0.04), 0.75 (0.12), respectively. Values in brackets represent average ratios for each ion in sea-water (Brewer, 1975). Larger ratios of ions indicate incorporation of non-marine constituents in aerosols. As to low mean Cl^-/Na^+ ratios than sea-water ratio indicates that a minor fraction of Na^+ may be contributed from other sources such as mineral dust. But also low ratio could be due to modifications of sea-salt fraction by non-marine constituents. Chloride loss may be explained by heterogeneous reaction of airborne sea-salt with acidic gases and aerosol species (Millero, 2006).

3.4. Time Series Of Pm Mass And Selected Aerosol Species

Time series of PM mass, selected acids and ions species in PM2.5 and PM10 fractions as a function of sampling time are shown in Figs. 2 and 3. Nss-SO_4^{2-} in Fig. 2 was hereby obtained by subtracting sea-salt contribution from measured SO_4^{2-} data. Sea-salt contribution of SO_4^{2-} was obtained as 0.252Na^+, whereby Na^+ is the measured concentration of Na^+ and 0.252 is SO_4^{2-}/Na^+ ratio in the bulk seawater composition given by Riley and Chester (1971). As can be observed in Fig. 2, selected species in both size fractions showed no clear trends that can be noted but showed slightly variation during sampling period especially for PM10 aerosols. The observed behaviour of the species could be resulted from variations in sources strengths and meteorological conditions, such as mixing height. Additionally, high relative humidity (mean: 73%) during the campaign could serve as removal mechanisms hence lead to a daily variation in carboxylates levels.

In this study, oxalate concentrations were high to a factor of 10 than those of formate during the sampling period and in both PM2.5 and PM10 aerosol particles. These results indicate that formate was mainly from photochemical oxidation, while oxalates might have other sources besides photochemical oxidation. On the other hand, the concentrations of acetate were high than those of oxalate. Acetic acid in the atmosphere has been reported to be produced by oxidations of longer-chain dicarboxylic acids (Kawamura et al., 1996). Therefore, the observed acetate levels suggest that longer-chain dicarboxylic acids were possibly available at our site. Unfortunately, no data for high molecular weight dicarboxylic acids measured for the campaign.

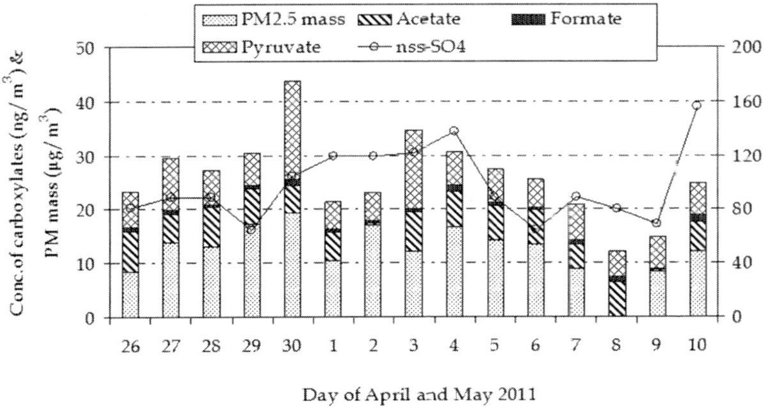

Figure 2. Time series of PM mass, carboxylates and nss-SO_4^{2-} in PM2.5fraction at Morogoro

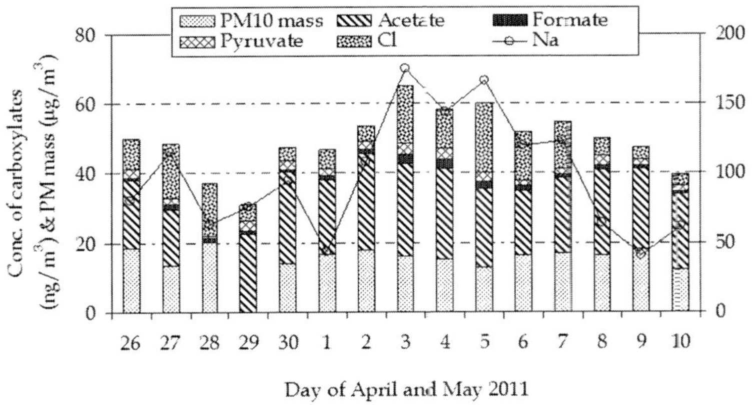

Figure 3. Time series of PM mass, carboxylates and Na^+ and Cl^- in PM10fraction at Morogoro

3.5. Pm2.5 To Pm10 Ratios

The average PM2.5 (fine) to PM10 (coarse) percentage ratios and associated standard deviations for PM mass, carboxylates and various water-soluble inorganic ions are shown in Fig. 4. The ratios were calculated on the basis of data for PM2.5 and PM10 samples taken in parallel and then averaged over all samples from the sampling period. The mean fine to coarse ratios for all species with exception of those for acetate, Na^+, Ca^{2+}, NO_3^- and Cl^- were predominantly associated with fine fraction (for more than 55%). High fine/coarse ratio for PM mass may be due to a less contribution from soil dust, which is known to be mostly associated with coarse particles. For carboxylates, high ratios (even larger than 70%) are considered to be attributed from secondary organic aerosols (SOA), biomass burning activities andhigh temperature (average: 26.8 °C during sampling period). Concentrations of oxalate in PM2.5 showed strong correlations ($r^2 = 0.70$) with those in PM10 aerosols. The slope of linear regression equations (PM2.5 = 2.48 x PM10) indicated that oxalate was mainly present in fine fraction during sampling period. On the other hand, different size distributions between oxalate and acetate could be related to their different physical characteristics. Acetate in PM2.5 fraction could easily volatilize (more volatile than oxalate) to gas phase, part of which could be absorbed on PM10 particles.

For water-soluble inorganic ions, as expected, sea-salt elements (Na, Cl) and indicator element for crustal matter (Ca) were predominantly (for more than 62%) associated with PM10 aerosols. $NH_4^+$and nss-SO_4^{2-} were mainly present in the fine aerosols suggesting that these species originated from high temperature sources and/or gas-to-particle conversion. The nss-SO_4^{2-} is due to oxidation of SO_2, which is predominantly from anthropogenic origin (e.g. biomass burning). A well-known indicator for biomass burning, K was associated with the fine particles (about 100%) suggesting vegetative emissions and crustal source could be an important source for K^+ aerosols at this site with small impact from biomass burning activities.

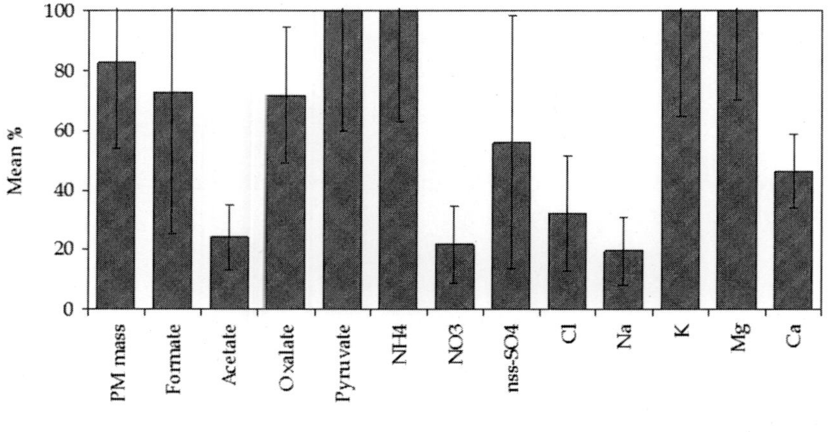

Figure 4. Mean contribution (%) of PM2.5 aerosolsmassand selected aerosol species in PM10 fractionduring the campaignin Morogoro.

3.6. Sources Of Carboxylates And Water-Soluble Ions In Aerosols

3.6.1. Correlation Analysis

Correlation coefficients of PM mass, carboxylates and source indicators, shown in Table 2 were performed in order to understand their possible sources and formation mechanisms. The selected source indicators include K^+ for biomass burning and vegetation emissions, Na^+ and $Cl^-$for sea spray or waste burning, and $SO_4^{2-}$for secondary formation of different mechanisms. Temperature and wind speed have been used as additional parameters to illustrate the atmospheric behaviours of carboxylic acids. In this study K^+ had good correlation with formate (r^2=53) and moderate to poorly correlations with other carboxylates in PM10. This indicates that formate could be originated from biomass and/or waste burning emissions but other carboxylates are considered to have other important sources than biomass burning. It can also be observed from Table 2 that there were possible similar sources for formate and other carboxylates (oxalate, Succinate and pyruvate) as verified by good correlation between them in PM2.5 and PM10 aerosols. Pyruvate shows good correlation with acetate (r^2 = 54) and succinate (r^2=51) in PM10 aerosols. These indicate a feature of photochemical decomposition of succinic acid (Yao et al., 2002).

Sea-salt derived aerosols have been reported to have particle with aerodynamic diameter between 1 and 5 μm (Kerminen et al., 2000). In this study, we found Na^+ has pronounced amount in PM10 aerosol and to slightly extent Cl^-, suggesting sea spray could be one of the contributing sources of the aerosol components at the site. But even though Na^+ correlate well (r^2=73) with Cl^-, the calculated Cl^- to Na^+mass ratio to sea-salt component in aerosol particles had mean value between 0.15 (PM2.5) and 0.10 (PM10). This suggest that continental contributions was most important than marine contribution since Cl^-/Na^+ ratio in marine aerosols varies between 1.0 and 1.7 (Chesselet et al., 1972).

Sulphate has been used as reference to investigate major formation routes of carboxylic acids (Yu et al., 2005). As shown in Table 2, formate and pyruvate showed good correlation with SO_4^{2-}in PM2.5, suggesting that in-cloud and heterogeneous formations play an important role in the formation of carboxylates. On the other hand, poor correlation of malonic with SO_4^{2-} in PM2.5 suggests that possibly the acid is volatile at ambient temperatures (Peng et al., 2001). Acetate and oxalateshowed poor correlations with SO_4^{2-} in both aerosol fractions, indicating that they mainly originated from primary emissions sources and/or the atmospheric processes different from those of SO_4^{2-}. This is contrary to what have been observed in other studies that in-cloud and heterogeneous formations can yield a good correlation between oxalate and SO_4^{2-} (Yu et al., 2005).

Wind speed poorly correlated with most carboxylates except with acetate (r^2=51) inPM10 aerosols. This indicates that in addition to secondary formation carboxylateswere mainly generated from local sources, while acetatemight be related to long range aerosols transport to the sampling site. It should also be noted that primary emissions are major sources of precursors for most carboxylic acids (Kawamura &Yasui, 2005).

Table 2. Correlation coefficients for PM mass, carboxylates, selected ions and wind speed in PM2.5 (upper diagonal triangle) and PM10 fractions (lower diagonal triangle) inMorogoro. Correlation coefficients that are larger than 0.50 are indicated in bold.

Species	FA	Ac	Oxa	Mal	Suc	Pyr	SO_4^{2-}	NO_3^-	Cl^-	Na^+	K^+	Ca^{2+}	WS
FA		0.26	0.10	0.19	–	**0.91**	**0.60**	0.31	0.10	–	0.38	–	-0.23
Ac	0.42		0.21	0.11	–	0.28	-0.04	0.45	0.27	–	-0.09	–	-0.21
Oxa	**0.77**	0.29		0.31	–	0.11	0.34	0.23	0.17	–	0.30	–	-0.07
Mal	–	–	–		–	0.17	0.17	**0.51**	0.33	–	0.13	–	0.08
Suc	**0.57**	0.42	-0.11	–		–	–	–	–	–	–	–	–
Pyr	0.48	**0.54**	0.13	–	**0.51**		**0.52**	0.21	-0.01	–	0.31	–	-0.20
SO_4^{2-}	–	0.26	0.30	–	0.31	-0.10		-0.16	-0.06	–	**0.93**	–	0.06
NO_3^-	**0.88**	0.17	**0.74**	–	**0.55**	0.30	0.11		0.78	–	-0.41	–	-0.03
Cl^-	**0.55**	-0.29	**0.56**	–	0.28	-0.11	0.05	**0.81**		–	-0.26	–	-0.12
Na^+	**0.81**	0.18	**0.74**	–	0.39	0.36	0.23	**0.87**	**0.73**		–	–	–
K^+	**0.53**	0.35	0.38	–	0.45	0.45	0.22	0.39	0.18	0.45		–	-0.02
Ca^{2+}	**0.65**	0.23	**0.52**	–	0.30	0.32	0.25	**0.58**	0.32	**0.58**	**0.51**		–
WS	-0.10	**0.51**	0.25	–	-0.44	-0.23	0.28	-0.06	-0.08	-0.19	0.11	0.26	

WS = Wind speed

3.6.2. Concentrations Ratios

The ratio of acetic to formic acid has been used as good indicator of contributions of primary (high ratio) and secondary sources (low ratio) to carboxylic acids (Talbot et al., 1988, 1990; Grosjean, 1992). As can be seen in Table 3, low acetate/formate ratios for both PM2.5 and PM10 aerosols particles indicate that secondary formation was an important contributing source of carboxylates at our site. This suggestion is supported by the fact that high mean average temperature during sampling period (mean: 26.8 °C) might be controlling factor in determining the contribution of primary and secondary sources to these carboxylates. However, there are various types of the atmospheric reactions forming carboxylic acids (e.g. formic, acetic, and oxalic) in urban and near urban atmospheres, which include oxidation of unsaturated fatty acids originating from cooking activities, ozone oxidation of olefins emitted from vehicular exhausts (Scheff&Wadden, 1993) and oxidation of aromatic hydrocarbons (Kawamura &Ikushima, 1993).

The ratio of oxalic acid to total dicarboxylic acid (for this study oxalic, malonic, succinic acids) can be used to evaluate aging process of organic aerosols (Kawamura &Sakaguchi, 1999), because diacid such as oxalic acid can be produced by oxidations of longer-chain dicarboxylic acids (Kawamura et al., 1996). In this study oxalate to total dicarboxylates ratios show low values in both aerosol fractions, indicating that aerosols emitted from various sources and transported to this site were less and equally aged. Since there relative humidity was high during the campaign (up to 73% on average), it is supposed that oxalate was also produced in aqueous phase. Aqueous phase chemistry in aerosol and/or cloud droplets is important in production of oxalic acid (Warneck, 2003). On the other hand, mean ratio of oxalate to K^+ in PM2.5 aerosols was 0.19±0.08, somewhat closeto or higher than those reported range (0.03−0.1) for flaming and smoldering phases in burning plumes (Yamasoe et al., 2000). This suggests that carboxylates might be originated from biogenic sources with contribution from biomass burning emissions.

Table 3. Mean ratios and ranges for carboxylates and K^+ in PM2.5 and PM10 aerosols in Morogoro.

Ratio	PM2.5		PM10	
	Mean±Stdev	Range	Mean±Stdev	Range
Acetate/Formate	0.24±0.31	0.07−1.07	0.05±0.03	0.03−0.10
Oxalate/Total dicarboxylates	0.33±0.08	0.23−0.49	0.33±0.17	0.15−0.83
Oxalate/K^+	0.19±0.08	0.07−0.36	−	−

4. CONCLUSION

PM2.5 and PM10 aerosols samples were collected from a rural site Morogoro, Tanzania and analysed for low molecular weight carboxylates and water-soluble inorganic ions. Oxalate and malonate were dominant species in PM2.5while acetate was most prominent species in PM10 aerosols followed by oxalate. Of the ionic components, SO_4^{2-}, K^+, and Mg^{2+} in PM2.5 and SO_4^{2-}, Na^+, and Mg^{2+} in PM10 made lager contribution to total water-soluble inorganic aerosol mass. Various ratios and correlations between carboxylates and ions used for possible source identification suggest that primary emissions, secondary formation, and to a slightly extent sea spray and biomass burning could be the sources for the aerosols at this site.The ratio of acetate to formate was used to distinguish primary and secondary sources of these carboxylates and was found to be close to reported value for secondary reactions, indicating dominance of secondary sources.Substantial concentration of carboxylates and water-soluble ions observed in the Morogoro atmosphere suggest that it was urgent to study the characteristics and sources of these species to better understand their roles in the Tanzania environment. However, more work is needed to determine longer-chain (high) molecular weight carboxylic acids and related organic compounds and their seasonal variations in other urban and rural sites in Tanzania.

ACKNOWLEDGEMENT

The authors are grateful to funding from Directorate of Research and Postgraduate Studies, Sokoine University of Agriculture (SUA)-Tanzania; Conselho Nacional de Desenvolvimento Científico e Tecnológico (CNPq)-Brazil; and Instituto Nacional de Ciência e Tecnologia, de Energia e Ambiente (INCT-E&A), Instituto de Química, Universidade Federal da Bahia (UFBA)-Brazil. We also wish to acknowledge Mr. Filbert T. Sogomba (SUA) for help in logistics and collection of aerosol samples and Cibele Cristina de AraújoSoares (UFBA) for PM mass measurements.

REFERENCES

1. S. G. Aggarwal, K. Kawamura, 2009Carbonaceous and inorganic composition in long-range transported aerosols over northern Japan: Implication for aging of water-soluble organic fractionAtmospheric Environment4316May 2009), 253225401352-2310
2. A. G. Allen, A. A. Cardoso, Rocha. G. O. da, 2004Influence of sugar cane burning on aerosol soluble ion composition in Southeastern BrazilAtmospheric Environment3830September 2004), 502550381352-2310
3. M. O. Andreae, P. Merlet, 2001Emission of trace gases and aerosols from biomass burningGlobal Biogeochemical Cycles154December 2001), 9559661944-9224
4. C. Bourotte, A. Curi-Amarante, P. , M. Forti, C. , A. Luiz, A.. Pereira, A. L. Braga, P. A. Lotufo, 2007Association between ionic composition of fine and coarse aerosol soluble fraction and peak expiratory flow of

asthmatic patients in São Paulo city (Brazil)Atmospheric Environment4110March 2007), 203620481352-2310

5. P. G. Brewer, 1975Minor elements in sea water, In: Chemical Oceanography, 1nd edition, J.P Riley and G. Skirrow (Ed.): 415-496, Academic Press, 0-12588-601-2UK
6. A. Chakraborty, T. Gupta, 2010Chemical Characterization and Source Apportionment of Submicron (PM1) Aerosols in Kanpur Region, India. Aerosol and Air Quality Research, 104December 2010), 4334452071-1409
7. R. Chesselet, M. Morelli, P. Buat-Menard, 1972Some aspects of the geochemistry of marine aerosols. In: Dyrssen, D., Jagner, D. (Eds.), The Changing Chemistry of the Oceans, Proceedings of Nobel Symposium 20, Wiley-Interscience, New York, 92
8. D. Dockery, A. Pope, 1996Epidemiology of acute health effects: Summary of time-series studies. In: Particles in Our Air: Concentration and Health Effects, J.D. Spengler, and R. Wilson, (Eds.): 123147Harvard University Press, 0-67424-077-4MA
9. D. Grosjean, 1992Formic and acetic acids: emissions, atmospheric formation and dry deposition at two southern California locations.Atmospheric Environment26A18December 1992), 327932 861352-2310
10. J. Gu, Z. Bai, A. Liu, L. Wu, Y. Xie, W. Li, H. Dong, X. Zhang, 2010Characterization of Atmospheric Organic Carbon and Element Carbon of PM2.5 and PM10 at Tianjin, China. Aerosol Aerosol and Air Quality Research, 102April), 1671762071-1409
11. R. M. Harrison, A. M. Jones, R. G. Lawrence, 2004Major component composition of PM10 and PM2.5 from roadside and urban background sites. Atmospheric Environment, 3827September 2004), 45311352-2310
12. R. Hitzenberger, A. Berner, Z. Galambos, W. Maenhaut, J. Cafmeyer, J. Schwarz, K. Miiller, G. Spindler, W. Wieprecht, K. Acker, R. . Hillamo, T. Makela, 2004Intercomparison of methods to measure the mass concentration of the atmospheric aerosol during INTERCOMP2000-influence of instrumentation and size cuts. Atmospheric Environment, 3838December 2004), 646764761352-2310
13. Intergovernmental Panel on Climate Change (IPCC),2007IPCC fourth assessment report (2007), Contribution of working group I. Cambridge University Press, London, 996
14. M. C. Jacobson, H. C. Hanson, K. J. . Noone, R. J. Charlson, 2000Organic atmospheric aerosols: review and state of the science. Reviews of Geophysics, 382May 2000), 2672941944-9208

15. A. D. Kappos, P. Bruckmann, T. Eikmann, N. Englert, U. Heinrich, P. Höppe, E. Koch, G. H. M. Krause, W. G. Kreyling, K. Rauchfuss, P. Rombout, V. S. Klemp, W. R. . Thiel, H. E. Wichmann, 2004Health effects of particles in ambient air. International Journal of Hygiene and Environmental Health, 2074August 2004), 3994071438-4639
16. S. . Karthikeyan, R. Balasubramanian, 2006Determination of water-soluble inorganic and organic species in atmospheric fine particulate matter. Microchemical Journal, 821January 2006), 49550002-6265X
17. K. . Kawamura, K. Ikushima, 1993Seasonal changes in the distribution of dicarboxylic acids in the urban atmosphere. Environmental Science and Technology, 2710September 1993), 222722331520-5851
18. K. Kawamura, H. Kasukabe, L. Barrie, 1996Source and reaction pathways of dicarboxylic acids, ketoacids and dicarbonyls in Arctic aerosols: one year of observations. Atmospheric Environment, 3010-11May 1996), 170917221352-2310
19. K. .Kawamura, F. Sakaguchi, 1999Molecular distribution of water soluble carboxylic acids in marine aerosols over the Pacific Ocean including tropics. Journal of Geophysical Research, 104D3February 1999), 350135092156-2202
20. V. Kerminen, M. , 2001Relative roles of secondary sulfate and organics in atmospheric cloud condensation nuclei production. Journal of Geophysical Research, 106D15August 1999), 17321173332156-2202
21. V. Kerminen, M. , C. Ojanen, T. Pakkanen, R. Hillamo, M. . Aurela, J. Merilaien, 2000Low-molecular-weight dicarboxylic acids in an urban and rural atmosphere. Journal of Aerosol Science, 313March 2000), 3493620021-8502
22. K. . Kawamura, O. Yasui, 2005Diurnal changes in the distribution of dicarboxylicacids, ketocarboxylic acids and dicarbonyls in the urban Tokyo atmosphere. Atmospheric Environment, 3910March 2005), 194519601352-2310
23. A. V. Kumar, R. S. . Patil, K. S. V. Nambi, 2001Source Apportionment of Ambient Particulate Matter at Two Traffic Junctions in Mumbai, India. Atmospheric Environment, 3525September 2001), 424542511352-2310
24. S. Kundu, K. Kawamura, T. W. Andreae, A. . Hoffer, M. O. Andreae, 2010Diurnal variation in the water-soluble inorganic ions, organic carbon and isotopic compositions of total carbon and nitrogen in biomass burning aerosols from the LBA-SMOCC campaign in Rondônia, Brazil. Journal of Aerosol Science, 411January 2010), 1181330021-8502

25. A. .Limbeck, H. Puxbaum, 1999Organic acids in continental background aerosols. Atmospheric Environment, 3312June 1999), 184718521352-2310
26. A. Limbeck, H. Puxbaum, L. Otter, M. C. Scholes, 2001Semi-volatile behavior of dicarboxylic acids and other polar organic species at a rural background site (Nylsvley, RSA). Atmospheric Environment, 3510April 2001), 185318621352-2310
27. M. T. Limon-Sanchez, J. L. Arriaga-Colina, S. Escalona-Segura, L. G. Ruíz-Suárez, 2002Observations of formic and acetic acids at three sites of Mexico City. Science of the Total Environment, 2873March 2002), 2032120048-9697
28. B. T. Mader, J. J. Schauer, J. H. Seinfeld, R. C. Flagan, J. Z. Yu, H. Yang, H. J. Lim, B. J. Turpin, J. T. Deminter, G. Heidemann, M. S. Bae, P. Quinn, T. Bates, D. J. Eatough, B. J. Huebert, T. Bertram, S. Howell, 2003Sampling methods used for the collection of particle-phase organic and elemental carbon during ACE-Asia. Atmospheric Environment, 3711April 2003), 143514491352-2310
29. W. Maenhaut, S. Nava, F. Lucarelli, W. Wang, X. . Chi, M. Kulmala, 2011Chemical composition, impact from biomass burning, and mass closure for PM2.5 and PM10 aerosols at Hyytiälä, Finland, in summer 2007. X-Ray Spectrometry, 403May/June 2011), 1681711097-4539
30. R. L. Mariani, W. Z. de Mello, 2007PM2.5-10, PM2.5 and associated water-soluble inorganic species at a coastal urban site in the metropolitan region of Rio de Janeiro. Atmospheric Environment, 4113April 2007), 288728921352-2310
31. F. J. Millero, 2006Chemical Oceanography (3rd ed.): 496, Taylor and Francis, CRC Press, Boca Raton, 0-84932-280-4
32. S. L. Mkoma, W. Wang, W. . Maenhaut, C. T. Tungaraza, 2010Seasonal Variation of Atmospheric Composition of Water-Soluble Inorganic Species at Rural Background Site in Tanzania, East Africa. Ethiopian Journal of Environmental Studies and Management, 32June 2010), 27381998-0507
33. S. L. Mkoma, W. Maenhaut, X. Chi, W. . Wang, N. Raes, 2009bChemical composition and mass closure for PM10 aerosols during the 2005 dry season at a rural site in Morogoro, Tanzania. X-Ray Spectrometry, 384July/August 2009), 2933001097-4539
34. S. L. Mkoma, X. Chi, W. Maenhaut, W. . Wang, N. Raes, 2009aCharacterisation of PM10 Atmospheric Aerosols for the Wet Season 2005 at Two Sites in East Africa. Atmospheric Environment, 433January 2009), 6316391352-2310

35. M. Mochida, A. Kawabata, K. Kawamura, H. Hatsushika, K. Yamazaki, 2003Seasonal variation and origins of dicarboxylic acids in the marine atmosphere over the western North Pacific. Journal of Geophysical Research, 108D6March 2003), 419342032156-2202
36. M. Narukawa, K. Kawamura, N. Takeuchi, T. Nakajima, 1999Distribution of dicarboxylic acids and carbon isopotic compositions in aerosols from 1997 Indonesian forest fires. Geophysical Research Letters, 2620October 1999) 310131041944-8007
37. J. F. Nicolas, N. Galindo, E. Yubero, C. Pastor, R. . Esclapez, J. Crespo, 2009Aerosol Inorganic Ions in a Semiarid Region on the Southeastern Spanish Mediterranean Coast. Water, Air and Soil Pollution, 2011-4July 2009), 1491591573-2932
38. D. Nyanganyura, W. Maenhaut, M. Mathuthu, A. . Makarau, F. X. Meixner, 2007The chemical composition of tropospheric aerosols and their contributing sources to a continental background site in northern Zimbabwe from 1994 to 2000. Atmospheric Environment, 4112April 2007), 264426591352-2310
39. C. M. Pavuluri, K. Kawamura, S. G. . Aggarwal, T. Swaminathan, 2011Characteristics, seasonality and sources of carbonaceous and ionic components in the tropical aerosols from Indian region. Atmospheric Chemistry and Physics, 1115August 2011), 821582301680-7316
40. C. Peng, M. N. Chan, C. K. Chan, 2001The hygroscopic properties of dicarboxylic and multifunctional acids: measurements and UNIFAC predictions. Environmental Science and Technology, 3522November 2001), 449545011520-5851
41. J. Putaud, P. , F. Raes, R. Van Dingenen, U. Baltensperger, E. Brüggemann, M. Facchini, C. , S. Decesari, S. Fuzzi, R. Gehrig, C. Hüglin, P. Laj, G. Lorbeer, W. Maenhaut, N. Mihalopoulos, K. Müller, X. Querol, S. Rodriguez, J. Schneider, G. Spindler, Brink. H. ten, K. . Tørseth, A. Wiedensohler, 2004A European aerosol phenomenology-2: chemical characteristics of particulate matter at kerbside, urban, rural and background sites in Europe. Atmospheric Environment, 3816May 2004), 257925951352-2310
42. K. Ram, M. M. Sarin, 2011Day and night variability of EC, OC, WSOC and inorganic ions in urban environment of Indo-Gangetic Plain: Implications to secondary aerosol formation. Atmospheric Environment, 452January 2011), 4604681352-2310
43. J. P. Riley, R. Chester, 1971Introduction to Marine Chemistry: 465, Academic Press, 0-12588-750-7

44. I. Salma, I. Balashazy, R. Winkler-Heil, W. . Hofmann, G. Zaray, 2002Effect of particle mass size distribution on the deposition of aerosols in the human respiratory system. Journal of Aerosol Science, 331January 2002), 1191320021-8502
45. P. . Scheff, R. A. Wadden, 1993Receptor modeling of volatile organic compounds: 1. Emission inventory and validation. Environmental Science and Technology, 274April 1993), 6176251520-5851
46. S. R. Souza, P. C. Vasconcellos, L. R. F. Carvalho, 1999Low molecular weight carboxylic acids in an urban atmosphere: winter measurements in São Paulo City, Brazil. Atmospheric Environment, 3316July 1999), 256325741352-2310
47. E. A. Stone, J. J. Schauer, B. B. Pradhan, P. M. Dangol, G. Habib, C. . Venkataraman, V. Ramanathan, 2010Characterization of emissions from South Asian biofuels and application to source apportionment of carbonaceous aerosol in the Himalayas. Journal of Geophysical Research, 115D06301doi:10.1029/2009JD011881March 2010), 2156-2202
48. D. G. Streets, K. F. Yarber, J. Woo, H. , G. R. Carmichael, 2003Biomass burning in Asia: Annual and seasonal estimates and atmospheric emissions. Global Biogeochemical Cycles, 174October 2003), 109911181944-9224
49. R. W. Talbot, M. O. Andreae, H. Berresheim, D. J. Jacob, K. M. Beecher, 1990Sources and sinks of formic, acetic and pyruvic acids over central Amazonia: 2. Wet deposition. Journal of Geophysical Research, 95D10January 1990), 167991681 12156-2202
50. Tanzania Bureau of Standards, TBS2006National Environmental Standards Compendium EMDC 6(1733). TZS 845: 2006 Air Quality Specification. 74
51. B. J. Turpin, P. Saxena, E. Andrews, 2000Measuring and simulating particulate organics in the atmosphere: problems and prospects. Atmospheric Environment, 3418July 2000), 298330131352-2310
52. R. Van Dingenen, F. Raes, J. Putaud, P. , U. Baltensperger, A. Charron, M. Facchini, C. , S. Decesari, S. Fuzzi, R. Gehrig, H. Hansson, C. , R. M. Harrison, C. Hüglin, A. M. Jones, P. Laj, G. Lorbeer, W. Maenhaut, F. Palmgren, X. Querol, S. Rodriguez, J. Schneider, Brink. H. ten, P. Tunved, K. Tørseth, B. Wehner, E. Weingartner, A. . Wiedensohler, P. Wåhlin, 2004A European aerosol phenomenology-1: physical characteristics of particulate matter at kerbside, urban, rural and background sites in Europe. Atmospheric Environment, 3816May), 256125771352-2310

53. P. Warneck, 2003In-cloud chemistry opens pathway to the formation of oxalic acid in the marine atmosphere. Atmospheric Environment, 3717June 2003), 242324271352-2310
54. M. A. Yamasoe, P. Artaxo, A. H. Miguel, A. G. Allen, 2000Chemical composition of aerosol particles from direct emissions of vegetation fires in the Amazon Basin: water-soluble species and trace elements. Atmospheric Environment, 3410July 2000), 164116531352-2310
55. X. Yao, M. Fang, C. K. Chan, 2002Size distributions and formation of dicarboxylic acids in atmospheric particles. Atmospheric Environment, 3613May 2002), 209921071352-2310
56. X. Yao, M. Fang, C. K. Chan, M. Hu, 2003Formation and size distribution characteristics of ionic species in atmospheric particulate matter in Beijing, China: (2) dicarboxylic acids. Atmospheric Environment, 3721July 2003), 300130071352-2310
57. J. Yu, X. Huang, J. Xu, M. Hu, 2005When aerosol sulfate goes up, so does oxalate: implication for the formation mechanisms of oxalate. Environmental Science and Technology, 391January 2005), 1281331520-5851

Index

A

C

D

G

H

I

L

M